CALCULUS 2

HONG-JONG KIM
HEASEUNG KWON

SNUPRESS
Seoul National University Press

CALCULUS 2

First edition published by Seoul National University Press in 2010
8th printing 2024

Seoul National University Press
1 Gwanak-ro, Gwanak-gu, Seoul 08826, Republic of Korea
Tel: +82-2-880-5252
Fax: +82-2-889-0785
Homepage: http://www.snupress.com
Facebook: @snupress1947
Instagram: @snupress
E-mail: snubook@snu.ac.kr

ISBN 978-89-521-1123-4 (94410)
978-89-521-1124-1 (set)

Preface

The book of nature is written in the language of mathematics
—Galileo Galilei (1564–1642)

Calculus is a branch of mathematics which consists of **differential calculus** and **integral calculus**. The differential calculus was founded by Isaac Newton (1643–1727) and Gottfried Wilhelm Leibniz (1646–1716), and the integral calculus was based on the **method of exhaustion** by an ancient Greek Eudoxus (408–355(?) B.C.) or Archimedes (287–212 B.C.) of Syracuse.[1] Calculus has widespread applications in science, engineering and economy and so it is indispensable for students in those fields. Also calculus gives the general public the wisdom of life.

It has been about 50 years since Seoul National University began to teach calculus. In 1988 Professor Sehie Park, who was the dean of math department then, suggested to make a textbook and Young-So Ko, Dohan Kim, Hong-Jong Kim started to write a textbook, which had been used since 1990. However, there were too many obstacles to produce a good book, such as SNU Press' out-of-date printing technique and lack of fresh contents. There were any valuable suggestions to improve it. However it was not easy to devote enough time to a new work. Even some said "There are plenty of good, expensive foreign textbooks for calculus. Why do we need spend several years to make a new textbook?"

"What should be taught in calculus?" Depending on people and their specialties, there are somewhat different opinions about it. Some think that "$\epsilon - \delta$" or the concept of **infinitesimal** should be emphasized, and some think that "abstract vector spaces" and "differential forms" should be taught more. Also there are other opinions: introducing basic differential equations, more computer-related education, showing more applications of calculus, giving more exercises and so on. Despite all those different viewpoints, everyone agrees on the following: "It should be fun to learn mathematics and anyone who learns math should feel that math is worth learning and useful."

[1]The calculus by Newton and Leibniz was not complete at all. Since the time of Leibniz and Newton, many mathematicians have contributed to the continuing development of calculus. "*Calculus*" in Latin means 'a small stone used for counting.'

In broad sense, math has a deep impact on the future of country[2]it is very important to teach math well. However it would be frustrating if young, fresh professors take too much time for making a new textbook or other chores to devote themselves to do their own pioneering research.

Nowadays there are fakes and nonsense everywhere in our society. I hope that the readers are not blinded by such things and contribute to establish "**healthy society**".

How to Read

This book is for students and lecturers who already know exponential functions, logarithmic functions, trigonometric functions and how to differentiate and integrate them.

This book consists of the following six parts:

I	Series and Taylor Expansion
II	Vectors and Matrices
III	Curves
IV	Multivarible Functions and Differentiation
V	Multiple Integrals and Green's Theorem
VI	Surface Integrals, Divergence Theorem and Stokes' Theorem

This text is intended for one-year calculus course: the first half for one semester and the other half for the the following semester.

We begin each chapter with the overview of the chapter in smaller fonts. Each chapter contains three types of exercises. There are "**Basic Exercises**" in the text and regular **Exercises** at the end of each section. Each chapter ends with **Advanced Exercises** They require quite a time to think. We recommend people who will read this book just once or less not to try Advanced Exercises. "Solving problems by oneself" is essential in appreciating mathematics. It won't be a desirable habit to look over hints or solutions for exercises in advance, which will appear time to time. Even though you may feel frustrated in search of answers, time spent to think will help you build up your ability and get though upcoming difficulties. When you get stuck to solve a problem, first think whether there are some parts in the text which you don't understand completely.

Terminologies which appear for the first time are usually written in **bold face**.

Some terminologies are explained in the [`Math Dictionary`] in Appendix.

The **logarithmic function**[3] used in this book is the "natural logarithm" function. The natural logarithm is often denoted by $\ln x$, but we will denote it by $\log x$.

---

[2]It is a very practical and concrete study that makes people think right and creatively, have sound judgment and understand others

[3]The theory of logarithmic function was made by a Scotish mathematician, John Napier (1550–1617).

The derivative of the natural logarithm function is $\frac{1}{x}$, while the derivative of the common logarithm function is $\frac{d}{dx}\log_{10} x = \frac{\log_{10} e}{x}$.[4] Mathematicians love beautiful equations. Even though the number

$$e = \lim_{n\to\infty}\left(1+\frac{1}{n}\right)^n$$

which is the base for the natural logarithm, is one of the most important numbers, it doesn't have any proper name.[5] The fact that the number e doesn't have its own symbolic notation also shows the "freedom of mathematics" well. Some denote the number by "e" instead of "e", but we will use the notation e even though it could be somewhat confusing. The **exponential function** means the exponential function with the base e, $\exp(x) := e^x$.

Let us introduce some notations in this book. $\mathbb{R}, \mathbb{Q}, \mathbb{Z}, \mathbb{N}, \mathbb{C}$ are the set of all Real numbers, RATIOnal numbers(Quotient of integers), integers(Zahlen), Natural numbers and Complex numbers, respectively. a/b is often used for $\dfrac{a}{b}$ to spare the space, and the notations $\leq$ and $\geq$ are used for $\leqq$ and $\geqq$ to save ink. You can find all notations and symbols used in this book in Appendix.

The identity matrix is denoted by I and the identity map whose domain X by id_X or 1_X. The function x is the identity map on the field of real numbers unless otherwise stated.

We tried to use tilt roman letters "$a, b, c, \ldots$" for scalars, bold faces "$\mathbf{a}, \mathbf{b}, \mathbf{c}, \ldots$" for vectors, capital letters "$A, B, C, \ldots$" for matrices and bold capitals "$\mathbf{A}, \mathbf{B}, \mathbf{C}, \ldots$" for vector fields.

Also "$a := b$" and "$b =: a$" means that "a is defined to be b." The logical symbols "$\forall a$" and "$\exists a$" are used to mean "for all a" and "there exists a" respectively.

The references are presented at the end of the book and indicated in [ ] with auther's name in the body. For example, if you are interested in how calculus has been developed for the last 100 years, you may refer to [Apostol et al.].

How To NOT Read

There are two types of footnotes: Ones that you don't have to read since it is not important, and the others which are very important but can be skipped if you are the first-time reader or you are lack of intellectual curiosity.

There are solutions or hints for some exercises but I recommend readers not to look over them in advance. Also to those who will read this book just once or less, I do not recommend to see **Advanced Exercises**

[4] It will verify that the number e is indispensable.

[5] The number e is often called "Euler's number" after Leonhard Euler who started to use the letter e for the constant in 1727. However Euler did so much more and there are other numbers called the very same name. Some call the number e "Napier's number" [Conway, Guy].

We put the symbol □ at the end of proof instead of the latin **Q.E.D.** (quod erat demonstrandum), which is traditionally used since Euclid's **Elements** (*Stoicheia*).[6] So, for those who don't want to know proofs, find □ to be free from proofs.

For Lecturers

Some introduce inverse functions, differentiation and integration of inverse functions and then define the exponential function $\exp x$ as the inverse of $\log x := \displaystyle\int_1^x \frac{dt}{t}$. However such a definition does not seem natural. Moreover we do not think starting a book with "inverse function" is proper. We introduce inverse function theorem for one-variable function while expanding the arctangent function into power series.

We gave a thought to the possibility of "considering only **continuously differentiable** functions rather than differentiable function" for multivariable functions, but didn't carry out the thought. We hope that you don't turn your back on a demand for "easy math."

We don't think that every "Basic Exercise" in the text should be mentioned during the lecture. Those exercises can be done to help students understand materials, while you can skep them if students understand materials well enough. You don't have to emphasize the contents in footnotes or put them in test.

For TAs

TAs who lead the practice session may keep in mind the followings to help students more effectively.

- Lead questions from students if they have difficulty to understand some materials.
- Have some office hour for students to ask questions freely.
- Provide additional explanation about the importance of exercises and introduce various methods to solve similar problems. Also emphasize how to write down answers in exam[7] and typical problems which often appear in exam.
- In the classroom, don't forget:
  - to say hello to students, to remember their names,
  - to speak loud enough, not too fast, not in monotone
  - to write big and clear enough in board, to clean the blackboard properly
  - to use proper chalks and colored ones
  - to arrange proper lighting
  - to keep yourself neat and trim
  - to have a sense of humor

[6] Euclid's Elements consists of 13 volumes.

[7] The answer must be written in a right, clear, beautiful and good way. The answer cannot obtain high score unless it shows the truthful procedure to deduce the conclusion.

Acknowledgement

I would like to thank many professors in math department in SNU for pointing out incorrect and unbeautiful writing, speaking out their thoughts. Professor Seok-jin Kang, Young-So Ko, Sung-Ki Kim, Changho Keem, Ok-Kyung Yoon, Jai-Han Yoon, In-Sok Lee, Hyun Koo Lee, Dong-Pyo Chi, Suhyoung Choi, Hyeong-In Choi, Jun-Muk Hwang read through the first version of this book though they were very busy. I would particularly like to thank Jae-Myung Chung, the chair of math department, without whose endless encouragement and suggestions this book wouldn't exist.

I would like to thank many lecturers and TAs who proofread exercises and point out errors, including Gyounghee Kim, Ilhyoung Kim, Dongjun Shin, Yongjin Yeom, Jongwon Oh, Yangdong Yoon, Sonam Choi, Insong Choi, Kil-Chan Ha. I also would like to thank various people who offered assistance in various ways, though I do not mention their names.

There are some expressions changed and some parts eliminated against our will. However I am responsible for any fault. If I may make an excuse, it was hard to struggle with frequently-changing system and policy and I have to submit this book on time even though I knew there were many things to improve. Maybe I should be thankful to have no more time left.

There are lots of pictures, which were made in various ways. I used TeX Cad, GnuPlot, MATHEMATICA 3.0, Maple V, Mdraw, etc. and sometimes I drew pictures by hand or used scanner. I would like to thank Professor Knuth who made TeX and KiHyoung Ko, JaeChoon Cha, Jin-Hwan Cho, Kwang-Hee Eun who provide Korean TeX.

I used Mercator-projection for the map of Korea while introducing double integral. This map was made as the polygon with 739 sides. Lastly, I would like to thank the one who waited till late at night, calculating coordinates for those points one by one....

January 1999, Kwanaksan

Preface for English Translation

Since 2005, Seoul National University has been offering a calculus course taught in English. For a couple of years a well-known English book was used as textbook. However most textbooks written in English are based on the curriculum for high school and university in the countries where those books are published, and the one used in SNU was not the exception. Its contents or level was quite different from the textbook used for the original calculus course. So we need to unify the contents taught and decide to translate the original textbook into English. It has been used for several years. Now we are delighted to publish it at SNU press.

We will be happy to to receive any suggestions and corrections from our reader. Please send your request to Heaseung Kwon (hkwon@snu.ac.kr).

Contents

Part IV

Functions of Several Variables and Differentiation

Chapter 10

Multivariable Functions

"What is the use of a book," thought Alice, "without pictures or conversations."
—in *Alice's Adventures in Wonderland* by Lewis Carroll (1832–1898)

So far we've been dealt with mostly the differential calculus of **one-variable functions**, that is, **functions of a single variable**. In this chapter, we study differential calculus of **multi-variable functions**, which depend on two or more variables, and we will see that the derivative at each point is given by the gradient vector. Also we study the chain rule to differentiate composite functions, and observe that the gradient vector is perpendicular to the level surface.

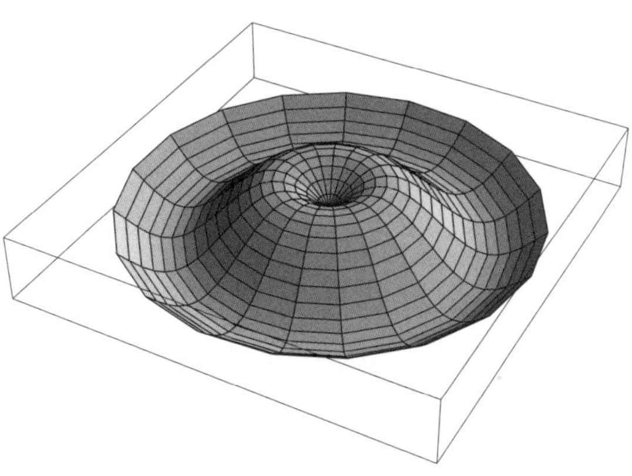

1 Graphs and Level Surfaces

1.1 Functions of Several Variables

The functions we've been studying so far were mostly functions whose domain is the field of real numbers or its subset. However, if we think of the function

$$d(x, y) := |x - y|$$

which measures the distance between two points x and y on the real line $\mathbb{R}$, the domain is the set of all pairs of real numbers (x, y)—that is, the two-dimensional plane $\mathbb{R}^2$. Like this, the function depending on two **variables** is called a **function of two variables** (or **two-variable function**).

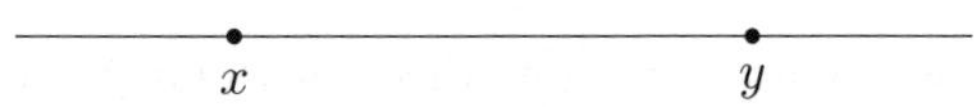

In this chapter, we will study the properties of functions with several variables. Suppose that a function

$$f : U \to \mathbb{R}$$

is defined on a subset U of $\mathbb{R}^n$. That is, f assigns to each point $X = (x_1, x_2, \ldots, x_n)$ in $U \subset \mathbb{R}^n$ a real number $f(X) = f(x_1, x_2, \ldots, x_n)$.[1] Then we call f a **function of n variables** (or an **n-variable function**) and also a **function of several variables** when $n \geq 2$.

1.1.1 Basic Exercise

1. Suppose that the function f assigns to three points in the plane the area of a triangle with vertices at those three points. How many variables does the function f have?

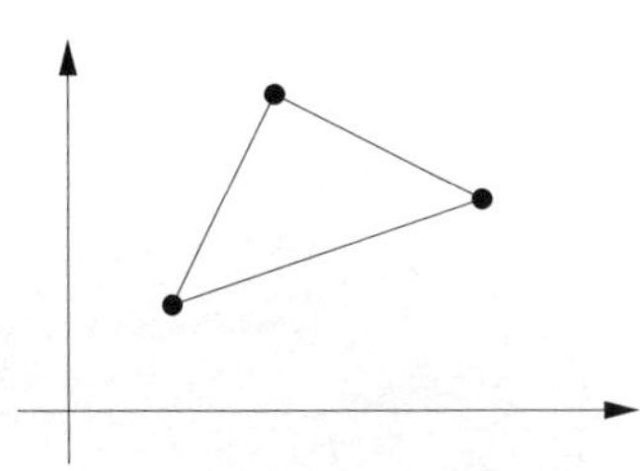

[1]It is very rare to denote a value of a function f at a point $(x_1, \ldots, x_n)$ by $f((x_1, \ldots, x_n))$. However we sometimes denote a value of a function f at a point X by fX. Some mathematicians think that it would be better to denote "the composite of two functions" by Xf rather than fX. The parenthesis in the notation $f(X)$ is not very important but the parenthesis appeared in $(x_1, \ldots, x_n)$ is the notation to denote an "ordered-set". In computer programming, the parenthesis like in $f(X)$ plays an important role.

2. When the function f assigns to each point in $\mathbb{R}^3$ the distance from the origin to the point, how many variables does the function f have?
3. When the function f assigns to two points in $\mathbb{R}^3$ the distance between those two points, how many variables does the function f have?

1.2 Graphs

One way to visualize a multivariable function f defined on a subset U in $\mathbb{R}^n$ is to consider its graph. The **graph** of f is defined by the set

$$\{(x_1, x_2, \ldots, x_n, z) \in \mathbb{R}^{n+1} \mid (x_1, x_2, \ldots, x_n) \in U,\ z = f(x_1, x_2, \ldots, x_n)\}.$$

1.2.1 Example

The graph of a two-variable function $f(x, y) = x^2 + y^2$ is the **paraboloid** given by the equation[2]

$$z = x^2 + y^2.$$

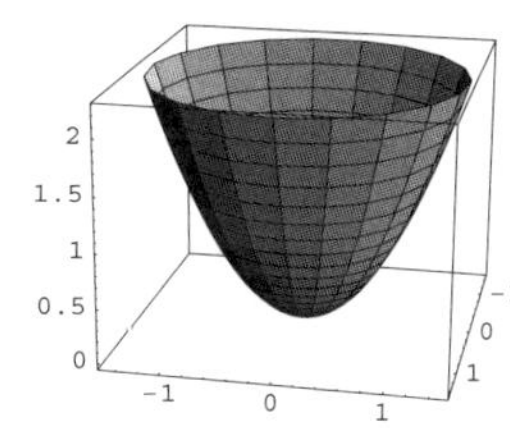

1.2.2 Example

The graph of the function

$$f(x, y) = 1 - x - y$$

is the plane in space given by the equation $z = 1 - x - y$.

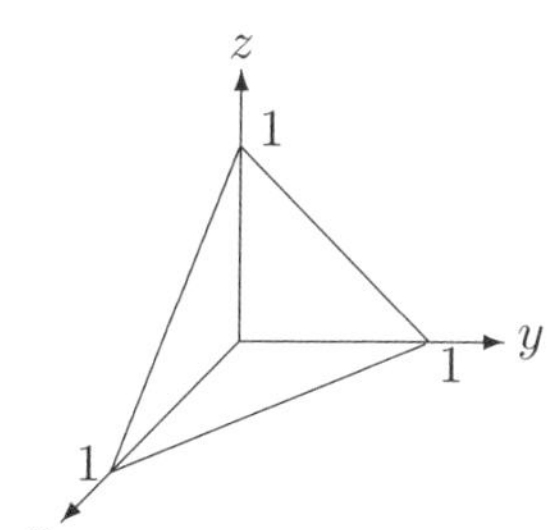

Picture: $x + y + z = 1$

[2]Since the intersection of the plane $z = c > 0$ and the graph of f is a circle with radius $\sqrt{c}$, the graph of f is a surface of revolution obtained by rotation about the z-axis. Also on the yz-plane given by the equation $x = 0$, the graph of f is the parabola $z = y^2$ and so the graph of f is a paraboloid obtained by rotating the parabola.

1.3 Slopes of Linear Functions

As a function of one variable, a **linear function** is of the form

$$y = ax + b,$$

and the **slope** of (the graph of) the function is[3]

$$\frac{\Delta y}{\Delta x} = \frac{(a(x+\Delta x)+b)-(ax+b)}{\Delta x} = a.$$

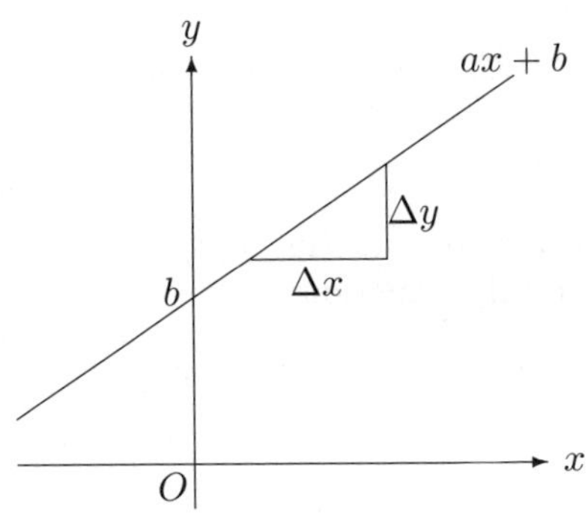

As a function of two variables, a linear function is expressed as

$$z = z(x,y) = ax + by + c.$$

If we investigate the graph of the function only along the line parallel to the x-axis on the xy-plane, then the slope is[4]

$$\frac{z(x+\Delta x, y) - z(x,y)}{\Delta x} = \frac{\big(a(x+\Delta x)+by+c\big)-(ax+by+c)}{\Delta x} = a.$$

Similarly, if we investigate the graph of the function along the line parallel to the y-axis, then the slope is

$$\frac{z(x, y+\Delta y) - z(x,y)}{\Delta y} = b.$$

[3]The slope of a linear function is constant, independent of x or Δx. In particular, we may let $\Delta x = 1$. In other words, the slope of a linear function means the change in values of the function while the (independent) variable increases by 1.

[4]Or it can be thought of as the increase in values of the function while the independent variable increases by 1 in the direction of the x-axis.

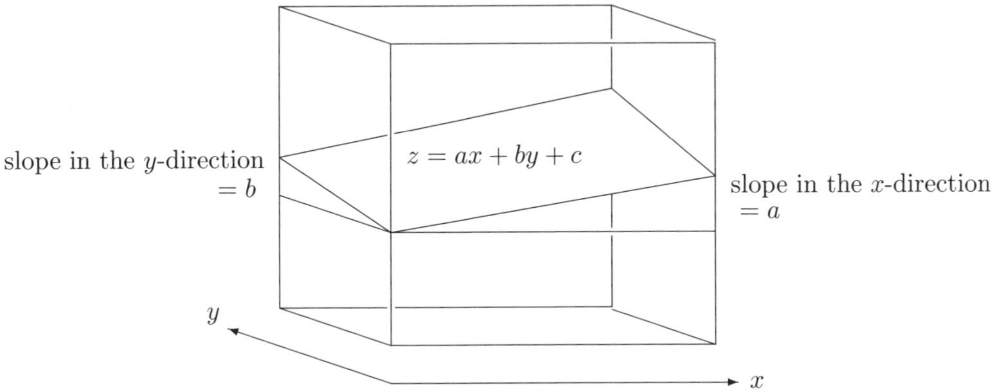

The **slope** of a linear function of two variables is the vector made up of the "slope in the x-direction" and "slope in the y-direction"

$$(a, b) = a\,\mathbf{i} + b\,\mathbf{j}.$$

This vector is called the **gradient vector**[5] of the linear function $z = ax + by + c$.

The slopes in the x- and y-directions for the function in the previous example

$$z = 1 - x - y$$

are both -1, and thus its gradient vector is $(-1, -1)$.

1.3.1 Basic Exercise

For the function $z = x + 2y + 3$, find the slopes in the x- and y-directions and explain their meanings. Also find its gradient vector.

A linear function of n variables is given by

$$z = z(\mathbf{x}) = \mathbf{a} \cdot \mathbf{x} + c = a_1 x_1 + \cdots + a_n x_n + c$$

where $\mathbf{a} = (a_1, \ldots, a_n)$ and $\mathbf{x} = (x_1, \ldots, x_n)$. The "slope in the x_1-direction" is

$$\frac{z(\mathbf{x} + t\mathbf{e}_1) - z(\mathbf{x})}{t} = \frac{(\mathbf{a} \cdot (\mathbf{x} + t\mathbf{e}_1) + c) - (\mathbf{a} \cdot \mathbf{x} + c)}{t} = \mathbf{a} \cdot \mathbf{e}_1 = a_1.$$

It is the slope of the graph along the line on the $x_1 z$-plane. Similarly, for $i = 2, \cdots, n$, the "slope in the x_i-direction" is a_i and this is the slope of the graph along the line on the $x_i z$-plane.

[5]The gradient of a general nonlinear function will appear in Section 3.

Theorem 1.3.2 *For a linear function* $z = a_1x_1 + \cdots + a_nx_n + c = \mathbf{a}\cdot\mathbf{x} + c$ *and a unit vector* $\mathbf{v} = (v_1, \ldots, v_n)$*, the slope of* z *in the* $\mathbf{v}$*-direction*[6] *is*

$$a_1v_1 + \cdots + a_nv_n = \mathbf{a}\cdot\mathbf{v}$$

Proof. Since

$$\frac{z(\mathbf{x}+t\mathbf{v}) - z(\mathbf{x})}{t} = \frac{(\mathbf{a}\cdot(\mathbf{x}+t\mathbf{v})+c) - (\mathbf{a}\cdot\mathbf{x}+c)}{t} = \mathbf{a}\cdot\mathbf{v},$$

we get the desired conclusion.[7] □

1.3.3 Basic Exercise

For the linear function

$$f(x, y) = x + 2y + 3,$$

find the gradient in the $\left(\frac{3}{5}, \frac{4}{5}\right)$-direction. how much does the value of the function increase when the variables change from (x, y) to $(x, y) + (3, 4)$?[8]

1.4 Level Surfaces

For a function $f : U \to \mathbb{R}$ of several variables defined on a set $U \subset \mathbb{R}^n$ and a real number c, the **inverse image** of c under f

$$f^{-1}(c) := \{X \in U \mid f(X) = c\}$$

is called the c-**level surface** or c-**isocline** of f. When $n = 2$, it is often called a **level curve**. If f represents the temperature, height or pressure, then it is called an **isotherm**, **contour** or **isobar**, respectively. A level surface is a set of points at which the function has a constant value.

---

[6] You may consider it as the change of values of the function as the variable changes by 1 in the $\mathbf{v}$-direction.

[7] "For a linear function", the slope in the $\mathbf{v}$-direction depends on neither the point $\mathbf{x}$ nor the amount of change t, but does depend only on the unit vector $\mathbf{v}$. Therefore

$$z(\mathbf{x}+\mathbf{v}) - z(\mathbf{x}), \qquad z(\mathbf{v}) - z(\mathbf{0})$$

are both the slope in the $\mathbf{v}$-direction.

[8] Answer: The slope in the $\left(\frac{3}{5}, \frac{4}{5}\right)$-direction is $(1, 2)\cdot\left(\frac{3}{5}, \frac{4}{5}\right) = 1\cdot\frac{3}{5} + 2\cdot\frac{4}{5} = \frac{11}{5}$ and the change in values of the function is $5\cdot\frac{11}{5} = 11$.

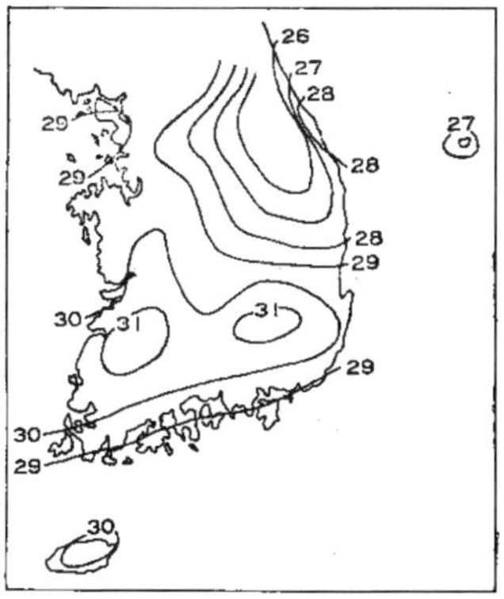

1.4.1 Example

For a function $f(x, y, z) = x^2 + y^2 + z^2$, the level surface $f^{-1}(c)$ is the sphere of radius $\sqrt{c}$ if $c > 0$, one point if $c = 0$, and the empty set if $c < 0$.

1.4.2 Basic Exercise

1. The following picture shows the level curves of the function $f(x, y) = x^2 - y^2$. In this picture, point out the part where the values of f are positive and the part where the values of f are negative.

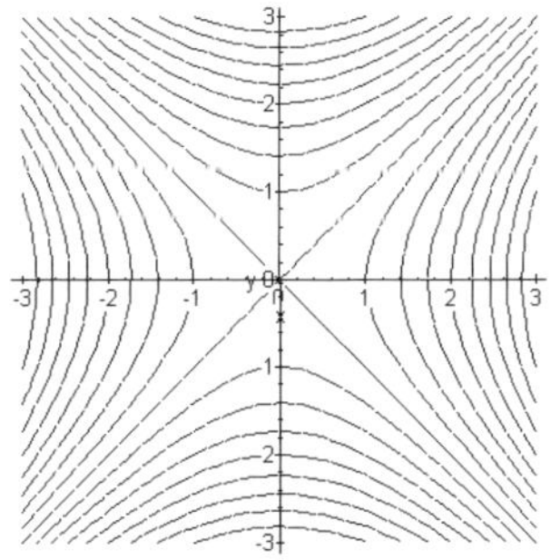

2. Sketch the level curves of the function $f(x, y) = x + y$. Also sketch the level surfaces of the function $g(x, y, z) = x^2 + y^2 - z^2$. (Since there are many level surfaces, you may sketch some of them for a few appropriate real values.)
3. Explain that the graph of the function $z = f(x, y)$ is equal to the 0-level surface of the function $F(x, y, z) = f(x, y) - z$.

While the graph of a function lies in the space of dimension one higher than the dimension of the domain, the level surfaces remains in the domain.

Exercise: Chapter 10 Section 1

1. When the function f assigns to three points in $\mathbb{R}^3$ the distance from the center of those points to the origin, how many variables does f have?

2. Sketch the graph of the following function by hand. Also use a computer program to graph the function.
   (1) $f(x, y) = \sqrt{x^2 + y^2}$ (2) $f(x, y) = \sqrt{1 + x^2 + y^2}$

3. Show that the surface obtained by rotating the graph of
$$z = f(y), \qquad y \geq 0$$
on the yz-plane around the z-axis in $\mathbb{R}^3$ is described by the equation $z = f\left(\sqrt{x^2 + y^2}\right)$.

4. Draw the graph of the following function.
   (1) $(\sin \pi r)/\pi r$ (Here, $r = \sqrt{x^2 + y^2}$) (2) $2(x^2 + y^2)e^{-x^2-y^2}$

5. For $c = -1, 0, 1, 2, 3$, sketch the c-level curves or surfaces of the function f.
   (1) $f(x, y) = x - y$ (2) $f(x, y) = xy$
   (3) $f(x, y) = x^2 + y^2$ (4) $f(x, y) = -x^2 + y^2$
   (5) $f(x, y, z) = x^2 + y^2 - z^2$ (6) $f(x, y, z) = x^2 - y^2 - z^2$

6. Sketch the level set of the function f containing the point P.
   (1) $f(x, y) = x + y$, $P = (1, 1)$
   (2) $f(x, y) = x^3 + x^2 - y^2$, $P = (0, 0)$
   (3) $f(x, y, z) = x^2 + y^2 - z^2$, $P = (1, 0, 0)$
   (4) $f(x, y, z) = x^2 - y^2 - z^2$, $P = (1, 0, 0)$
   (5) $f(x, y, z) = x^2 + 4y^2 + 9z^2$, $P = \left(\frac{1}{\sqrt{3}}, \frac{1}{2\sqrt{3}}, \frac{1}{3\sqrt{3}}\right)$

7. For a unit vector $\mathbf{v} = (1, 1, 1)/\sqrt{3}$, find the gradient of the linear function $f(x, y, z) = x + 2y + 3z + 4$ in the $\mathbf{v}$-direction.

2 Limits and Continuous Functions

For a function $f : U \to \mathbb{R}$ defined on an open set $U \subset \mathbb{R}^n$ and a point P in U, f is called **continuous** at P if

$$\lim_{X \to P} f(X) = f(P). \tag{10.1}$$

If we think over the meaning of the above equation (10.1), it means that

If X in U is sufficiently close to P, then $f(X)$ is also close to $f(P)$.[9]

To show that the equation (10.1) is true, it is not sufficient to show that

$$\lim_{i \to \infty} f(X_i) = f(P) \tag{10.2}$$

for some sequence (X_i) which converges to P, we have to show that (10.2) is true for every sequence (X_i) which converges to P.

2.0.1 Basic Exercise

For a function

$$f(x,y) = \begin{cases} \dfrac{x^2 - y^2}{x^2 + y^2} & , \quad (x,y) \neq (0,0) \\ 0 & , \quad (x,y) = (0,0), \end{cases}$$

find the following limit if exists. Also determine whether the function $f(x,y)$ is continuous at the origin.[10]

1. $\lim_{x \to 0} f(x,y)$
2. $\lim_{y \to 0} f(x,y)$
3. $\lim_{y \to 0} \left(\lim_{x \to 0} f(x,y) \right)$
4. $\lim_{x \to 0} \left(\lim_{y \to 0} f(x,y) \right)$
5. $\lim_{(x,y) \to (0,0)} f(x,y)$

---

[9]In other words,

$$\lim_{X \to P} |f(X) - f(P)| = 0.$$

In this equation, the "notation of absolute value" is not very important, but it is helpful to use the notation of absolute value when f is a vector-valued function. "A point X is close to a point P" means that "the real number $|X - P|$ is close to 0." It is obvious that $X = (x_1, \dots, x_n)$ **converges** to $P = (p_1, \dots, p_n)$ if and only if each component x_i of X converges to the corresponding component p_i of P.

[10]$\lim_{t \to 0} g(t) = l$ means that if the absolute value of t is very small but $t \neq 0$ then the difference between $g(t)$ and l is arbitrarily small. Answer: 1. -1 if $y \neq 0$ and 1 if $y = 0$; 2. 1 if $x \neq 0$ and -1 if $x = 0$; 3. -1 ; 4. 1 ; 5. The limit doesn't exist. Therefore $f(x,y)$ is not continuous at the origin.

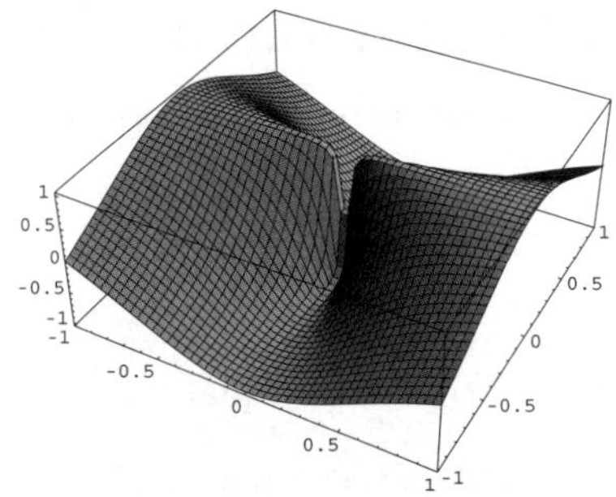

Picture: The graph[11] of $f(x,y) = \dfrac{x^2 - y^2}{x^2 + y^2}$

2.0.2 Continuity on a Set

If a function $f : U \to \mathbb{R}$ is continuous at every point in U, then f is called **continuous** on the set U.

2.0.3 Example

For a fixed point Q in $\mathbb{R}^n$, the function f measures the distance from Q to an arbitrary point X,

$$f(X) := |X - Q|.$$

Show that f is continuous.

For any point P,

$$|f(X) - f(P)| = \big||X - Q| - |P - Q|\big| \le |X - P|$$

and so

$$\lim_{X \to P} |f(X) - f(P)| = 0,$$

that is,

$$\lim_{X \to P} f(X) = f(P).$$

Therefore f is continuous.

[11]Since $f(x,y)$ satisfies $f(cx,cy) = f(x,y)$ for any nonzero real number c, the graph near the origin looks similar, no matter how closely we zoom in the graph.

2.0.4 The Set of All Continuous Functions

Often in mathematics, once a certain concept is introduced, in order to understand the concept, it is very useful to consider not only a single object with the concept but also the set of all such objects and the properties of that set.

Just as for functions of a single variable, for functions of several variables, *constant functions are continuous, and the sums and products of continuous functions are continuous. The composition of two continuous functions is continuous too.* In particular if f, g are both continuous and g is never zero, then $f/g = f \cdot \frac{1}{g}$ is continuous.[12]

2.0.5 Basic Exercise

1. Show that the orthogonal projections

$$(x_1, x_2, \ldots, x_n) \mapsto x_k \qquad (k = 1, 2, \ldots, n)$$

are all continuous. Also show that the **linear function**

$$(x_1, x_2, \ldots, x_n) \mapsto a_1 x_1 + a_2 x_2 + \cdots + a_n x_n$$

is continuous.

2. Show that the following function defined on $\mathbb{R}^2$ is continuous.[13]

$$f(x, y) = \begin{cases} \dfrac{xy}{\sqrt{x^2 + y^2}} & , \quad (x, y) \neq (0, 0) \\ 0 & , \quad (x, y) = (0, 0) \end{cases}$$

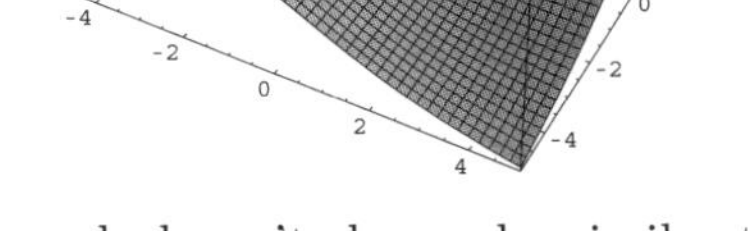

Picture: The graph doesn't change by similar transformations

3. Show that the function defined on $\mathbb{R}^2$ is not continuous.[14]

[12] $1/g$ is the composition of a function g and the "continuous function" $x \mapsto x^{-1}$.

[13] Hint: Use the equation for the arithmetic and geometric means, $|xy| \leq \dfrac{1}{2}(x^2 + y^2)$.

[14] It suffices to show that if (x, y) gets close to the origin along the line $y = tx$, the value of $f(x, y)$ converges to $\dfrac{t}{1 + t^2}$.

$$f(x,y) = \begin{cases} \dfrac{xy}{x^2+y^2} & , \quad (x,y) \neq (0,0) \\ 0 & , \quad (x,y) = (0,0) \end{cases}$$

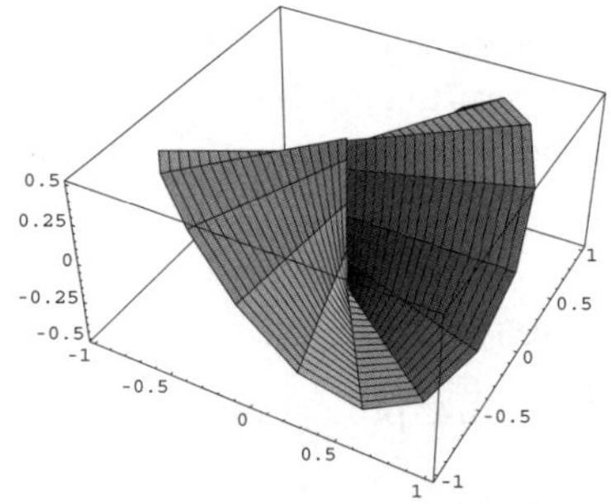

2.1 Extreme Value Theorem

Among interesting problems which appear in nature, social phenomena, economics or physiology, there are lots of problems about finding maximum or minimum values. For those problems, it is important to know the following:

EXISTENCE: Does the given problem have a solution?

UNIQUENESS: If there exists a solution, how many solutions are there?

THE METHOD TO FIND SOLUTIONS: How can we find the concrete solutions efficiently?

The "Extreme Value Theorem" is an important theorem that *guarantees the existence of solutions.*[15] It would be vain to try to solve problems for which the **existence** of solutions is not guaranteed, such as "trisecting an arbitrary angle", "finding the formula for roots of quintic polynomials".[16]

[15]Even though the theorem guarantees the existence of solutions, it is nothing to do with the practical method to find solutions.

[16]The formula for roots of cubic polynomials was discovered by Italian Girolamo Cardano (1501–1576), and the formula for roots of quartic polynomials was discovered by his student Ferrari (1522–1565). However, Norwegian Niels H. Abel (1802–1829) and French Évaliste Galois (1811–1832) showed that there is no general formula for roots of polynomials of degree five or higher.

2.1.1 Closed, Bounded Sets

Weierstrass' Extreme Value Theorem states that any continuous function defined on a "good domain" attains the maximum and minimum values — that is, the existence of solutions.

Here let's investigate what kind of sets are "good domains" .

We will introduce the terms "open ball" and "closed ball". An **open ball** of radius $r > 0$ with the center at P in $\mathbb{R}^n$ is defined to be

$$\mathbf{B}^n(P, r) := \{X \in \mathbb{R}^n : |X - P| < r\}.$$

Also a **closed ball** of radius $r > 0$ with the center at P in $\mathbb{R}^n$ is defined to be

$$\overline{\mathbf{B}}^n(P, r) := \{X \in \mathbb{R}^n : |X - P| \leq r\}.$$

An open or closed ball is called a **ball**.

Of course, a 1-dimensional ball is an **interval**, a 2 -dimensional ball is a **disk**.

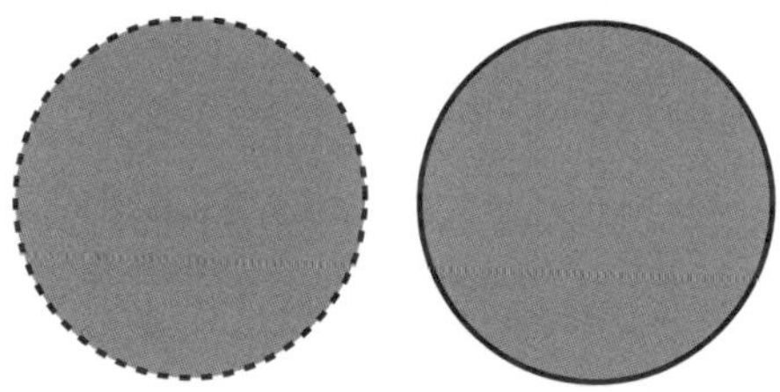

Picture: An open disk $\mathbf{B}^2$ and a closed disk $\overline{\mathbf{B}}^2$

A subset $\mathcal{A}$ of $\mathbb{R}^n$ is called **bounded** if there exists a ball which contains $\mathcal{A}$.[17]

And a subset of $\mathbb{R}^n$ is called **closed** if it contains the **boundary**.[18] A closed interval in the field of real numbers is an example of closed set.

In the Extreme Value Theorem, a "good domain" means a "closed, bounded set".

Theorem 2.1.2 (Extreme Value Theorem) *A continuous function defined on a closed, bounded set D in $\mathbb{R}^n$ must attain the absolute maximum and minimum values in D.*

[17] A subset $\mathcal{A}$ of $\mathbb{R}^n$ is bounded if and only if

$$\mathcal{A} \to \mathbb{R}, \quad X \mapsto |X|$$

is a **bounded function**.

[18] You may refer to the last section for closed set and boundary.

This theorem is the extension of the one-dimensional case,[19] and it can be proved in a similar way, too.[20]

The sketch of proof of the Extreme Value Theorem can be found in [`Math Dictionary`].

[19]A continuous function defined on a closed, bounded interval attains the absolute maximum and minimum values.

[20]If we use this theorem, we can prove the **fundamental theorem of algebra** saying that *"Any nonconstant polynomial (with complex coefficients) has a zero (in the field of complex numbers)"*. Its proof can be found in Appendix 1 [`Math Dictionary`].

Exercise: Chapter 10 Section 2

1. Find the following limit if it exists.

$$(1)\quad \lim_{(x,y)\to(0,0)} \frac{1}{1-\log(x^2+y^2)} \qquad (2)\quad \lim_{(x,y)\to(0,0)} \frac{e^{-1/(x^2+y^2)}}{x^2+y^2}$$

2. Investigate the continuity of the function.

$$(1)\ f(x,y) = \begin{cases} xy^2/(x^2+y^4), & (x,y) \neq (0,0) \\ 0, & (x,y) = (0,0) \end{cases}$$

$$(2)\ f(x,y) = \begin{cases} (x^2-y^2)/(x^2+y^2), & (x,y) \neq (0,0) \\ 0, & (x,y) = (0,0) \end{cases}$$

$$(3)\ f(x,y) = \begin{cases} (x^2y-xy^2)/(x^2+y^2), & (x,y) \neq (0,0) \\ 0, & (x,y) = (0,0) \end{cases}$$

3. Show that the function[21]

$$f(x,y) := \begin{cases} \dfrac{x^2y}{x^4+y^2}, & (x,y) \neq (0,0) \\ 0, & (x,y) = (0,0) \end{cases}$$

is continuous if it is restricted on any straight line through the origin but not continuous when it is restricted on the parabola $y = x^2$.

4. For the function

$$f(x,y) := \frac{ax+by}{\sqrt{x^2+y^2}}$$

defined on $\mathbb{R}^2 \setminus \{(0,0)\}$, find all real values of a, b for which $f(x,y)$ is continuous if we assign a proper value of $f(x,y)$ at the origin.

5. What does it mean to say that a subset in $\mathbb{R}^n$ is "bounded"?

6. Describe when a continuous function defined on a subset in $\mathbb{R}^n$ does not attain the absolute maximum.

[21] Therefore $f(x,y)$ is not continuous at the origin.

3 Directional and Partial Derivatives

3.1 Directional Derivatives

In order to find the rate of change for multi-variable functions, we have to determine the direction in which the variable changes. For functions of a single variable, there is essentially only one direction[22] in which the variable can change, but for functions of several variables , there are infinitely many directions for change. Any nonzero vector determines a **direction**.

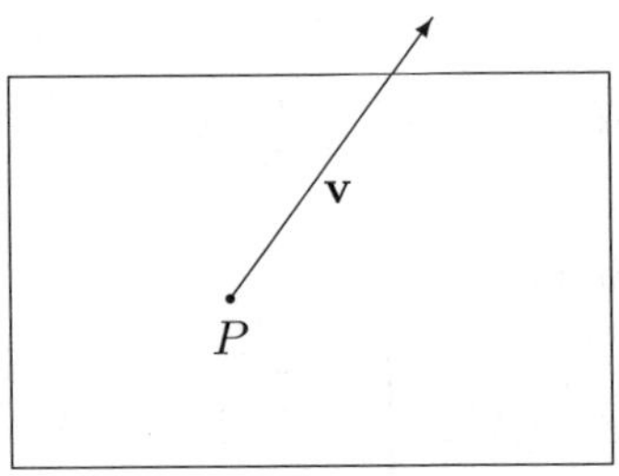

On the other hand, to investigate the change of a function, the function need to be defined not only at one point but also on a **neighborhood** of the point. Therefore when we differentiate a function, we require its domain to be a "set which contains a small neighborhood of every point in it". Such a set is called an **open set**. We will discuss a little more about open set in the last section. An open set is a "set which doesn't contain the boundary."

Now for a function

$$f : U \to \mathbb{R}$$

defined on an open set U in $\mathbb{R}^n$ and for a point P in U and a vector[23] $\mathbf{v}$, if the limit

$$D_{\mathbf{v}} f(P) := \lim_{t \to 0} \frac{f(P + t\mathbf{v}) - f(P)}{t} = \left. \frac{d}{dt} \right|_0 f(P + t\mathbf{v})$$

---

[22] There is only one direction in which the variable increases.

[23] Here a vector means a vector with initial point P. Such a vector is often called a **tangent vector** at the point P. This vector deduces the way to differentiate a function with respect to that direction. In modern **differential geometry** a vector is identified with a "differential operator." Likewise, many concepts are important on their own, but they can be defined by the relations with other concepts, for example, matrices as linear transformations, basis of spaces as coordinate systems.

exists, then it is called the **directional derivative** of f at P in the direction of $\mathbf{v}$ or directional **instantaneous rate of change** in the direction of $\mathbf{v}$.[24] For a unit vector $\mathbf{v}$, geometrically, the directional derivative in the direction of $\mathbf{v}$ means the **slope** of the graph of the function f at the point $(P, f(P))$ when we restrict the graph of f along the line $P + t\mathbf{v}$.

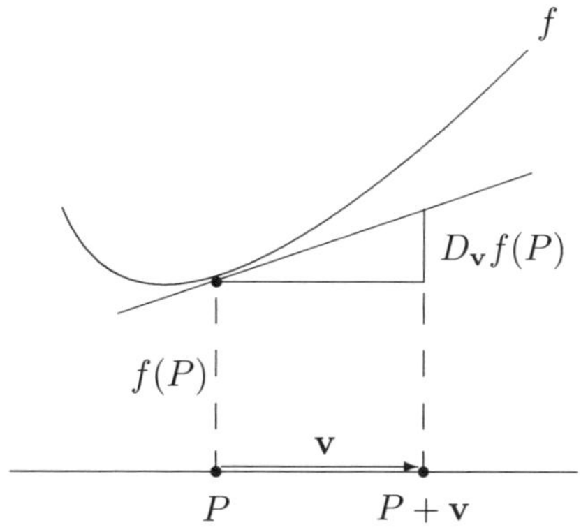

3.1.1 Example

For a function $f(x, y) = x^2 + y^2$, the directional derivative at $P = (0, 1)$ in the direction of $(2, 3)$ is

$$\left.\frac{d}{dt}\right|_0 f\big((0,1)+t(2,3)\big) = \left.\frac{d}{dt}\right|_0 f(2t, 1+3t) = \left.\frac{d}{dt}\right|_0 \big((2t)^2 + (1+3t)^2\big) = 6.$$

3.1.2 Basic Exercise

Suppose that for two functions f and g defined on an open set U in $\mathbb{R}^n$, there exist directional derivatives at the point $P \in U$ in the direction of $\mathbf{v}$. Show that[25]

$$\begin{aligned} D_{c\mathbf{v}}f(P) = cD_{\mathbf{v}}f(P) = D_{\mathbf{v}}(cf)(P), \quad \forall c \in \mathbb{R} \\ D_{\mathbf{v}}(f+g)(P) = D_{\mathbf{v}}f(P) + D_{\mathbf{v}}g(P) \\ D_{\mathbf{v}}(fg)(P) = D_{\mathbf{v}}f(P) \cdot g(P) + f(P) \cdot D_{\mathbf{v}}g(P). \end{aligned}$$

[24] In fact, it would be appropriate to call a unit vector a **direction**, but we will use the term 'direction' for 'direction with magnitude'. Some books use the term "directional derivatives" only for unit vectors.

[25] For example, the first equation means that there exists the " $c\mathbf{v}$-directional derivative of f, and it is c times $\mathbf{v}$-directional derivative". In short, we can write

$$D_{c\mathbf{v}}f = cD_{\mathbf{v}}f = D_{\mathbf{v}}(cf), \quad D_{\mathbf{v}}(f+g) = D_{\mathbf{v}}f + D_{\mathbf{v}}g, \quad D_{\mathbf{v}}(fg) = (D_{\mathbf{v}}f) \cdot g + f \cdot (D_{\mathbf{v}}g).$$

3.2 Partial Derivatives

There are n special directions in $\mathbb{R}^n$, that is, n unit vectors pointing the direction of the axes. They form standard basis vectors

$$\mathbf{e}_1 := (1, 0, \ldots, 0), \quad \ldots \quad , \quad \mathbf{e}_n := (0, 0, \ldots, 1).$$

From now on, at a point $P = (p_1, \ldots, p_n)$, the directional derivative $D_{\mathbf{e}_k} f(P)$ of a function $f : U \to \mathbb{R}$ in the direction of $\mathbf{e}_k$ will be written simply as

$$D_k f(P),$$

and called the kth **partial derivative** of f at P. Using coordinates, it is expressed as

$$D_k f(P) = \left.\frac{d}{dt}\right|_0 f(p_1, \ldots, p_k + t, \ldots, p_n)$$

and it is the rate of change when we let the kth component of $P = (p_1, \ldots, p_n)$ vary. Thus we may denote [26]

$$\frac{\partial f}{\partial x_k} := D_k f.$$

The kth partial derivative $\dfrac{\partial f}{\partial x_k}$ is a derivative obtained when we consider only x_k as the variable and all other components as constants.

3.2.1 Example

At $P = (1, 2, 3)$, the partial derivatives of $f(x, y, z) = x^2y^3 + z$ are[27]

$$\begin{aligned} D_1 f(P) &= \frac{\partial f}{\partial x}(P) = 2xy^3(P) = 16, \\ D_2 f(P) &= \frac{\partial f}{\partial y}(P) = 3x^2y^2(P) = 12, \\ D_3 f(P) &= \frac{\partial f}{\partial z}(P) = 1. \end{aligned}$$

[26] We use the notation $\dfrac{df}{dx}$ when the function f changes depending on the variable x. However when the function f depends not only the variable x but also some other variables, we usually use the "round d" ∂ instead of d and write $\dfrac{\partial f}{\partial x}$. For example, to compute $\dfrac{\partial f}{\partial x}$ for a two-variable function $f(x, y)$, we have to fix the variable y. In other words, the notation $\dfrac{\partial}{\partial x}$ depends not only x but also the "coordinate system" (x, y).

[27] Usually, we denote the orthogonal projections onto the x-, y-, z-axis in $\mathbb{R}^3$ by

$$x, y, z : \mathbb{R}^3 \to \mathbb{R}$$

respectively. Refer Chapter 4 Section 4.

If a function f has the kth partial derivative at every point in the domain U, we obtain the function

$$D_k f : U \to \mathbb{R}, \qquad P \mapsto D_k f(P),$$

and we call it the kth **partial derivative** of f.

3.2.2 Basic Exercise

Find all partial derivatives of the function.

1. $f(x,y,z) = \sqrt{x^2+y^2+z^2}$
2. $f(x,y,z) = xe^{-yz}$
3. $f(x,y,z) = x^{yz}$
4. $f(x,y,z) = \sin(xyz)$

3.2.3 Example

By the **fundamental theorem of calculus**, for a continuous function $g(t)$,

$$\left.\frac{d}{dx}\right|_{x_0} \int_a^x g(t)\,dt = g(x_0), \qquad \left.\frac{d}{dy}\right|_{y_0} \int_y^b g(t)\,dt = -g(y_0).$$

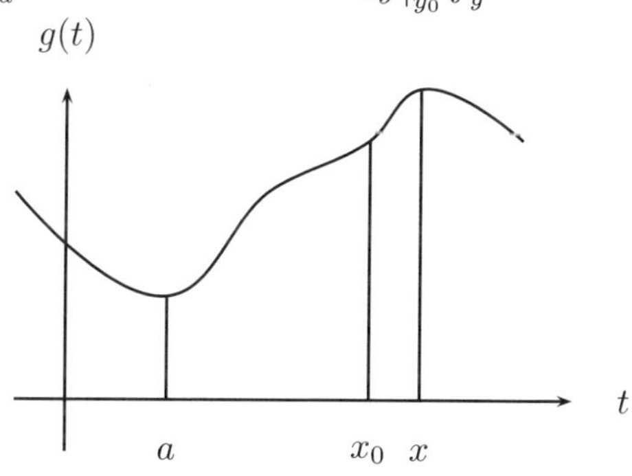

If we let

$$f(x,y) := \int_y^x e^{-t^2}\,dt$$

be the area on the interval $[y,x]$ for the "normal distribution", we get

$$\frac{\partial f}{\partial x} = e^{-x^2}, \qquad \frac{\partial f}{\partial y} = -e^{-y^2}.$$

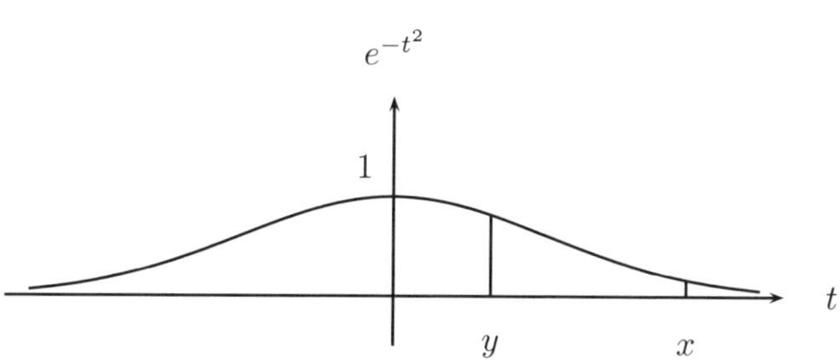

3.3 The Gradient Vector

When a function f of n variables has all partial derivatives at a point P, f is called **partially differentiable** at P. And the vector

$$\operatorname{grad} f(P) := (D_1 f(P), \ldots, D_n f(P))$$

is called the **gradient vector** of f at P.[28]

As the slope of the tangent line for a single-variable function $f(x)$ is $f'(x)$, the slope of the "tangent plane" for a multi-variable function $f(X)$ is nothing but $\operatorname{grad} f(P)$.

3.3.1 Example

At the origin $O = (0, 0, 0)$, find the gradient vector of

$$f(x, y, z) = \exp(x + yz).$$

$$\begin{aligned} D_1 f(x, y, z) &= f(x, y, z), \\ D_2 f(x, y, z) &= z f(x, y, z), \\ D_3 f(x, y, z) &= y f(x, y, z), \end{aligned}$$

and thus $D_1 f(O) = 1$, $D_2 f(O) = 0$, $D_3 f(O) = 0$. Therefore

$$\operatorname{grad} f(O) = (1, 0, 0).$$

3.3.2 Example

For a point $A = (a_1, \ldots, a_n)$ in $\mathbb{R}^n$, find the gradient vector of the function

$$f(X) = |X - A|^2 = (x_1 - a_1)^2 + \cdots + (x_n - a_n)^2 \qquad (X \in \mathbb{R}^n).$$

Since

$$\frac{\partial f}{\partial x_1} = 2(x_1 - a_1), \ \ldots \ , \frac{\partial f}{\partial x_n} = 2(x_n - a_n),$$

we get

$$\operatorname{grad} f(X) = 2(x_1 - a_1, \ldots, x_n - a_n) = 2(X - A).$$

We will soon see the important properties of gradient vector.

---

[28] Just as the "slope" of the "graph" of a single-variable function $y = ax + b$ is a, the "gradient vector" in the "graph" of a multi-variable function $y = a_1 x_1 + \cdots + a_n x_n + b$ is $(a_1, \ldots, a_n)$. $\operatorname{grad} f$ is often written as ∇f. That is,

$$\nabla = (D_1, \ldots, D_n) = \left(\frac{\partial}{\partial x_1}, \ldots, \frac{\partial}{\partial x_n} \right).$$

We read the notation ∇ "del" or "nabla". $\operatorname{grad} f(P)$ always means $(\operatorname{grad} f)(P)$.

Exercise: Chapter 10 Section 3

1. For $f(x, y, z) = x^2y^3 + z$, find $\operatorname{grad} f(1, 2, 3)$.

2. Find the directional derivative of the following function at a point P in the direction of $\mathbf{v}$.

   (1) $f(x, y) = x^2 + y^3$, $P = (-1, 3)$, $\mathbf{v} = (1, 2)$

   (2) $f(x, y) = 2x^2 + xy - y^2$, $P = (3, -2)$, $\mathbf{v} = (1, -1)$

   (3) $f(x, y) = e^x \log y$, $P = (0, e)$, $\mathbf{v} = (2, 1)$

   (4) $f(x, y) = e^x \sin y$, $P = (0, \pi/4)$, $\mathbf{v} = (2, 1)$

   (5) $f(x, y) = \log \sqrt{x^2 + y^2}$, $P = (1, 1)$, $\mathbf{v} = (2, 1)$

3. For **partially differentiable functions** f and g defined on an open set U in $\mathbb{R}^n$, show that

$$\begin{aligned}\operatorname{grad}(f + g) &= \operatorname{grad} f + \operatorname{grad} g \\ \operatorname{grad}(cf) &= c \operatorname{grad} f, \qquad c \in \mathbb{R} \\ \operatorname{grad}(fg) &= f \operatorname{grad} g + g \operatorname{grad} f.\end{aligned}$$

   Also show that if $h : \mathbb{R} \to \mathbb{R}$ is differentiable, then for any $P \in U$,

$$\operatorname{grad}(h \circ f)(P) = h'(f(P)) \operatorname{grad} f(P).$$

4. For the identity map $X = (x_1, \ldots, x_n) : \mathbb{R}^n \to \mathbb{R}^n$ and

$$|X| := \sqrt{x_1^2 + \cdots + x_n^2},$$

   show that

$$\operatorname{grad} |X| = \frac{X}{|X|}.$$

   Also for any given point P in $\mathbb{R}^n$, show that

$$\operatorname{grad} |X - P| = \frac{X - P}{|X - P|}.$$

5. In $\mathbb{R}^n$, show that[29]

$$\operatorname{grad} \frac{1}{|X|} = -\frac{X}{|X|^3}.$$

   Also for any real number k, show that

$$\operatorname{grad} \frac{1}{|X|^k} = -\frac{kX}{|X|^{k+2}}.$$

[29] Like the gravitational force is in inverse proportion to the square of the distance.

6. When we let $r = |X|$ in $\mathbb{R}^n$, fine the gradient vector of the following function at a point X.

   (1) $f(X) = \log r$ (2) $f(X) = e^{-r^2}$ (3) $f(X) = re^{-r}$

7. Explain that for the function $f(x, y) = xy^2$,
$$\frac{\partial f}{\partial x}(1,1) = D_1 f(1,1) = 1$$
   but for the function $g(y, x) = yx^2$,
$$\frac{\partial g}{\partial x}(1,1) = D_2 g(1,1) = 2.$$
   Also check that $f = g$.

8. In polar coordinates (r, θ), show that from $r = \sqrt{x^2 + y^2}$,
$$\frac{\partial r}{\partial x} = \frac{x}{r} = \cos\theta,$$
   and from $x = r\cos\theta$,
$$\frac{\partial x}{\partial r} = \cos\theta.$$
   Use those to show that[30]
$$\frac{\partial r}{\partial x} \cdot \frac{\partial x}{\partial r} \neq 1.$$

9. Doctors (in particular, pediatricians), often need to know the body surface area of patients. For example, the body surface area is used to analyze the result of a physical exam for a certain function of kidney. According to statistic data, for a person with weight W kg and height H cm, the body surface area A m^2 is approximated by
$$A = 0.007\, W^{0.425} H^{0.725}.$$
   When $W = 54$ and $H = 165$, find $\dfrac{\partial A}{\partial W}$ and $\dfrac{\partial A}{\partial H}$, and explain their meaning.

   (Here, $54^{0.425} \approx 5.4$, $54^{-0.575} \approx 0.10$, $165^{0.725} \approx 40.5$, $165^{-0.275} \approx 0.25$.)

[30]In order to show that two functions are not equal, it suffices to show that there is a point in the domain at which their values are not equal.

4 Differentiable Functions

In this section, we will introduce the concept of "differentiability" of a function f defined on an open set in $\mathbb{R}^n$. In the previous section, we have defined directional derivatives of a function. You might love to think that if a function f has directional derivatives of all directions at a point P then f must be differentiable at P. But it is NOT the way how we define differentiability.

In Advanced Exercise in this chapter, we introduce a function such that its directional derivatives exist in all directions but it is not continuous. Once we properly define "differentiable function" later on, we will see that a differentiable function is always continuous.[31]

4.1 Differentiablity

First, recall what "a function $f(x)$ being differentiable at a point p" means when $n = 1$. Geometrically it means that the graph of the function f has the **tangent line** at the point $(p, f(p))$.

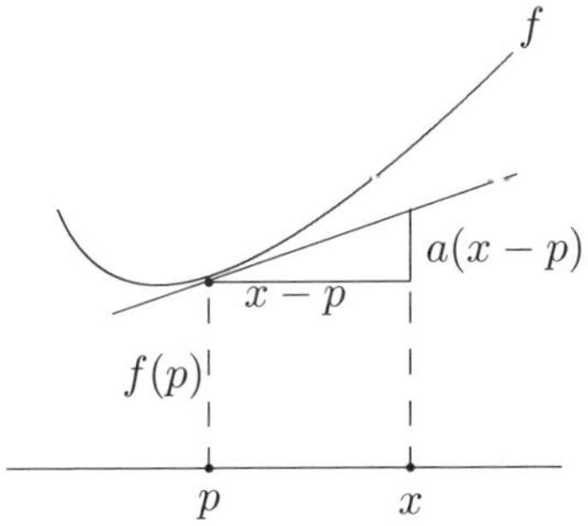

The equation of the straight line through $(p, f(p))$ is given by[32]

$$y = f(p) + a(x - p). \tag{10.3}$$

The line will be **tangent** to the curve $y = f(x)$ when $x = p$ if the difference between $f(x)$ and $f(p) + a(x - p)$ approaches 0 much faster than x approaches p, that is,[33]

$$\lim_{x \to p} \frac{f(x) - f(p) - a(x - p)}{x - p} = 0.$$

[31] As a definition makes a property, a property makes a definition.

[32] When we consider the differentiability, we exclude the straight line parallel to the y-axis, that is, the line whose slope is the infinity. The function with the instantaneous rate of change ∞ should be treated carefully.

[33] If we let $x = p + v$,

$$\lim_{v \to 0} \frac{1}{v} (f(p + v) - f(p) - av) = 0,$$

that is, $f(p + v) - f(p) = av + o(v)$.

Or we may write

$$f'(p) := \lim_{x \to p} \frac{f(x) - f(p)}{x - p} = a.$$

That is, the slope a of the tangent line is $f'(p)$.

Likewise, in $\mathbb{R}^n$ we will say that a function $y = f(X)$ is **differentiable** at a point P if the graph of f has the **tangent plane** at the point $(P, f(P)) \in \mathbb{R}^n \times \mathbb{R} = \mathbb{R}^{n+1}$. The equation of the plane[34] through $(P, f(P))$ is of the form

$$y = f(P) + \mathbf{a} \cdot (X - P).$$

Now we can say that this plane is tangent to $y = f(X)$ at the point $(P, f(P))$ if

$$\lim_{X \to P} \frac{f(X) - f(P) - \mathbf{a} \cdot (X - P)}{|X - P|} = 0.$$

Here if we let $X = P + \mathbf{v}$, we have

$$\lim_{\mathbf{v} \to \mathbf{0}} \frac{|f(P + \mathbf{v}) - f(P) - \mathbf{a} \cdot \mathbf{v}|}{|\mathbf{v}|} = 0. \tag{10.4}$$

This equation may be written simply as

$$f(P + \mathbf{v}) - f(P) - \mathbf{a} \cdot \mathbf{v} = o(\mathbf{v})$$

or[35]

$$f(P + \mathbf{v}) = f(P) + \mathbf{a} \cdot \mathbf{v} + o(\mathbf{v}). \tag{10.5}$$

We may summarize what we said above as follows.

---

[34]Here the plane means the plane of higher dimension— i.e., **hyperplane**. In general, the equation of the hyperplane through a point $(P, y_0) \in \mathbb{R}^n \times \mathbb{R}$ (with finite slope) is of the form

$$y = y_0 + \mathbf{a} \cdot (X - P).$$

Here the vector $\mathbf{a} = (a_1, \ldots, a_n)$ denotes the **slope** of the hyperplane.

[35]For two functions $g(\mathbf{v}), h(\mathbf{v})$ defined near the origin in $\mathbb{R}^n$, $g(\mathbf{v}) = o(h(\mathbf{v}))$ means the following: For any given $\epsilon > 0$, if $\mathbf{v}$ is very close to the origin, then

$$|g(\mathbf{v})| \le \epsilon |h(\mathbf{v})|.$$

The definition of this notation "$g(\mathbf{v}) = o(h(\mathbf{v}))$" can be adapted to "vector-valued functions" g and h, and $g = o(h)$ is equivalent to $|g| = o(|h|)$. For instance, for the identity map $\mathbf{v} : \mathbb{R}^n \to \mathbb{R}^n$, $g(\mathbf{v}) = o(\mathbf{v})$ means that for any given $\epsilon > 0$,

$$|\mathbf{v}| \ll 1 \quad \Rightarrow \quad |g(\mathbf{v})| \le \epsilon |\mathbf{v}|.$$

4.1.1 Definition

A function $f : U \to \mathbb{R}$ defined in an open set U in $\mathbb{R}^n$ is called **differentiable** at a point $P \in U$ if there exists a vector $\mathbf{a}$ which satisfies the equation (10.4) at P. If f is differentiable at every point in U, then f is called **differentiable** in U.

Theorem 4.1.2 *If a function $f : U \to \mathbb{R}$ defined on an open set U in $\mathbb{R}^n$ is differentiable at a point $P \in U$, then*

(i) *f is continuous at P.*

(ii) *f has the directional derivatives in all directions at P and*

$$D_{\mathbf{v}} f(P) = \operatorname{grad} f(P) \cdot \mathbf{v} \qquad (\mathbf{v} \in \mathbb{R}^n).$$

(iii) *The vector* $\mathbf{a}$ *which satisfies the equation* (10.4) *is* $\operatorname{grad} f(P)$.

Proof. (i) From the equation (10.5), we get

$$\lim_{\mathbf{v}\to\mathbf{0}} f(P+\mathbf{v}) = \lim_{\mathbf{v}\to 0} \big(f(P) + \mathbf{a}\cdot\mathbf{v} + o(\mathbf{v})\big) = f(P).$$

(ii), (iii) According to the equation (10.4), for a unit vector $\mathbf{v}$,

$$0 = \lim_{|t\mathbf{v}|\to 0} \frac{f(P+t\mathbf{v}) - f(P) - \mathbf{a}\cdot t\mathbf{v}}{|t\mathbf{v}|},$$

that is,

$$0 = \lim_{t\to 0} \frac{f(P+t\mathbf{v}) - f(P) - \mathbf{a}\cdot t\mathbf{v}}{t} = \left(\lim_{t\to 0} \frac{f(P+t\mathbf{v}) - f(P)}{t}\right) - \mathbf{a}\cdot\mathbf{v}.$$

Therefore the directional derivative $D_{\mathbf{v}} f(P)$ exists and

$$D_{\mathbf{v}} f(P) = \lim_{t\to 0} \frac{f(P+t\mathbf{v}) - f(P)}{t} = \mathbf{a}\cdot\mathbf{v}.$$

If we let $\mathbf{a} = (a_1, \ldots, a_n)$, then

$$D_i f(P) = D_{\mathbf{e}_i} f(P) = \mathbf{a}\cdot\mathbf{e}_i = a_i$$

and thus

$$\operatorname{grad} f(P) = \big(D_1 f(P), \ldots, D_n(P)\big) = (a_1, \ldots, a_n) = \mathbf{a}.$$

Therefore

$$D_{\mathbf{v}} f(P) = \mathbf{a}\cdot\mathbf{v} = \operatorname{grad} f(P)\cdot\mathbf{v}.$$

□

It follows from the above theorem that f is differentiable at P if and only if

(i) $\mathrm{grad} f(P)$ exists and

(ii) $f(P+\mathbf{v}) - f(P) \equiv \mathrm{grad} f(P) \cdot \mathbf{v} + o(\mathbf{v})$.

Thus the gradient vector grad $f(P)$ can be called the "**derivative** of f at a point P", and written as $f'(P)$ or $Df(P)$:

$$\mathrm{grad}\, f(P) = f'(P) = Df(P).$$

The derivative of a function of several variables is not a real number any longer but a vector.[36]

For a differentiable function, once we know the gradient vector (hence the slope of the function in the direction of each axis), it is very easy to find any directional derivative, because[37]

$$D_{\mathbf{v}} f(P) = \mathrm{grad} f(P) \cdot \mathbf{v}. \tag{10.6}$$

From this equation, we know that

$$\boxed{D_{\mathbf{v}+\mathbf{w}} f(P) = D_{\mathbf{v}} f(P) + D_{\mathbf{w}} f(P)}$$

4.1.3 Basic Exercise

1. Suppose that at the origin a function $f(x, y)$ is differentiable at the origin and its directional derivatives in the direction of $(1,1)$ and $(1,2)$ are 1 and 2 respectively. Find the its directional derivative in the direction of $(2,3)$.[38]

2. For a vector $\mathbf{a} \in \mathbb{R}^n$, show that the **linear function $\mathbf{a}^*$** defined by

$$\mathbf{a}^*(\mathbf{x}) := \mathbf{a} \cdot \mathbf{x} \qquad (\mathbf{x} \in \mathbb{R}^n)$$

is differentiable. Also show that

$$\mathrm{grad}\ \mathbf{a}^*(\mathbf{x}) = \mathbf{a}, \qquad (\mathbf{x} \in \mathbb{R}^n).$$

---

[36] In fact, it would be better to consider the derivative of a function of several variables grad $f(P)$ as a linear map

$$\mathbf{v} \mapsto \mathrm{grad}\, f(P) \cdot \mathbf{v}$$

rather than a vector.

[37] Using coordinates, this equation is written as $D_{\mathbf{v}} f = v_1 \dfrac{\partial f}{\partial x_1} + \cdots + v_n \dfrac{\partial f}{\partial x_n}$ or simply

$$D_{\mathbf{v}} = v_1 \frac{\partial}{\partial x_1} + \cdots + v_n \frac{\partial}{\partial x_n}.$$

[38] Once you know $(2,3) = (1,1) + (1,2)$, it is obvious that the answer is 3.

3. Show that the set of all differentiable functions defined on an open set in $\mathbb{R}^n$ is closed under the addition and multiplication of functions[39]

$$f + g, \qquad f \cdot g.$$

4. Use two exercises above to show that every **polynomial function** is differentiable.

4.2 Gradient Vectors and the Rate of Change of a Function

In the equation (10.6), if we let $\mathbf{v}$ be a unit vector and θ be the angle between $\mathbf{v}$ and grad$f(P)$, we get

$$D_{\mathbf{v}} f(P) = \operatorname{grad} f(P) \cdot \mathbf{v} = |\operatorname{grad} f(P)| \cos\theta.$$

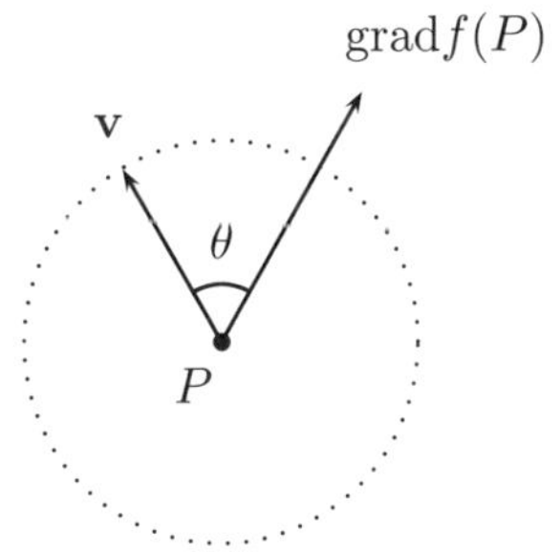

So the unit vector $\mathbf{v}$ which maximizes $D_{\mathbf{v}} f(P)$ is the vector with $\cos\theta = 1$ — that is, the unit vector which has the same direction as grad$f(P)$, and the magnitude of grad $f(P)$ is the rate of change in the direction of $\mathbf{v}$.[40]

Also the unit vector $\mathbf{v}$ which minimizes $D_{\mathbf{v}} f(P)$ is the vector with $\cos\theta = -1$ — that is, the unit vector which has the same direction as $-$grad$f(P)$.

In the same reason, the instantaneous rate of change in the direction perpendicular to grad$f(P)$ is 0.

---

[39] We regard a real numbers as a **constant function**. Therefore the scalar multiple of a differentiable function is still differentiable.

[40] From this principle, we know that

$$\operatorname{grad} f(P) = \left(\frac{\partial f}{\partial x_1}(P), \ldots, \frac{\partial f}{\partial x_n}(P) \right)$$

is independent of the choice of coordinate system.

4.2.1 Example

Suppose that the temperature distribution in plane is given by[41]

$$f(x,y) = 273\,e^{-2x^2-3y^2} \qquad (^\circ\mathrm{K}).$$

Find the direction in which the temperature increases most rapidly at $(0,1)$? How fast will the temperature increase in that direction? (In this problem, a "direction" means a "unit vector".)

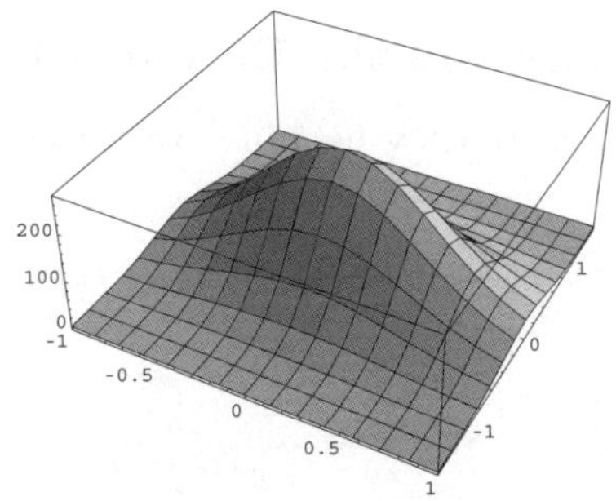

Solution. The gradient vector of the function f is

$$\operatorname{grad} f(x,y) = 273\,e^{-2x^2-3y^2}(-4x,-6y),$$

and thus

$$\operatorname{grad} f(0,1) = 273\,e^{-3}(0,-6).$$

Therefore, at the given point $(0,1)$, the direction in which the temperature increases fastest is the "warm south" —i.e. the $(0,-1)$-direction and the rate of change in temperature is

$$|\operatorname{grad} f(0,1)| = 273\,e^{-3}\cdot 6 \approx 81.6.$$

4.2.2 Basic Exercise

When a differentiable function f defined near a point P in $\mathbb{R}^n$ has the constant directional derivative $D_{\mathbf{v}}f(P)$ for any unit vector $\mathbf{v}$, find the value.

4.3 Continuously Differentiable Functions

We will introduce a theorem which gives an easy way to determine the differentiability of functions. First let's introduce a terminology.

[41]Kelvin is the unit for absolute temperature. If you want to know about Sir Kelvin, look up William Thomson in Chapter 18.

4.3.1 Definition

For a function $f(x_1, \ldots, x_n)$ defined on an open set in $\mathbb{R}^n$, if its partial derivatives

$$\frac{\partial f}{\partial x_1}, \quad \ldots, \quad \frac{\partial f}{\partial x_n}$$

are all continuous, then f is called a **continuously differentiable function** or $\mathcal{C}^1$ **function**.

Theorem 4.3.2 *A continuously differentiable function defined on an open set in $\mathbb{R}^n$ is differentiable.*

The proof of this theorem is not difficult, but we will leave its proof in the last section.[42]

Given a function, it is often quite hard to determine whether the function satisfies the definition of differentiability. The theorem may give an easier way to check the differentiability of the given function.

4.3.3 Example

The function $e^{xy} \sin z + \log(1 + x^2 + y^2)$ is $\mathcal{C}^1$ and hence differentiable.

42

The one who loves practice without theory is like the sailor who boards ship without a rudder and compass and never knows where he may cast.

Leonardo da Vinci (1452–1519)

Exercise: Chapter 10 Section 4

1. Suppose that the temperature distribution in the plane is given by the function

$$f(x,y) = x^2y + xy^2 \qquad (°\mathrm{C})$$

and Kildong is currently at $(2,3)$.

   (1) In what direction should he move in order to get cool fastest? And how fast will the temperature decrease?

   (2) In what direction does the temperature increase fastest? And how fast will the temperature increase?

   (3) Find the direction in which the instantaneous rate of change in temperature is 0.

2. Suppose that for a differentiable function $f(x,y)$ in $\mathbb{R}^2$, the directional derivative at the origin in the direction $(1,2)$ is 1 and the directional derivative at the origin in the direction $(2,3)$ is 2. Find the directional derivative of f in the direction $(3,5)$ at the origin. Also find the gradient vector $\operatorname{grad} f(O)$.

3. Is a **homogeneous** polynomial function

$$a_0x^n + a_1x^{n-1}y + \cdots + a_{n-1}xy^{n-1} + a_ny^n$$

continuously differentiable? Also is it differentiable? Is any polynomial function is continuously differentiable? Is any **rational function** given by a ratio of two polynomials is continuously differentiable (in its domain)?

4. Explain that if differentiable functions $f(x,y)$, $g(x,y)$ defined on a rectangle $(a,b)\times(c,d)\subset\mathbb{R}^2$ satisfy

$$\frac{\partial f}{\partial x} = \frac{\partial g}{\partial x},$$

there is a differentiable function $h:(c,d)\to\mathbb{R}$ which satisfies

$$f(x,y) = g(x,y) + h(y).$$

5. Suppose that for a function $f(x,y)$ defined near the origin in the plane, $D_1f(O) = 1$ and $D_2f(O) = -2$. Find the direction in which the values of the function hardly change.

5 The Chain Rule

The **chain rule** gives the rule for differentiating a composite function. For two differentiable single-variable functions $f : (a, b) \to (c, d)$ and $g : (c, d) \to \mathbb{R}$, the derivative of their composite

$$g \circ f : (a, b) \to \mathbb{R}$$

is

$$\boxed{(g \circ f)'(x) = g'(f(x)) \cdot f'(x)}$$

That is, the derivative of the composite function $g \circ f$ is the product of the derivatives of g and f.

5.0.1 Basic Exercise

Find the derivative $f'(x)$ for the function

$$f(x) := \int_0^{x^2} \sin e^t \, dt.$$

Also for a continuous function $h(t)$ and a differentiable function $g(x)$, show that

$$\frac{d}{dx} \int_0^{g(x)} h(t) \, dt = h(g(x)) g'(x).$$

The chain rule for functions of a single variable is extended to the chain rule for functions of several variables similarly.

Theorem 5.0.2 (Chain Rule) *For a differentiable function f defined on an open set U in $\mathbb{R}^n$ and a differentiable curve $X : I \to U$ defined on an interval I, $f \circ X$ is differentiable and*

$$\frac{d}{dt}(f \circ X)(t) = \operatorname{grad} f(X(t)) \cdot X'(t), \quad \forall t \in I.$$

Proof. Fix $t \in I$. For a real number Δt with very small absolute value,

$$\begin{aligned}
& f\big(X(t + \Delta t)\big) - f\big(X(t)\big) \\
= \; & \operatorname{grad} f\big(X(t)\big) \cdot \big(X(t + \Delta t) - X(t)\big) + o(|X(t + \Delta t) - X(t)|) \\
= \; & \operatorname{grad} f\big(X(t)\big) \cdot \big(X'(t)\Delta t + o(\Delta t)\big) + o(\Delta t) \\
= \; & \operatorname{grad} f\big(X(t)\big) \cdot X'(t)\Delta t + o(\Delta t).
\end{aligned}$$

as we are required to prove.[43] □

In order to find the net rate of change of a function f at a point P in the domain in the direction of $\mathbf{v}$, it suffices to find the net rate of change of f along a **straight line** $X(t)$ which satisfies

$$X(0) = P, \qquad X'(0) = \mathbf{v} :$$

$$D_{\mathbf{v}} f(P) = \left.\frac{d}{dt}\right|_0 f(X(t)). \tag{10.7}$$

The chain rule shows that the equation (10.7) holds when $X(t)$ is not only a straight line but also a curve.

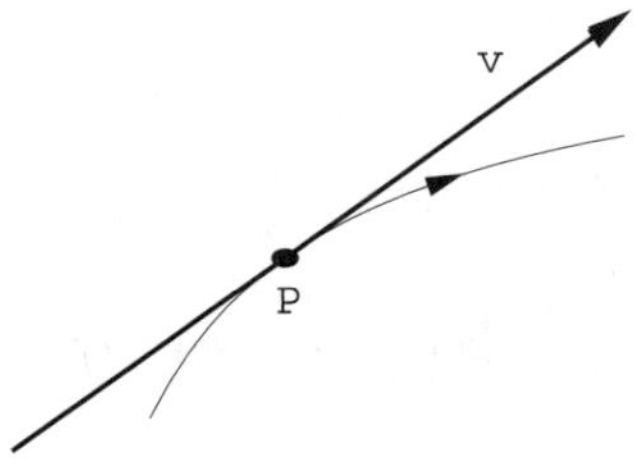

With coordinates, the chain rule can be expressed as[44]

$$\boxed{\frac{d}{dt} f\big(x_1(t), \ldots, x_n(t)\big) = \frac{\partial f}{\partial x_1}\frac{dx_1}{dt} + \cdots + \frac{\partial f}{\partial x_n}\frac{dx_n}{dt}}$$

5.0.3 Example

A function f defined on the plane is usually denoted by $f(x, y)$, using the standard rectangular coordinates (x, y). if we express it as $f(r, \theta)$ using the polar coordinates (r, θ),[45] then by the chain rule,

---

[43] If a function $h(\Delta t)$ is $o(|X(t + \Delta t) - X(t)|)$, then

$$\lim_{\Delta t \to 0} \frac{h(\Delta t)}{\Delta t} = \lim_{\Delta t \to 0} \frac{h(\Delta t)}{|X(t+\Delta t) - X(t)|} \frac{|X(t+\Delta t) - X(t)|}{\Delta t} = 0 \cdot |X'(t)| = 0.$$

(If $X'(t) \neq \mathbf{0}$, $|X(t+\Delta t) - X(t)| \neq 0$ when $0 < |\Delta t| \ll 1$. So the above equation holds. When $X'(t) = \mathbf{0}$, we can also show that $\lim_{\Delta t \to 0} h(\Delta t)/\Delta t = 0$.) Therefore $h(\Delta t) = o(\Delta t)$.

[44] The name of the chain rule is originated from the fact that the formula is linked like chains.

[45] In a strict sense, for a function $f : \mathbb{E}^2 \to \mathbb{R}$ and coordinate systems $(x, y) : \mathbb{E}^2 \to \mathbb{R}^2$ and $(r, \theta) : \mathbb{E}^2 \to \mathbb{R}^2$, we think of them as

$$f(x, y) := f \circ (x, y)^{-1}, \qquad f(r, \theta) := f \circ (r, \theta)^{-1}.$$

$$\frac{\partial f}{\partial r} = \frac{\partial f}{\partial x}\frac{\partial x}{\partial r} + \frac{\partial f}{\partial y}\frac{\partial y}{\partial r} = \cos\theta\,\frac{\partial f}{\partial x} + \sin\theta\,\frac{\partial f}{\partial y},$$

$$\frac{\partial f}{\partial \theta} = \frac{\partial f}{\partial x}\frac{\partial x}{\partial \theta} + \frac{\partial f}{\partial y}\frac{\partial y}{\partial \theta} = -r\sin\theta\,\frac{\partial f}{\partial x} + r\cos\theta\,\frac{\partial f}{\partial y}.$$

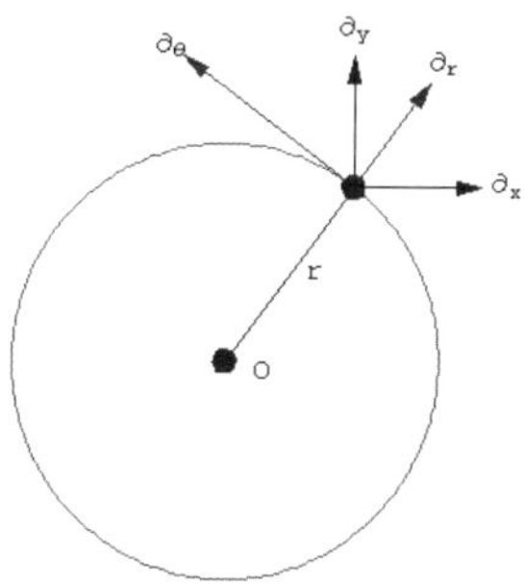

To simplify the notation, if we let

$$\partial_x := \frac{\partial}{\partial x}, \qquad \partial_y := \frac{\partial}{\partial y},$$
$$\partial_r := \frac{\partial}{\partial r}, \qquad \partial_\theta := \frac{\partial}{\partial \theta}$$

and omit f in common, we get[46]

$$\partial_r = \cos\theta\,\partial_x + \sin\theta\,\partial_y$$
$$\partial_\theta = r\,(-\sin\theta\,\partial_x + \cos\theta\,\partial_y)\,.$$

5.0.4 Basic Exercise

Suppose that we have a device to show the thickness of the fog near the point $(x_0, y_0) = (3, 4)$ and $f(x, y)$ is the scale that the needle indicates.

Suppose that we know $D_1 f(x_0, y_0) = -0.1$, $D_2 f(x_0, y_0) = 0.2$. When a person passes through a point (x_0, y_0) while driving a car along a circle

$$x(t) = 5\cos t, \qquad y(t) = 5\sin t,$$

find the net rate of change of f. Also find the net rate of change of f when a person drive a car along a straight line[47]

$$x(t) = tx_0, \qquad y(t) = ty_0.$$

---

[46] As already mentioned, directional derivatives are nothing but vectors. In other words, ∂_x means the vector (field) **i** and ∂_y means the vector (field) **j**. The above equations express the vector (fields) ∂_r and ∂_θ as linear combinations of the vector fields **i** and **j**.

[47] Along a circle, $\frac{d}{dt}\big|_{t_0} f(x(t), y(t)) = r_0(-\sin t_0, \cos t_0) \cdot \left(\frac{\partial f}{\partial x}, \frac{\partial f}{\partial y}\right) = 5(-\frac{4}{5}, \frac{3}{5}) \cdot (-0.1, 0.2) = 5(0.08 + 0.12) = 1$. And the instantaneous rate of change along a straight line is $(3, 4) \cdot (-0.1, 0.2) = -0.3 + 0.8 = 0.5$.

Exercise: Chapter 10 Section 5

1. For the rectangular coordinates (x, y) and polar coordinates (r, θ) in $\mathbb{R}^2$, show that

$$\begin{aligned}\partial_x &= (\cos\theta)\partial_r - \frac{\sin\theta}{r}\partial_\theta \\ \partial_y &= (\sin\theta)\partial_r + \frac{\cos\theta}{r}\partial_\theta.\end{aligned}$$

Also use it to derive

$$|\text{grad } f|^2 = \left(\frac{\partial f}{\partial r}\right)^2 + \frac{1}{r^2}\left(\frac{\partial f}{\partial \theta}\right)^2.$$

2. (**Euler's formula**) For a given natural number k, show that if a function $f(x_1, \ldots, x_n)$ satisfies[48]

$$f(tx_1, \ldots, tx_n) = t^k f(x_1, \ldots, x_n), \qquad \forall t, x_i \in \mathbb{R},$$

then

$$x_1 D_1 f(x_1, \ldots, x_n) + \cdots + x_n D_n f(x_1, \ldots, x_n) = k f(x_1, \ldots, x_n).$$

3. Show that if a differentiable function $f(\mathbf{x})$ satisfies $f(t\mathbf{x}) = tf(\mathbf{x})$ for any real number t and any $\mathbf{x} \in \mathbb{R}^n$ then there exists a vector $\mathbf{a}$ such that

$$f(\mathbf{x}) = \mathbf{a} \cdot \mathbf{x}.$$

4. For a differentiable function f, when we let $z = f\left(\frac{x-y}{y}\right)$, show that

$$x\frac{\partial z}{\partial x} + y\frac{\partial z}{\partial y} = 0.$$

5. For a function

$$f(x, y) := \int_{xy}^{x^2+y^2} e^{-(t-1)^2}\, dt,$$

find the partial derivatives $\frac{\partial f}{\partial x}(1, 0)$ and $\frac{\partial f}{\partial y}(0, 1)$.

---

[48] One example of such a function is homogeneous polynomial

$$f(x_1, \ldots, x_n) = a_1 x_1^k + \cdots + a_n x_n^k.$$

6. For a continuous function $g(t)$ and $\mathcal{C}^1$ functions $a(x, y)$, $b(x, y)$, find the partial derivatives $D_1 f(x, y)$ and $D_2 f(x, y)$ of

$$f(x, y) := \int_{a(x,y)}^{b(x,y)} g(t)\, dt.$$

7. When y is a function of a variable x and it satisfies the equation

$$f(x, y) = 0,$$

show that

$$D_1 f(x, y) + D_2 f(x, y) \frac{dy}{dx} = 0.$$

(In this problem, we assume that functions y and f are both differentiable.)

8. When $x^2 + y^2 = 1$ and $y \neq 0$, show that $dy/dx = -x/y$.

9. When $x^3 + x^2 y + xy^2 + y^3 = 1$, find $\frac{dy}{dx}(1)$.

10. Let (x, y) be the (longitude, latitude)-coordinates near Dokdo. Suppose that on some day, the seawater temperature nearby is given by $f(x, y)$ (°C), and at the point (x_0, y_0) the net rate of change of temperature is $+1$ (°C/rad) to the east and -1 (°C/rad) to the north. When a ship is sailing near Dokdo along the curve

$$x(t) = x_0 + 2(t - t_0), \qquad y(t) = y_0 + 3\sin(t - t_0),$$

find the net rate of change of temperature when the ship passes through the point (x_0, y_0)

11. Find the partial derivatives of the function with respect to x and y.

(1) $f(u, v) = (u - v)e^u, \quad u = xy, \quad v = x^2 - y^2$

(2) $f(u, v) = u \log v + v \log u, \quad u = \dfrac{y}{2} + \dfrac{2}{y}, \quad v = xe^y$

(3) $f(u, v) = u \log v, \quad u = x \sin y + y \sin x, \quad v = x \cos y + y \cos x$

(4) $f(u, v) = (u + v)/(1 - uv), \quad u = \sin 2x, \quad v = \cos(3x - y)$

12. For $w = f(x - y, y - x)$, show that $\dfrac{\partial w}{\partial x} + \dfrac{\partial w}{\partial y} = 0$.

13. For $w = x^3 f(y/x, z/x)$, show that $x\dfrac{\partial w}{\partial x} + y\dfrac{\partial w}{\partial y} + z\dfrac{\partial w}{\partial z} = 3w$.

14. For a function $g(x, y) = f(x + y, x - y)$, if we let $x + y = u$ and $x - y = v$, show that

$$\frac{\partial g}{\partial x} \cdot \frac{\partial g}{\partial y} = \left(\frac{\partial f}{\partial u}\right)^2 - \left(\frac{\partial f}{\partial v}\right)^2.$$

15. For a constant θ, let $x = u\cos\theta - v\sin\theta$ and $y = u\sin\theta + v\cos\theta$. For $f(x,y) = g(u,v)$, show that

$$\left(\frac{\partial g}{\partial u}\right)^2 + \left(\frac{\partial g}{\partial v}\right)^2 = \left(\frac{\partial f}{\partial x}\right)^2 + \left(\frac{\partial f}{\partial y}\right)^2.$$

16. Suppose that a differentiable function $g(r)$ of a single variable has the derivative which is never zero. Show that for a function

$$f(X) := g(|X|) \qquad (X \in \mathbb{R}^3 - \{O\})$$

the gradient $\operatorname{grad} f(X)$ is parallel to X.

17. If the gradient of a differentiable function $f(X)$ is parallel to X, there exists a function $g(x)$ such that

$$\operatorname{grad} f(X) = g(X)X.$$

Then show that $f(X)$ is constant on any sphere with the center at the origin.

6 Gradient Vectors and Level Surfaces

For a $\mathcal{C}^1$ function f, the geometrical property of the gradient vector $\operatorname{grad} f$ is the following:

> The gradient vector at a point P, $\operatorname{grad} f(P)$ is normal to the level surface $S := \{X \mid f(X) = f(P)\}$ containing the point P.[49]

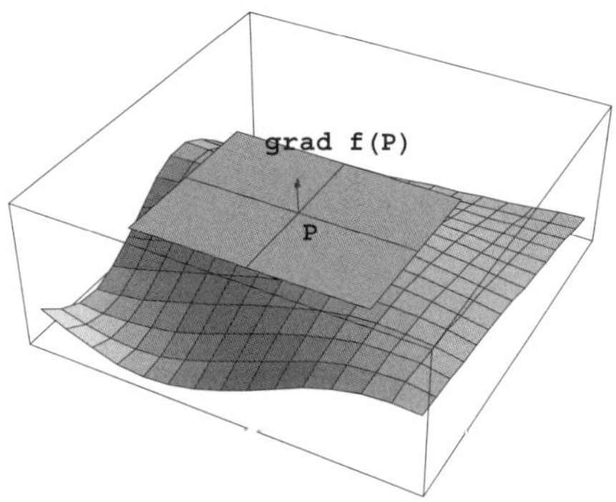

It can be shown as follows: First, observe that the tangent plane to the level surface S at P consists of all possible tangent vectors at P to curves that pass through P and lie on S. In other words, for any differentiable curve

$$X : \mathbb{R} \to S, \quad X(0) = P$$

the velocity vector $X'(0)$ is contained in the tangent plane to S at P, and conversely every element in the tangent plane can be obtained in such a way.[50]

Therefore it suffices to show that for any differentiable curve $X : \mathbb{R} \to S$ with $X(0) = P$,

$$\operatorname{grad} f(P) \cdot X'(0) = 0.$$

Since $X(t) \in S$,

$$f(X(t)) = f(P).$$

Thus if we differentiate both sides with respect to t, by the chain rule, we get $\operatorname{grad} f(X(t)) \cdot X'(t) = 0$. If we let $t = 0$, we get the desired conclusion.

---

[49] We say that a vector is perpendicular to a surface if the vector is perpendicular to the tangent plane of the surface. A vector perpendicular to a surface is called a **normal vector**. Since it is not interesting when the gradient $\operatorname{grad} f(P)$ is $\mathbf{0}$, we assume that $\operatorname{grad} f(P) \neq \mathbf{0}$.

[50] I'd like to have an opportunity to explain this as a result of the **inverse function theorem** in Chapter 12....

6.0.1 Basic Exercise

Explain which direction is the steepest direction on a map full of contour lines. Also find the direction to which the wind blows on a map with isobars. Explain that the magnitude of the gradient vector is large and the wind is strong at a point near which isobars are dense.

6.0.2 Example

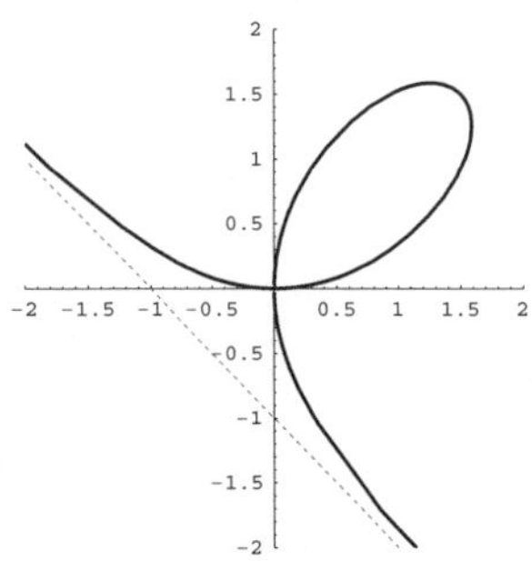

Find the tangent line of the folium of Descartes[51]

$$x^3 + y^3 = 3xy$$

at the point $P = (3/2, 3/2)$.

The folium of Descartes is the 0-level curve of the function

$$f(x, y) = x^3 + y^3 - 3xy.$$

Since the gradient vector of $f(x, y)$ is

$$\operatorname{grad} f(x, y) = 3(x^2 - y, y^2 - x),$$

the gradient vector at the point P is parallel to $(1, 1)$. Therefore the tangent line is perpendicular to $(1, 1)$ and the equation of the tangent line is $1(x - 3/2) + 1(y - 3/2) = 0$. This equation can be simplified by

$$x + y = 3.$$

6.0.3 Example(Ellipsoid)

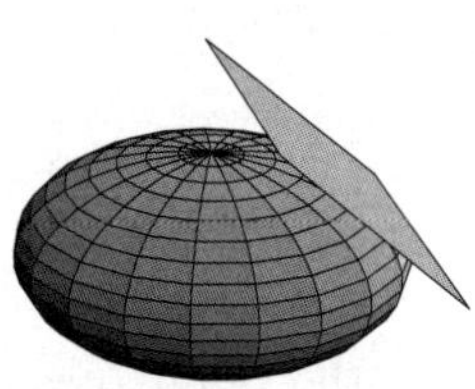

Find the equation of the tangent plane of

$$\frac{x^2}{9} + \frac{y^2}{8} + z^2 = 1$$

at a point $P\left(\frac{3}{5}, \frac{8}{5}, \frac{4}{5}\right)$.

We know that the given ellipsoid is a 1-level surface of the function

$$f(x, y, z) = \frac{x^2}{9} + \frac{y^2}{8} + z^2.$$

[51]Refer to Chapter 9 Section 1.

At the point P, a normal vector to this level surface is

$$\operatorname{grad} f\left(\frac{3}{5},\frac{8}{5},\frac{4}{5}\right)=2\left(\frac{\frac{3}{5}}{9},\frac{\frac{8}{5}}{8},\frac{4}{5}\right)=\frac{2}{15}\left(1,3,12\right).$$

Therefore the tangent plane at P is normal to the vector $(1,3,12)$ and its equation is given by

$$x+3y+12z=\frac{3}{5}+3\cdot\frac{8}{5}+12\cdot\frac{4}{5}=\frac{75}{5}=15.$$

6.0.4 Basic Exercise

Show that at a point (x_0,y_0,z_0) on an ellipsoid

$$\frac{x^2}{a^2}+\frac{y^2}{b^2}+\frac{z^2}{c^2}=1,$$

the equation of the tangent plane is

$$\frac{x_0x}{a^2}+\frac{y_0y}{b^2}+\frac{z_0z}{c^2}=1.$$

6.0.5 Example

When a light is emanated from a point $(0,0,1)$ in the direction of $\mathbf{v}=(1,6,-7)$, find the direction in which this light is reflected on an ellipsoid

$$x^2+2y^2+3z^2=1.$$

Solution.

First, let us find the point P at which the light emanated from $(0,0,1)$ toughes the surface. The point P is given by

$$P=(0,0,1)+t(1,6,-7)=(t,6t,1-7t).$$

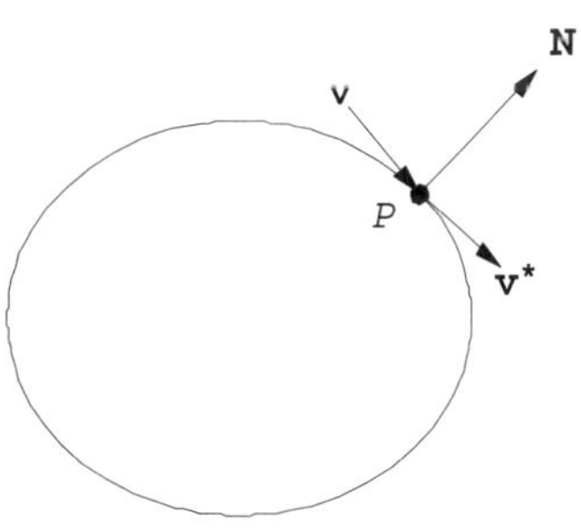

On the other hand, since P is a point on the surface,

$$1=t^2+2\cdot(6t)^2+3(1-7t)^2=220t^2-42t+3$$

and so

$$0=2(110t^2-21t+1)=2(10t-1)(11t-1).$$

Among values of t which satisfies the above equation, we have to pick the smaller one.[52] Thus we get $t=1/11$ and

$$P=\left(\frac{1}{11},\frac{6}{11},\frac{4}{11}\right).$$

[52]The larger value of t represents the point at which the light meets the ellipsoid again after it penetrates the ellipsoid.

On the other hand, a normal vector to the surface at the point P is

$$\mathbf{N} := 11 \cdot \left(\frac{1}{11}, 2 \cdot \frac{6}{11}, 3 \cdot \frac{4}{11}\right) = (1, 12, 12)$$

and so the direction $\mathbf{v}^*$ in which the light goes after reflection[53] is

$$\begin{aligned}
\mathbf{v}^* &= \mathbf{v} - 2 \cdot \frac{\mathbf{v} \cdot \mathbf{N}}{\mathbf{N} \cdot \mathbf{N}} \mathbf{N} \\
&= (1, 6, -7) - 2 \cdot \frac{1 + 72 - 84}{1 + 144 + 144} (1, 12, 12) \\
&= (1, 6, -7) + \frac{22}{289} (1, 12, 12) \\
&\approx (1.1, 6.9, -6.1).
\end{aligned}$$

6.0.6 Basic Exercise

At a point $(1, 1, 1)$, find the direction to which the light emanated in the direction $\mathbf{v} := (-1, -1, -1)/\sqrt{3}$ is reflected on the ellipsoid $x^2 + y^2 + 2z^2 = 1$. (Give the answer as a unit vector.[54])

---

[53] The vector $\mathbf{v}$ is the sum of the component perpendicular to the tangent plane $\frac{\mathbf{v} \cdot \mathbf{N}}{\mathbf{N} \cdot \mathbf{N}} \mathbf{N}$ and component parallel to the tangent plane $\mathbf{v} - \frac{\mathbf{v} \cdot \mathbf{N}}{\mathbf{N} \cdot \mathbf{N}} \mathbf{N}$. When the vector is reflected, the perpendicular component goes back but the parallel component goes ahead. Therefore the reflecting direction is

$$\mathbf{v}^* = \left(-\frac{\mathbf{v} \cdot \mathbf{N}}{\mathbf{N} \cdot \mathbf{N}} \mathbf{N}\right) + \left(\mathbf{v} - \frac{\mathbf{v} \cdot \mathbf{N}}{\mathbf{N} \cdot \mathbf{N}} \mathbf{N}\right) = \mathbf{v} - 2 \cdot \frac{\mathbf{v} \cdot \mathbf{N}}{\mathbf{N} \cdot \mathbf{N}} \mathbf{N}.$$

[54] The point where the light meets the ellipsoid is $P = (\frac{1}{2}, \frac{1}{2}, \frac{1}{2})$ and the direction perpendicular to the ellipsoid at P is $\mathbf{N} = (1, 1, 2)$. Therefore

$$\mathbf{v}^* = \mathbf{v} - 2 \frac{\mathbf{v} \cdot \mathbf{N}}{\mathbf{N} \cdot \mathbf{N}} \mathbf{N} = (1, 1, 5)/3\sqrt{3}.$$

Exercise: Chapter 10 Section 6

1. Show that at the point (x_0, y_0) on an ellipse

$$\frac{x^2}{a^2} + \frac{y^2}{b^2} = 1,$$

the equation of the tangent line is

$$\frac{x_0 x}{a^2} + \frac{y_0 y}{b^2} = 1.$$

2. Show that at the point (x_0, y_0) on a hyperbola

$$\frac{x^2}{a^2} - \frac{y^2}{b^2} = 1,$$

the equation of the tangent line is

$$\frac{x_0 x}{a^2} - \frac{y_0 y}{b^2} = 1.$$

3. Find the equation of the tangent plane of the surface given by the following equation at the point P.

(1) $8x^2 + y^2 + 8z^2 = 16,\qquad P = (1, 2, 1/\sqrt{2})$

(2) $x^2 + y^2 - z^2 = 4,\qquad P = (2, 1, 1)$

(3) $x^2 + \frac{1}{4}y^2 - \frac{1}{9}z^2 = 1,\qquad P = (1, 2, 3)$

(4) $\sin xy + \sin yz + \sin zx = 1,\qquad P = (1, \pi/2, 0)$

(5) $z - e^x \sin y = 0,\qquad P = (\log 3, 3\pi/2, -3)$

(6) $2y - z^3 - 3xz = 0,\qquad P = (1, 7, 2)$

(7) $x^3 - 2y^2 + z^2 = 0,\qquad P = (1, 1, 1)$

4. Find the the equation of the tangent plane of a hyperboloid

$$\frac{x^2}{a^2} + \frac{y^2}{b^2} - \frac{z^2}{c^2} = 1$$

at a point (x_0, y_0, z_0). Also find the equation of the tangent plane of a hyperboloid

$$\frac{x^2}{a^2} - \frac{y^2}{b^2} - \frac{z^2}{c^2} = 1$$

at a point (x_0, y_0, z_0).

7 Continuously Differentiable functions, Open sets, Closed sets, Bounded sets

7.1 Continuously Differentiable functions and Differentiability

In this section, we will prove the following theorem.

Theorem 7.1.1 *A continuously differentiable function defined on an open set in $\mathbb{R}^n$ is differentiable.*

Proof. First, consider the proof when $n = 2$.

Let f be a given $\mathcal{C}^1$ function. Fix a point $X = (x, y)$ in the domain of f. Then for a very small vector $\Delta X = (\Delta x, \Delta y)$, bythe mean value theorem, there exist $x^* \in [x, x + \Delta x]$ and $y^* \in [y, y + \Delta y]$ such that[55]

$$\begin{aligned}
& f(X + \Delta X) - f(X) \\
= \;& f(x + \Delta x, y + \Delta y) - f(x, y) \\
= \;& \big(f(x + \Delta x, y + \Delta y) - f(x, y + \Delta y)\big) + \big(f(x, y + \Delta y) - f(x, y)\big) \\
= \;& \Delta x \frac{\partial f}{\partial x}(x^*, y + \Delta y) + \Delta y \frac{\partial f}{\partial y}(x, y*).
\end{aligned}$$

Since $\frac{\partial f}{\partial x}$ and $\frac{\partial f}{\partial y}$ are continuous, we obtain

$$\begin{aligned}
& f(X + \Delta X) - f(X) - \mathrm{grad} f(X) \cdot \Delta X \\
= \;& \left(\Delta x \frac{\partial f}{\partial x}(x^*, y + \Delta y) + \Delta y \frac{\partial f}{\partial y}(x, y*)\right) - \left(\Delta x \frac{\partial f}{\partial x}(x, y) + \Delta y \frac{\partial f}{\partial y}(x, y)\right) \\
= \;& \Delta x \left(\frac{\partial f}{\partial x}(x^*, y + \Delta y) - \frac{\partial f}{\partial x}(x, y)\right) + \Delta y \left(\frac{\partial f}{\partial y}(x, y^*) - \frac{\partial f}{\partial y}(x, y)\right) \\
= \;& o(\Delta X).
\end{aligned}$$

It completes the proof when $n = 2$.

---

[55]We use the notation $[a, b]$ to denote the interval usually when $a < b$, but we understand it as $[a, b] := [b, a]$ when $a > b$ and a **singleton** when $a = b$.

For general n, we transform

$$\begin{aligned}
& f(X+\Delta X)-f(X) \\
= \; & f(x_1+\Delta x_1, x_2+\Delta x_2, \ldots, x_n+\Delta x_n)-f(x_1, x_2, \ldots, x_n) \\
= \; & f(x_1+\Delta x_1, x_2+\Delta x_2, \ldots, x_n+\Delta x_n)-f(x_1, x_2+\Delta x_2, \ldots, x_n+\Delta x_n) \\
& +f(x_1, x_2+\Delta x_2, \ldots, x_n+\Delta x_n)-f(x_1, x_2, x_3+\Delta x_3, \ldots, x_n+\Delta x_n) \\
& +\cdots \\
& +f(x_1, \ldots, x_{n-1}, x_n+\Delta x_n)-f(x_1, \ldots, x_{n-1}, x_n),
\end{aligned}$$

and if we apply the mean value theorem in each step, we may prove it in a similar way. □

7.1.2 Basic Exercise

1. For a function f defined on an open set U in $\mathbb{R}^n$ and linearly independent vectors $\mathbf{v}_1, \ldots, \mathbf{v}_n$ in $\mathbb{R}^n$, show that if the directional derivatives

$$D_{\mathbf{v}_1} f, \; \ldots, \; D_{\mathbf{v}_n} f \; : U \to \mathbb{R}$$

are all continuous then f is a $\mathcal{C}^1$ function.

2. Suppose that $S \subset \mathbb{R}^3$ is a connected surface and the gradient vector of a $\mathcal{C}^1$ function $f(x, y, z)$ is perpendicular to the tangent plane of S at each point on S. Then show that f is a constant function on S.

7.2 Open sets, Closed sets, Bounded sets

To differentiate a function, we need to measure the rate of change of the function, for which we do not need only the value of the function at one point in the domain but the values of the function near the point. A domain with such a property is called "open".

On the other hand, to answer the question of "when does a continuous function have the maximum or minimum?" or to integrate a continuous function, it is convenient that the domain of the function is "closed".

7.2.1 Interiors, Exteriors, Boundaries

A subset $\mathcal{A}$ of $\mathbb{R}^n$ classifies a point P in $\mathbb{R}^n$ into the following three kinds.

(i) If there exists a ball with center P which is contained in $\mathcal{A}$, then P is called an **interior point** of $\mathcal{A}$.

(ii) If there exists a ball with center P which doesn't intersect with $\mathcal{A}$ at all, then P is called an **exterior point** of $\mathcal{A}$.

(iii) If every ball with center P meets both $\mathcal{A}$ and its **complement**, $\mathcal{A}^{\complement} := \mathbb{R}^n - \mathcal{A}$, then P is called a **frontier point**, or **boundary point**. That is, a point which is neither an interior point nor an exterior point of $\mathcal{A}$ is a boundary point of $\mathcal{A}$.

interior point boundary point exterior point

From now on, for a subset $\mathcal{A}$ of $\mathbb{R}^n$, the set of all interior points of $\mathcal{A}$ is called the **interior** of $\mathcal{A}$, the set of all exterior points of $\mathcal{A}$ is called the **exterior** of $\mathcal{A}$ and the set of all boundary points of $\mathcal{A}$ is called the **frontier, boundary** of $\mathcal{A}$.

7.2.2 Example

In plane, the interior of the unit disk

$$\mathbf{B}^2 := \{X \in \mathbb{R}^2 : |X| < 1\}$$

is $\mathbf{B}^2$ itself and its exterior is

$$\{X \in \mathbb{R}^2 : |X| > 1\}.$$

The boundary is the (standard) unit circle

$$\mathbf{S}^1 := \{X \in \mathbb{R}^2 : |X| = 1\}.$$

7.2.3 Basic Exercise

Find the interior, exterior and boundary of the closed unit disk

$$\overline{\mathbf{B}}^2 := \{X \in \mathbb{R}^2 : |X| \leq 1\}.$$

In general, the boundary of an n-dimensional ball of radius r with center at P in $\mathbb{R}^n$ is an $(n-1)$-dimensional **sphere**[56]

$$\mathbf{S}^{n-1}(P, r) := \{X \in \mathbb{R}^n : |X - P| = r\}.$$

[56] A 1-dimensional sphere is called a **circle**.

7.2.4 Basic Exercise

Find the interior, exterior and boundary of the following two squares

$$S := \{(x, y) \mid \max\{|x|, |y|\} < 1\}, \qquad \overline{S} := \{(x, y) \mid \max\{|x|, |y|\} \leq 1\},$$

respectively.

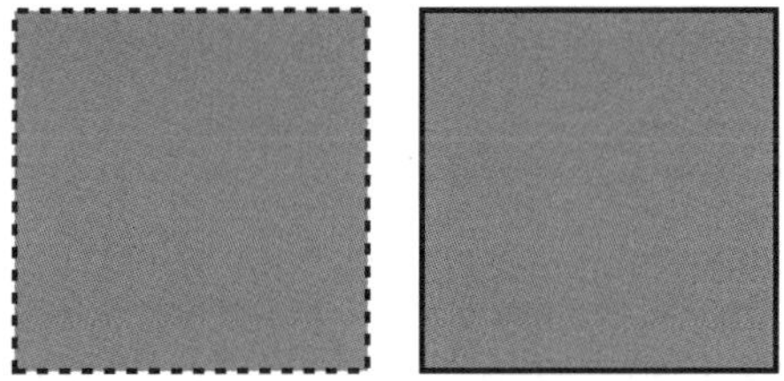

7.2.5 Neighborhoods

If a point P in $\mathbb{R}^n$ is an interior point of a subset $\mathcal{A}$ of $\mathbb{R}^n$, $\mathcal{A}$ is called a **neighborhood** of P.

7.2.6 Basic Exercise

Determine whether the following set $\mathcal{A}$ in the field of real numbers is a neighborhood of the origin.
(1) $\mathcal{A} - \{x \mid x > 0\}$ (2) $\mathcal{A} = \{x \mid x \geq 0\}$ (3) $\mathcal{A} = \mathbb{R}$

7.2.7 Open Sets

If every point in a set $\mathcal{A}$ is an interior point of $\mathcal{A}$, then $\mathcal{A}$ is called an **open set** (or **open subset**). In other words,, a subset $\mathcal{A}$ of $\mathbb{R}^n$ is called open if "for any point in $\mathcal{A}$, not only the point but also its small **neighborhood** is contained in $\mathcal{A}$."[57] If $\mathcal{A}$ is open, then a boundary point of $\mathcal{A}$ is not contained in $\mathcal{A}$, and conversely a set which contains no boundary points is open.

7.2.8 Basic Exercise

Show that an open disk in the plane is open.

---

[57]The set of all open subsets of a space is called the **topology** of the space and the study of topologies of spaces is called **topology**.

7.2.9 Closed Sets

If every boundary point of a subset $\mathcal{A}$ of $\mathbb{R}^n$ is contained in $\mathcal{A}$, $\mathcal{A}$ is called **closed** in $\mathbb{R}^n$.

7.2.10 Basic Exercise

Show that a closed disk in the plane is closed.

7.2.11 Openness and Closedness

The concept of 'openness' and 'closedness' is **relative**. For example, an open interval $I := (0, 1)$ is open as a subset of $\mathbb{R}$. However, it is not open as a subset of $\mathbb{R}^2$ which contains "$\mathbb{R}$ as the x-axis".[58]

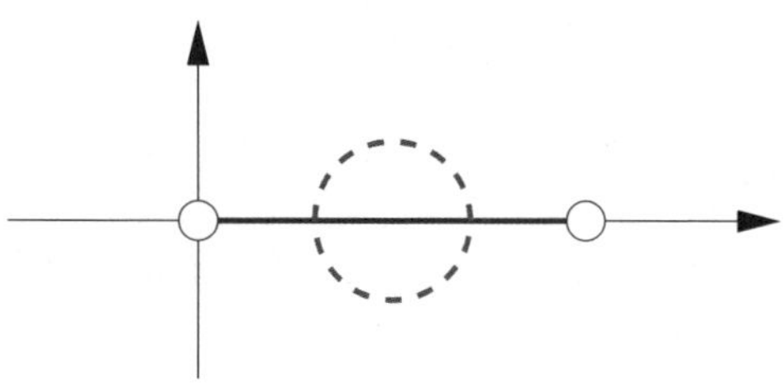

A subset of $\mathbb{R}^n$ can be neither open nor closed. For example, the interval $[0, 1)$ is neither open nor closed in $\mathbb{R}$.

The whole space itself is both open and closed in itself.

Theorem 7.2.12 *Every level surface of a continuous function* $f : \mathbb{R}^n \to \mathbb{R}$ *is a closed set.*

Proof. Fix a constant c. Then we need to show that the c-level surface of f

$$f^{-1}(c) = \{X \in \mathbb{R}^n \mid f(X) = c\}$$

[58]In such a viewpoint, the terms "open subset" or "closed subset" are more logical than the terms "open set" or "closed set". Nevertheless, we will use illogical terms. Once you know it, you are free from the logic.

is a closed set in $\mathbb{R}^n$. Therefore it suffices to show that if a point P in $\mathbb{R}^n$ is a boundary point of the c-level surface, P is still a point in the c-level surface.

Since P is a boundary point, there exists a sequence $(X_i \in f^{-1}(c) \mid i = 1, 2, 3, \dots)$ such that[59]

$$\lim_{i\to\infty} X_i = P.$$

From the continuity of f, we get

$$f(P) = \lim_{i\to\infty} f(X_i) = c$$

and thus $P \in f^{-1}(c)$. □

7.2.13 Basic Exercise

In the three-dimensional space, determine whether the set of all points satisfying the equation

$$x^2 + 3y^4 + 4z^6 = 1$$

is closed.

7.2.14 Bounded Sets

The **diameter** of a subset $\mathcal{A}$ in $\mathbb{R}^n$ is the supremum of distances between any two points in the set, and it is denoted by $\operatorname{diam}\mathcal{A}$. That is, we define[60]

$$\operatorname{diam}\mathcal{A} := \sup\{|P-Q| : P \in \mathcal{A},\ Q \in \mathcal{A}\} \in [0, \infty].$$

From now on, a set with finite diameter is called a **bounded set**. A bounded set is contained in a big(?) ball and conversely a set which is contained in some ball is bounded.

7.2.15 Basic Exercise

1. Show that the diameter of a ball in $\mathbb{R}^n$ is twice its radius.
2. If two subsets $\mathcal{A}, \mathcal{B}$ of $\mathbb{R}^n$ satisfy the inclusion $\mathcal{A} \subset \mathcal{B}$, show that

$$\operatorname{diam}\mathcal{A} \le \operatorname{diam}\mathcal{B}.$$

[59] Refer to Exercise 2in this section.

[60] Without any special remark, we consider nonempty sets. We think of the diameter of the empty set as 0.

7.2.16 Example

Find the diameter of the set

$$\mathcal{A} := \{(x, y) \in \mathbb{R}^2 \mid |x| + |y| \leq 1\}.$$

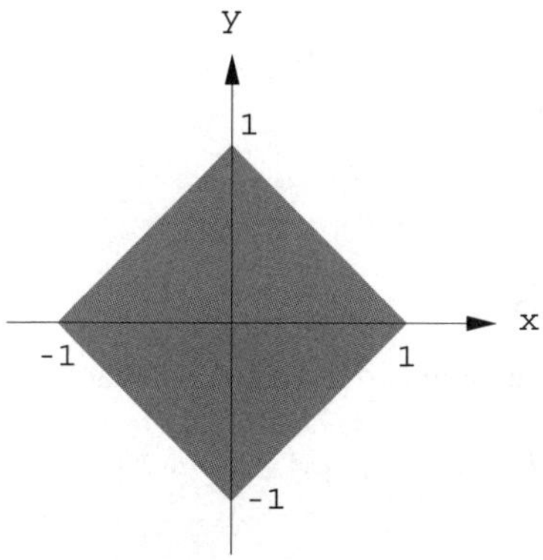

The set $\mathcal{A}$ is contained in the unit disk

$$\overline{\mathbf{B}}^2 := \{(x, y) : x^2 + y^2 \leq 1\}.$$

Therefore

$$\operatorname{diam} \mathcal{A} \leq \operatorname{diam} \overline{\mathbf{B}}^2 = 2.$$

On the other hand, since two points $(1, 0)$ and $(-1, 0)$ in $\mathcal{A}$ have the distance 2, we get

$$\operatorname{diam} \mathcal{A} \geq 2.$$

From two inequalities above, we have

$$\operatorname{diam} \mathcal{A} = 2.$$

7.2.17 Example

In the plane, the **first quadrant**

$$\{(x, y) \in \mathbb{R}^2 \mid x > 0, \ y > 0\}$$

is not bounded.

Exercise: Chapter 10 Section 7

1. In the field of real numbers $\mathbb{R}$, find the interior, exterior and boundary of the field of rational numbers [61] $\mathbb{Q}$.

2. Explain that if a point P is a boundary point of a set $\mathcal{A}$ in $\mathbb{R}^n$, there exists a sequence[62] $(X_i \in \mathcal{A} \mid i = 1, 2, 3, \dots)$ such that[63]

$$\lim_{i \to \infty} X_i = P.$$

3. Show that a subset $\mathcal{A}$ of $\mathbb{R}^n$ is open if and only if its compliment $\mathcal{A}^{\mathsf{C}} = \mathbb{R}^n - \mathcal{A}$ is closed.

4. Show that a subset $\mathcal{A}$ of $\mathbb{R}^n$ is closed if and only if every point which is not contained in $\mathcal{A}$ is an exterior point of $\mathcal{A}$.

5. Show that the union and intersection of two open sets are both open.

6. Show that the union and intersection of two closed sets are both closed.

7. Find the diameter of the following set in the plane.
   (1) $\{(x, y) \mid |x|^{1/3} + |y|^{1/3} = 1\}$ (2) $\{(x, y) \mid |x|^{1/3} + |y|^{1/3} \leq 1\}$

8. Describe the boundary of the set given by the following equation in the plane, and examine whether it is open, closed or bounded. Also find its diameter.
   (1) $|x| + |y| < 1$ (2) $\max\{|x|, |y|\} < 1$
   (3) $\min\{x + y, x - y\} \geq 1$ (4) $-\sqrt{1 - x^2} < y \leq \sqrt{1 - x^2}$

---

[61] A **field** means a set which is closed under the addition, subtraction, multiplication and division.

[62] It would be correct to call it a **sequence of points**. However we call the "sequence of functions", "sequence of sets", "sequence of spaces", "sequence of matrices", etc, simply a **sequence**.

[63] That is, $\lim_{i \to \infty} |X_i - P| = 0$.

Chapter 10 Advanced Exercise

1. Show that the function

$$h(x,y)=\begin{cases}\dfrac{xy}{x^2+y^2}, & (x,y)\neq(0,0)\\ 0, & (x,y)=(0,0)\end{cases}$$

is not continuous. Also show that there exist the partial derivatives of $h(x,y)$, $D_1h(x,y)$ and $D_2h(x,y)$ at every point. Is h differentiable?

2. Is the function

$$f(x,y)=\begin{cases}\dfrac{y^3}{x^2+y^2}, & (x,y)\neq(0,0)\\ 0, & (x,y)=(0,0)\end{cases}$$

continuous? Does f have directional derivatives in all directions at the origin? Is f differentiable?

3. Show that the function

$$f(x,y)=\begin{cases}\dfrac{xy}{x^2-y}, & x^2\neq y\\ 0, & x^2=y\end{cases}$$

is not continuous but has directional derivatives in all directions at the origin.

4. Show that the function

$$f(x,y)=\begin{cases}\dfrac{x^2y\sqrt{x^2+y^2}}{x^4+y^2}, & (x,y)\neq(0,0)\\ 0, & (x,y)=(0,0)\end{cases}$$

is continuous and every directional derivative is 0 at the origin, but it is not differentiable at the origin.

5. Investigate the differentiability of the following function.

(1) $$f(x,y)=\begin{cases}\left(x^3+y^3\right)/\left(x^2+y^2\right), & (x,y)\neq(0,0)\\ 0, & (x,y)=(0,0)\end{cases}$$

(2) $$f(x,y,z)=\begin{cases}\dfrac{x^3+y^3+z^2}{x^2+y^2+z^2}, & (x,y,z)\neq(0,0,0)\\ 0, & (x,y,z)=(0,0,0)\end{cases}$$

6. For an orthonormal basis $\mathbf{v}_1,\mathbf{v}_2,\ldots,\mathbf{v}_n$ of $\mathbb{R}^n$ and a differentiable function $f(x_1,\ldots,x_n)$, show that

$$\mathrm{grad}f(P)=\sum_{i=1}^{n}D_{\mathbf{v}_i}f(P)\,\mathbf{v}_i.$$

7. In cylindrical coordinates

$$(x, y, z) = G(r, \theta, z) = (r\cos\theta, r\sin\theta, z),$$

let

$$\mathbf{e}_r := \frac{\partial G}{\partial r}\Big/\left|\frac{\partial G}{\partial r}\right|, \qquad \mathbf{e}_\theta := \frac{\partial G}{\partial \theta}\Big/\left|\frac{\partial G}{\partial \theta}\right|, \qquad \mathbf{e}_z := \frac{\partial G}{\partial z}\Big/\left|\frac{\partial G}{\partial z}\right|.$$

Show that for any C^1 function $f(x, y, z)$,

$$\operatorname{grad} f = \frac{\partial f}{\partial r}\,\mathbf{e}_r + \frac{1}{r}\frac{\partial f}{\partial \theta}\,\mathbf{e}_\theta + \frac{\partial f}{\partial z}\,\mathbf{e}_z.$$

8. In spherical coordinates

$$(x, y, z) = G(\rho, \varphi, \theta) = (\rho\sin\varphi\cos\theta, \rho\sin\varphi\sin\theta, \rho\cos\varphi),$$

let

$$\mathbf{e}_\rho := \frac{\partial G}{\partial \rho}\Big/\left|\frac{\partial G}{\partial \rho}\right|, \qquad \mathbf{e}_\varphi := \frac{\partial G}{\partial \varphi}\Big/\left|\frac{\partial G}{\partial \varphi}\right|, \qquad \mathbf{e}_\theta := \frac{\partial G}{\partial \theta}\Big/\left|\frac{\partial G}{\partial \theta}\right|.$$

Show that for any C^1 function $f(x, y, z)$,

$$\operatorname{grad} f = \frac{\partial f}{\partial \rho}\,\mathbf{e}_\rho + \frac{1}{\rho}\frac{\partial f}{\partial \varphi}\,\mathbf{e}_\varphi + \frac{1}{\rho\sin\varphi}\frac{\partial f}{\partial \theta}\,\mathbf{e}_\theta.$$

9. Show that if a continuous function $f : A \to \mathbb{R}$ defined on a closed set A in $\mathbb{R}^n$ satisfies

$$\lim_{P\to\infty} f(P) = \infty,$$

then f attains the minimum.[64]

[64] "$P \to \infty$" means "$|P| \to \infty$".

Chapter 11

Maxima, Minima and Higher-order Derivatives

Differentiate!

S.-S. Chern, (1911–2004)

If we differentiate a function once, then we get the first order derivatives, if twice then the second order derivatives, if several times then the higher order derivatives. In this chapter, we investigate how to use dierivatives to find the maximum or minimum and study the Lagrange multiplier method to find the maximum or minimum for functions with constraints. Also we study the approximation theory using Taylor expansion and investigate how to use the second order derivatives to find the extreme values for functions of several variables.

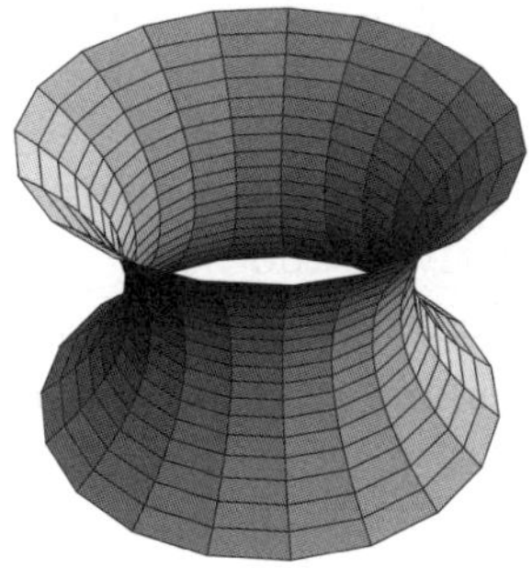

Picture: The Euler-Lagrange equation tells us that the catenary is a minimal surface.

1 Differentiation under the Integral Sign

Leibniz found out that for a $\mathcal{C}^1$ function $f(x, y)$ defined on a rectangle $[a, b] \times [c, d]$, there holds

$$\boxed{\frac{d}{dx} \int_c^d f(x,y)\,dy = \int_c^d \frac{\partial f}{\partial x}(x,y)\,dy}$$

We will call it **Leibniz's rule** (for differentiation under the integral sign).[1]

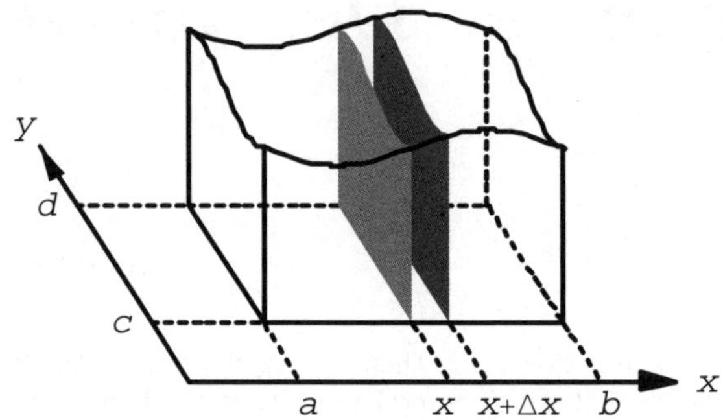

From the following, we can see that this rule is quite natural. The definite integral is the limit of a finite sum:

$$\int_c^d f(x,y)\,dy \sim \sum_{i=1}^{n} f(x, y_i)\,\Delta y_i \qquad (c = y_0 < \cdots < y_n = d).$$

Meanwhile, the differentiation is a linear operator — that is, the derivative of sum is the sum of derivatives, and so it is a very natural phenomenon that differentiation and definite integral commute:

We leave the proof of Leibniz's rule in the last section.

---

[1] A famous physicist, R. Feynman(1918–1988) emphasized several times that this formula is useful: "So because I was self-taught using that book, I had peculiar methods for doing integrals, $\cdots$. I used that one damn tool again and again." [Feynman, p.72, p.93]. Often the product rule, that is, the rule for differentiating the product of two functions,

$$(f \cdot g)' = f' \cdot g + f \cdot g'$$

is called **Leibniz's rule**, but this is nothing to do with differentiation of integrals.

1.0.1 Example

For positive numbers c, d and x, consider the definite integral

$$\int_c^d \frac{1}{y} e^{-xy^2}\, dy.$$

The value varies depending on x, but it is well-known that its indefinite integral is impossible to find. In other words, it is impossible to express the antiderivative of the function $\frac{1}{y}e^{-y^2}$ as an **elementary function**.[2] Elementary functions are polynomials, rational functions, exponential functions, logarithmic functions, trigonometric functions, inverse trigonometric functions, or functions expressed as their sum, product, composition or inverse. Liouville showed that there exists an elementary function whose antiderivative is not an elementary function [`Math Dictionary`]. One example for such a function is the function we have here.

However if we use Leibniz's rule, we get[3]

$$\begin{aligned} \frac{d}{dx}\int_c^d \frac{1}{y} e^{-xy^2}\, dy &= \int_c^d \frac{\partial}{\partial x}\left(\frac{1}{y}e^{-xy^2}\right) dy = \int_c^d -ye^{-xy^2}\, dy \\ &= \left[\frac{1}{2x}e^{-xy^2}\right]_c^d \\ &= \frac{1}{2x}\left(e^{-xd^2} - e^{-xc^2}\right). \end{aligned}$$

1.0.2 Basic Exercise

Compute the following derivative.

$$\frac{d}{dy}\int_1^{\pi} \frac{\sin xy}{x}\, dx.$$

---

[2] I hope that the reader won't misunderstand that the function $\frac{1}{y}e^{-y^2}$ is not integrable. Every continuous function is integrable.

[3] In this equation, if $x > 0$, we can get

$$\frac{d}{dx}\int_1^{\infty} \frac{1}{y} e^{-xy^2}\, dy = -\frac{1}{2x}e^{-x}.$$

Exercise: Chapter 11 Section 1

1. Find the following derivative when $x > 0$.

$$\frac{d}{dx}\int_{x^2}^{x^3} \frac{1}{y} e^{xy^2}\, dy.$$

2. Show that for $\mathcal{C}^1$ functions $a(x)$, $b(x)$, $g(x,y)$,

$$\frac{d}{dx}\int_{a(x)}^{b(x)} g(x,y)\, dy = g(x,b(x))b'(x) - g(x,a(x))a'(x) + \int_{a(x)}^{b(x)} D_1 g(x,y)\, dy.$$

3. When $k < 1$, show that the function

$$f(k) := \int_0^{\pi/2} \frac{dx}{\sqrt{1 - k\cos^2 x}}$$

is an increasing function for k. At $k = 0$, find the derivative of f.

4. For a $\mathcal{C}^2$ curve $C(s)$ ($s_0 \le s \le s_1$) parameterized by arc length in the plane, if we denote the unit tangent vector by

$$\mathbf{t}(s) := C'(s),$$

then at each point on C the **curvature** is given by

$$\kappa(s) := |\mathbf{t}'(s)|.$$

If we translate the curve $C(s)$ by u in the direction of a unit normal vector[4]

$$\mathbf{n}(s) := -\mathbf{t}'(s)/\kappa(s),$$

then we get a new curve

$$C_u(s) := C(s) + u\,\mathbf{n}(s), \qquad s_0 \le s \le s_1.$$

Let

$$L(u) := \int_{s_0}^{s_1} |C_u'(s)|\, ds$$

be the length of the curve C_u.

(1) Show that $\mathbf{n}'(s) = \kappa(s)\mathbf{t}(s)$.

---

[4]Since $|\mathbf{t}| = 1$, $\mathbf{t} \perp \mathbf{t}'$. For convenience, we assume $\kappa(s) \neq 0$ in this problem. This assumption is not very important, since we may define $\mathbf{n}(s)$ by "the vector obtained by rotating $\mathbf{t}(s)$ by 90°", instead of defining it by the above equation. If $\mathbf{t}$ is $\mathcal{C}^1$, then $\mathbf{n}$ obtained by rotating $\mathbf{t}$ by 90° is also $\mathcal{C}^1$.

(2) Show that if $|u| \ll 1$, that is, if the absolute value of u is very small, then the derivative of $L(u)$ is the total curvature of the original curve, that is,

$$L'(u) = \int_{s_0}^{s_1} \kappa(s)\, ds.$$

(3) When the original curve C is an ellipse, find the value of $L'(0)$.

2 Second Order Derivatives

For a $\mathcal{C}^1$ function f defined on an open set in $\mathbb{R}^n$, if its first order partial derivatives

$$D_1 f, \ \ldots, \ D_n f \ : U \to \mathbb{R}$$

are all

(i) differentiable functions, then f is called **twice differentiable.**[5]

(ii) $\mathcal{C}^1$ functions, then f is called a $\mathcal{C}^2$ **function.**[6]

Of course, a $\mathcal{C}^2$ function f is twice differentiable. At any point $P \in U$, a twice differentiable function f has n^2 **second-order partial derivatives**

$$D_i D_j f(P) = \frac{\partial^2 f}{\partial x_i \partial x_j}(P) \qquad (1 \le i, j \le n).$$

We denote ${D_i}^2 f := D_i D_i f$.

2.0.1 Example

For the function $f(x, y) := \sin(xy)$,

$$D_1 f(x, y) = y\cos(xy), \qquad D_2 f(x, y) = x\cos(xy)$$

and

$$\begin{aligned} {D_1}^2 f(x, y) &= D_1\big(y\cos(xy)\big) = -y^2\sin(xy) \\ D_2 D_1 f(x, y) &= D_2\big(y\cos(xy)\big) = \cos(xy) - xy\sin(xy) \\ D_1 D_2 f(x, y) &= D_1\big(x\cos(xy)\big) = \cos(xy) - xy\sin(xy) \\ {D_2}^2 f(x, y) &= D_2\big(x\cos(xy)\big) = -x^2\sin(xy). \end{aligned}$$

Here we can observe that

$$D_1 D_2 f(x, y) = D_2 D_1 f(x, y).$$

Indeed, it is a general phenomenon.

---

[5] In general, if all first order partial derivatives of a function are k-times differentiable, then the function is called $(k+1)$-times differentiable.

[6] In general, if all first order partial derivatives of a function are $\mathcal{C}^k$, then the function is called a $\mathcal{C}^{k+1}$ function. Also if a function is $\mathcal{C}^k$ for every natural number k, it is called an **infinitely differentiable** function or a $\mathcal{C}^\infty$ function.

Euler found out the following fact while he was working on thermodynamics.

Theorem 2.0.2 (Commutativity of Partial Derivatives) *For a C^2 function f defined on an open set U in $\mathbb{R}^n$ and any point P in the domain U,*[7]

$$D_iD_jf(P) = D_jD_if(P) \qquad (1 \le i, j \le n).$$

Proof. Since we want to examine the relation between the ith variable and jth variable, it suffices to show the theorem for the function of two variables $f(x, y)$.

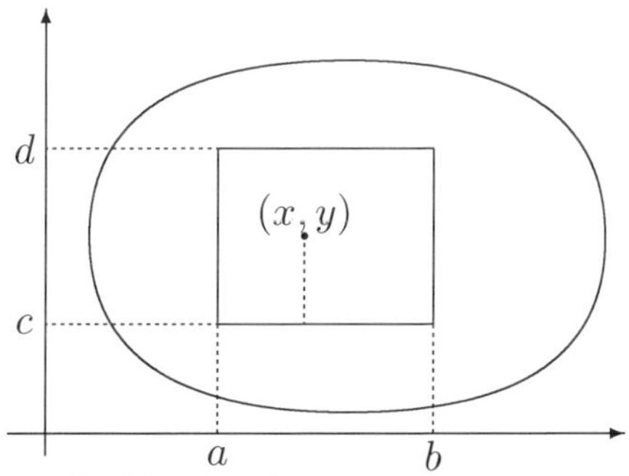

In a small rectangle $[a, b] \times [c, d]$ contained in U, we get

$$\begin{aligned}\int_c^y D_1D_2f(x,t)\,dt &= \frac{\partial}{\partial x}\int_c^y D_2f(x,t)\,dt \\ &= \frac{\partial}{\partial x}\big(f(x,y) - f(x,c)\big) \\ &= D_1f(x,y) - D_1f(x,c).\end{aligned}$$

Since the left-hand side of this equation is differentiable with respect to y, the right-hand side is also differentiable with respect to y and thus if we differentiate both sides with respect to y, we get

$$D_1D_2f(x,y) = D_2D_1f(x,y).$$

□

[7]From Euler's "commutativity of partial derivatives", we may derive Leibniz's "differentiation under the integral sign".

2.0.3 Basic Exercise

For the function $f(x, y, z) = e^{xy} + z\cos x$, investigate

$$\frac{\partial^2 f}{\partial x \partial y} = \frac{\partial^2 f}{\partial y \partial x}, \qquad \frac{\partial^2 f}{\partial x \partial z} = \frac{\partial^2 f}{\partial z \partial x}, \qquad \frac{\partial^2 f}{\partial y \partial z} = \frac{\partial^2 f}{\partial z \partial y}.$$

If $f(x, y)$ is $\mathcal{C}^3$, then $D_2 f$ is $\mathcal{C}^2$ and thus from Euler's commutativity of partial derivatives, $D_1 D_2(D_2 f) = D_2 D_1(D_2 f)$. On the other hand, since $D_2 D_1(D_2 f) = D_2(D_1 D_2 f) = D_2(D_2 D_1 f)$, we know that

$$D_1 {D_2}^2 f, \qquad D_2 D_1 D_2 f, \qquad {D_2}^2 D_1 f$$

are all equal.

Similarly, if $f(x, y, z)$ is $\mathcal{C}^4$, then the followings are all equal.

$${D_1}^2 D_2 D_3 f, \quad D_2 {D_1}^2 D_3 f, \quad D_2 D_3 {D_1}^2 f, \quad \ldots, \quad D_1 D_2 D_3 D_1 f.$$

2.0.4 Basic Exercise

For the following function, investigate that

$$\frac{\partial^3 f}{\partial x \partial y \partial z} := \frac{\partial}{\partial x}\frac{\partial}{\partial y}\frac{\partial}{\partial z} f \quad \text{and} \quad \frac{\partial^3 f}{\partial z \partial y \partial x} := \frac{\partial}{\partial z}\frac{\partial}{\partial y}\frac{\partial}{\partial x} f$$

are equal. Also show that

$$\frac{\partial^3 f}{\partial x^2 \partial y} := \left(\frac{\partial}{\partial x}\right)^2 \frac{\partial}{\partial y} f \quad \text{and} \quad \frac{\partial^3 f}{\partial y \partial x^2} := \frac{\partial}{\partial y}\left(\frac{\partial}{\partial x}\right)^2 f$$

are equal.

1. $f(x, y, z) = x^2 yz$
2. $f(x, y, z) = \sin(x + y + z)$
3. $f(x, y, z) = \log(x^2 - yz) + e^{xyz}$
4. $f(x, y, z) = 1/(x^2 + y^2 + z^2)$

In general, we get the following theorem.

Corollary 2.0.5 *For a $\mathcal{C}^k$ function f defined on an open set in $\mathbb{R}^n$ and $\{i_1, i_2, \ldots, i_k\} \subset \{1, 2, \ldots, n\}$, the higher order partial derivatives of f*

$$D_{i_1} D_{i_2} \cdots D_{i_k} f$$

remain the same regardless of the order of $D_{i_1}, D_{i_2}, \ldots, D_{i_k}$.

2.1 Directional Differential Operators

For a $\mathcal{C}^1$ function $f(x_1, \ldots, x_n)$ and a vector $\mathbf{v} = v_1\mathbf{e}_1 + \cdots + v_n\mathbf{e}_n$, we know that

$$D_{\mathbf{v}}f = \mathbf{v} \cdot \operatorname{grad} f = (v_1D_1 + \cdots + v_nD_n)f.$$

We will denote this equation as[8]

$$\boxed{D_{\mathbf{v}} = v_1D_1 + \cdots + v_nD_n}$$

For a $\mathcal{C}^2$ function f,

$$D_{\mathbf{v}}{}^2 f := D_{\mathbf{v}}\left(D_{\mathbf{v}}f\right) = \sum_{i=1}^{n} v_iD_i\left(\sum_{j=1}^{n} v_jD_jf\right) = \sum_{i,j=1}^{n} v_iv_jD_iD_jf.$$

That is,[9]

$$D_{\mathbf{v}}{}^2 = \left(\sum_{i=1}^{n} v_iD_i\right)^2 = \sum_{i,j=1}^{n} v_iv_jD_iD_j.$$

In $\mathbb{R}^2$, for a $\mathcal{C}^2$ function $f(x,y)$ and a vector $\mathbf{v} = a\,\mathbf{i}+b\,\mathbf{j} = (a,b)$, we get

$$\boxed{D_{(a,b)}{}^2 f = a^2D_1{}^2f + 2abD_1D_2f + b^2D_2{}^2f}$$

For example, if $f(x,y) = x^2y^3$, for a vector $\mathbf{v} = (a,b)$,

$$D_{\mathbf{v}}{}^2 f(x,y) = 2a^2\,y^3 + 12ab\,xy^2 + 6b^2\,x^2y,$$

and in particular,

$$D_{(1,2)}{}^2 f(3,1) = 2 + 12 \cdot 2 \cdot 3 + 6 \cdot 2^2 \cdot 3^2 = 2 + 72 + 216 = 290.$$

While a **first-order partial derivative** represents the instantaneous rate of change or "slope", a **second-order partial derivative** tells us the **convexity** of the function.[10]

---

[8]The directional differential operator $D_{\mathbf{v}}$ maps a $\mathcal{C}^1$ function to a continuous function, a $\mathcal{C}^2$ function to a $\mathcal{C}^1$ function, and a $\mathcal{C}^k$ function to a $\mathcal{C}^{k-1}$ function in general. A function defined between "spaces of functions " is called an "operator". A directional differential operator maps a $\mathcal{C}^\infty$ function to a $\mathcal{C}^\infty$ function.

[9]$D_{\mathbf{v}}{}^2$ is the composition of a directional differential operator $D_{\mathbf{v}}$ and itself.

[10]For a function of a single variable $f(x)$, if $f''(x_0) > 0$ then this function is **convex** (upward) at $x = x_0$, and if $f''(x_0) < 0$ then this function is **concave** at $x = x_0$. In the

For a function of several variables, in order to investigate the convexity we still need the direction and the **second-order directional derivative** in the direction of **v** is

$$D_{\mathbf{v}}{}^2 f(P).$$

Lemma 2.1.1 *A C^2 function f defined on an open set U in $\mathbb{R}^n$ satisfies*

$$\left.\frac{d^2}{dt^2}\right|_0 f(P+t\mathbf{v}) = D_{\mathbf{v}}{}^2 f(P)$$

for any point P in U and any vector $\mathbf{v}$.[11]

Proof. From the chain rule, we know that

$$\frac{d}{dt} f(P+t\mathbf{v}) = \operatorname{grad} f(P+t\mathbf{v}) \cdot \mathbf{v} = D_{\mathbf{v}} f(P+t\mathbf{v}).$$

Therefore

$$\frac{d^2}{dt^2} f(P+t\mathbf{v}) = \frac{d}{dt}\left(D_{\mathbf{v}} f\right)(P+t\mathbf{v}) = D_{\mathbf{v}}\left(D_{\mathbf{v}} f\right)(P+t\mathbf{v}).$$

If we let $t = 0$ then we can get the conclusion we want. □

It follows from the theorem that

- if $D_{\mathbf{v}}{}^2 f(P) > 0$, f is **convex** in the direction of **v** at P.

picture below, the first three pictures are graphs of convex functions and the remaining three pictures are graphs of concave functions.

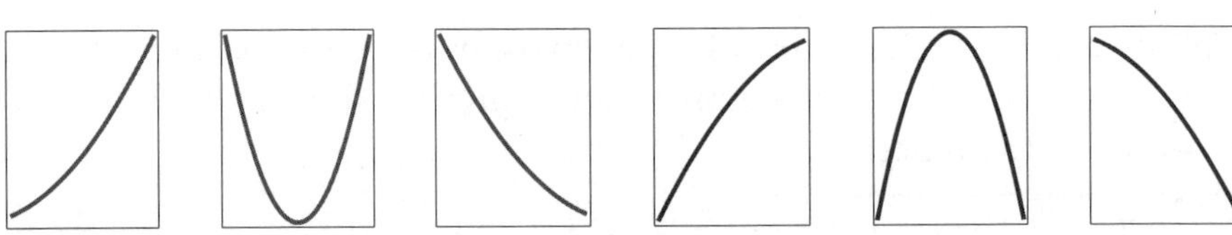

A convex function defined on a closed interval achieves its maximum always on the boundary, a concave function achieves its minimum always on the boundary. Such a principle is called the **maximum principle**. We leave the precise definition of convex function on [Math Dictionary].

[11]The second derivative of a function with a variable t: $\left(\frac{d}{dt}\right)^2 = \frac{d^2}{dt^2}$.

- if ${D_{\mathbf{v}}}^2 f(P) < 0$, f is **concave** in the direction of $\mathbf{v}$ at P.

For example, for a function $f(x, y) = xy^2$ and a vector $\mathbf{v} = a\mathbf{i} + b\mathbf{j}$,

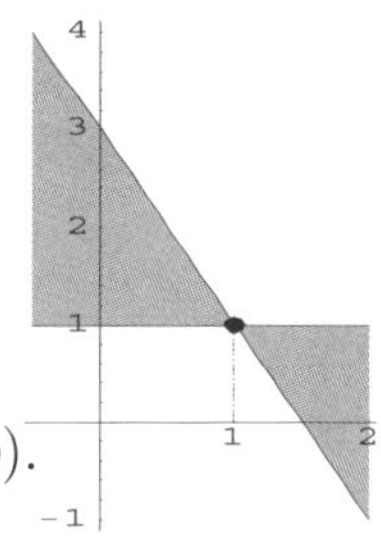

$${D_{\mathbf{v}}}^2 f(x, y) = 4aby + 2b^2 x$$

and thus

$${D_{\mathbf{v}}}^2 f(1, 1) = 4ab + 2b^2 = 2b(2a + b).$$

Therefore at $(1, 1)$, f is concave in the direction of the vector (a, b) satisfying $b(2a + b) < 0$. The shaded region denotes directions in which f is concave at $(1, 1)$.

2.1.2 Higher Order Derivatives

For a $\mathcal{C}^3$ function $f(x, y)$ and a vector $\mathbf{v} = a\,\mathbf{i} + b\,\mathbf{j}$,

$$\begin{aligned} {D_{\mathbf{v}}}^3 f &:= D_{\mathbf{v}}\left({D_{\mathbf{v}}}^2 f\right) \\ &= (aD_1 + bD_2)\left(a^2 {D_1}^2 f + 2abD_1 D_2 f + b^2 {D_2}^2 f\right) \\ &= a^3\, {D_1}^3 f + 3a^2 b\, {D_1}^2 D_2 f + 3ab^2\, D_1 {D_2}^2 f + b^3\, {D_2}^3 f. \end{aligned}$$

2.1.3 Basic Exercise

1. For a function $f(x, y) = xy^3$, find ${D_{(a,b)}}^3 f(x, y)$.
2. For a $\mathcal{C}^3$ function f, show that

$$\left.\frac{d^3}{dt^3}\right|_0 f(P + t\mathbf{v}) = {D_{\mathbf{v}}}^3 f(P)$$

for any point P in the domain and a vector $\mathbf{v}$.

In general, if a function $f(x, y)$ is $\mathcal{C}^k$, then[12]

$${D_{(a,b)}}^k f = \sum_{i+j=k} \binom{k}{i} a^i b^j {D_1}^i {D_2}^j f.$$

[12]Here the coefficients of the binomial expansion $\binom{k}{i} = {}_k\mathrm{C}_i = \dfrac{k!}{(k-i)!i!}$ represents the number of ways of picking i unordered outcomes from k possibilities.

2.1.4 Basic Exercise

For a function $f(x, y) = xy^3 + \sin x + \cos y$, find ${D_{(a,b)}}^4 f(x, y)$.

Exercise: Chapter 11 Section 2

1. For the following function, examine if $D_1D_2f = D_2D_1f$ holds.
   (1) $2x^2y^3 - x^3y^5$ (2) $\cos^7 x - \sin^2 y$
   (3) $\arctan(xy)$ (4) $\log \dfrac{x-1}{y-1}$

2. Find all second order partial derivatives of $f(x,y) = xy^2 + ye^{-x}\sin(x-y)$.

3. For $z = \sin(x+ct) + \cos(2x+2ct)$, show that[13]

$$\frac{\partial^2 z}{\partial t^2} = c^2 \frac{\partial^2 z}{\partial x^2}.$$

4. For a $\mathcal{C}^2$ function $f(x,y)$ and any vector $\mathbf{v}$, is the second order directional derivative in the direction of $\mathbf{v}$ equal to the second order directional derivative in the direction of -$\mathbf{v}$ at each point in the domain?

5. Suppose that for a $\mathcal{C}^3$ function $f(x,y)$ defined near the origin the second order directional derivative in the direction of $\mathbf{v}$ at the origin is 1. Find the second order directional derivative in the direction of $2\mathbf{v}$. Also when the third order directional derivative of f in the direction of $\mathbf{v}$ at the origin is 1, find the third order directional derivative in the direction of $2\mathbf{v}$.

6. For a curve $(x_1(t), \ldots, x_n(t))$ and a function $f(x_1, \ldots, x_n)$, show that

$$\frac{d^2}{dt^2} f(x_1, \ldots, x_n) = \sum_{i,j=1}^{n} \frac{\partial^2 f}{\partial x_i \partial x_j} \frac{dx_i}{dt} \frac{dx_j}{dt} + \sum_{k=1}^{n} \frac{\partial f}{\partial x_k} \frac{d^2 x_k}{dt^2}.$$

7. In $\mathbb{R}^n$, the differential operator

$$\nabla^2 := {D_1}^2 + \cdots + {D_n}^2$$

is called the **Laplace operator** or **Laplacian.**[14] Show that we may express the Laplacian in $\mathbb{R}^2$ using polar coordinates as follows:

$$\nabla^2 = {\partial_x}^2 + {\partial_y}^2 = {\partial_r}^2 + \frac{1}{r}\partial_r + \frac{1}{r^2}{\partial_\theta}^2.$$

8. A function which satisfies **Laplace's equation** $\nabla^2 f = 0$ is called a **harmonic function**. Show that the following function is harmonic.
   (1) $f(x,y) = \log(x^2+y^2)$ (2) $f(x,y) = e^x \sin y$
   (3) $f(x,y,z) = x^2 + y^2 - 2z^2$ (4) $f(x,y,z) = 1/\sqrt{x^2+y^2+z^2}$

---

[13]This differential equation is called the **wave equation**.

[14]Laplace studied solutions of Laplace's equation while working on the gravity since 1782. Laplace's equation first appeared on the paper of Euler in 1752. The Laplace operator ∇^2 is often denoted as Δ.

9. Show that $\left(\dfrac{\partial}{\partial x}+x\dfrac{\partial}{\partial y}\right)^2$ and $\left(\dfrac{\partial}{\partial x}\right)^2+2x\dfrac{\partial^2}{\partial x\partial y}+x^2\left(\dfrac{\partial}{\partial y}\right)^2$ are not equal.

10. For $f(x,y)=\log(1+xy)$, find ${D_{\mathbf{v}}}^2 f$ and ${D_{\mathbf{v}}}^3 f$. And investigate whether it is convex or concave at $(1,1)$ in the direction of $\mathbf{v}=(1,2)$.

11. For a continuous function

$$f(x,y)=\begin{cases}\dfrac{xy(x^2-y^2)}{x^2+y^2}, & (x,y)\neq(0,0)\\ 0, & (x,y)=(0,0),\end{cases}$$

show that $D_{\mathbf{v}}f(0,0)=0$ for any vector $\mathbf{v}$.[15] Also show that at the origin, there exist $\dfrac{\partial^2 f}{\partial x\partial y}$ and $\dfrac{\partial^2 f}{\partial y\partial x}$, but

$$\frac{\partial^2 f}{\partial x\partial y}(0,0)\neq\frac{\partial^2 f}{\partial y\partial x}(0,0).$$

Explain why Euler's commutativity of partial derivatives (2.0.2) doesn't hold.

12. Let $\mathcal{C}^\infty(U)$ be the set of all infinitely differentiable functions defined on an open set U in $\mathbb{R}^n$. Show that for any two vectors $\mathbf{v},\mathbf{w}$, the directional differential operators

$$D_{\mathbf{v}},\ D_{\mathbf{w}}\ :\ \mathcal{C}^\infty(U)\to\mathcal{C}^\infty(U)$$

are commutative, that is,[16]

$$D_{\mathbf{v}}\circ D_{\mathbf{w}}=D_{\mathbf{w}}\circ D_{\mathbf{v}}.$$

13. If a function $f(x_1,\ldots,x_n)$ is $\mathcal{C}^k$, show that for any vector $\mathbf{v}=\sum_{i=1}^n v_i\,\mathbf{e}_i$,

$${D_{\mathbf{v}}}^k f=\sum_{i_1,\ldots,i_k=1}^{n} v_{i_1}\cdots v_{i_k}D_{i_1}\cdots D_{i_k}f.$$

[15] Here the origin is called a **critical point**.

[16] As it is a very important problem to know when two matrices are commutative, it is also a very important problem to know when two operators are commutative. An operation is called **commutative** if the order of operations doesn't matter. In general, "taking a pill after meal" is quite different from "having meal after a pill" .

3 Taylor Expansion and Approximation

Suppose that a function $f(x_1, \ldots, x_n)$ is differentiable at a point P in the domain. If a vector $\mathbf{v}$ has a small norm, then

$$f(P+\mathbf{v}) \approx f(P) + \operatorname{grad} f(P) \cdot \mathbf{v}.$$

Here

$$f(P) + \operatorname{grad} f(P) \cdot \mathbf{v}$$

is called the **linear approximation** of $f(P+\mathbf{v})$ at P.

3.0.1 Example

Find an approximation of $\left(0.99\, e^{0.02}\right)^8$.

Solution. If we let $f(x,y) := (xe^y)^8 = x^8 e^{8y}$, then

$$f(1,0) = 1, \qquad \operatorname{grad} f(x,y) = 8x^7 e^{8y}(1,x), \qquad \operatorname{grad} f(1,0) = (8,8).$$

Therefore[17]

$$\left(0.99\, e^{0.02}\right)^8 = f(0.99, 0.02) \approx f(1,0) + (8,8)\cdot(-0.01, 0.02) = 1 + 0.08 = 1.08.$$

3.0.2 Basic Exercise

1. Explain that if a point P varies by $\mathbf{v}$ in the domain of a differentiable function f, the rate of change of f with respect to the original value $f(P)$, called the relative rate of change of f, $\dfrac{f(P+\mathbf{v}) - f(P)}{f(P)}$, is approximately $\dfrac{\operatorname{grad} f(P) \cdot \mathbf{v}}{f(P)}$.
2. When there are x skilled workers and y unskilled workers in a factory, the output per day is given by $f(x,y) = 10x^2 y$. Suppose that there are 20 skilled workers and 40 unskilled workers in this factory now. If 1 more skilled worker comes, approximately by what percentage would the output increase? If 1 more unskilled worker comes, approximately by what percentage would the output increase? If 1 more skilled worker and 1 more unskilled worker come, approximately by what percentage would the output increase? (Answer[18])

[17] Indeed, $\left(0.99\, e^{0.02}\right)^8 \approx 1.082850933$.

[18] This problem require to find

$$\frac{\operatorname{grad} f(20,40) \cdot (n,m)}{f(20,40)}.$$

Taylor's theorem which we will introduce next is often used to find the approximate value of a function and the error bound.

Theorem 3.0.3 (Taylor Expansion) *Suppose that f is a k-times differentiable function defined on an open set U in $\mathbb{R}^n$, and that for a point P in U and a vector $\mathbf{v}$, the line segment $[P, P+\mathbf{v}] := \{P + t\mathbf{v} \mid 0 \le t \le 1\}$ is contained in U. Then there exists t^* between 0 and 1 such that*

$$\begin{aligned} f(P+\mathbf{v}) &= f(P) + D_{\mathbf{v}}f(P) + \frac{1}{2!}{D_{\mathbf{v}}}^2 f(P) + \cdots \\ &\quad + \frac{1}{(k-1)!}{D_{\mathbf{v}}}^{k-1} f(P) + \frac{1}{k!}{D_{\mathbf{v}}}^k f(P + t^*\mathbf{v}). \end{aligned}$$

Proof. For a function

$$g(t) := f(P + t\mathbf{v}),$$

we will use mathematical induction to prove

$$g^{(i)}(t) = {D_{\mathbf{v}}}^i f(P + t\mathbf{v}), \qquad i = 0, 1, \ldots, k.$$

First, when $i = 0$ it is obvious. By the hypothesis of induction,

$$g^{(i+1)}(t) = \frac{d}{dt}(g^{(i)}(t)) = \frac{d}{dt}({D_{\mathbf{v}}}^i f)(P + t\mathbf{v}) = \left(D_{\mathbf{v}}({D_{\mathbf{v}}}^i f)\right)(P + t\mathbf{v})$$

and therefore we complete the indcution. Now if we apply the Taylor expansion for a function of single variable to $g(t)$, we know that there exists $t^* \in (0, t)$ such that

$$g(t) = g(0) + g'(0)t + \cdots + \frac{g^{(k-1)}(0)}{(k-1)!}t^{k-1} + \frac{g^{(k)}(t^*)}{k!}t^k.$$

If we let $t = 1$ here, we are done. □

The approximate value is 10%, 2.5%, 10% + 2.5% = 12.5%, respectively. The real value is 10.25%, 2.5%, 13.00625%, respectively.

For example, if $f(x,y)$ is twice differentiable and $\mathbf{v} = a\,\mathbf{i} + b\,\mathbf{j}$, then there exists $(x^*, y^*) := (x_0 + t^*a, y_0 + t^*b)$ such that

$$\begin{aligned} & f(x_0 + a, y_0 + b) \\ = \;& f(x_0, y_0) + D_{\mathbf{v}} f(x_0, y_0) + \frac{1}{2} {D_{\mathbf{v}}}^2 f(x^*, y^*) \\ = \;& f(x_0, y_0) + \left(a \frac{\partial f}{\partial x}(x_0, y_0) + b \frac{\partial f}{\partial y}(x_0, y_0) \right) \\ & + \frac{1}{2} \left(a^2 \frac{\partial^2 f}{\partial x^2}(x^*, y^*) + 2ab \frac{\partial^2 f}{\partial x \partial y}(x^*, y^*) + b^2 \frac{\partial^2 f}{\partial y^2}(x^*, y^*) \right). \end{aligned}$$

More concretely, if

$$f(x,y) = \frac{1}{6}(x^3 + y^3), \qquad (x_0, y_0) = (0, 0),$$

at the origin the values of the function and its first derivatives are all zero and

$$\frac{\partial^2 f}{\partial x^2} = x, \quad \frac{\partial^2 f}{\partial x \partial y} = 0, \quad \frac{\partial^2 f}{\partial y^2} = y.$$

Therefore there exists $t^* = \frac{1}{3} \in (0,1)$ such that

$$\frac{1}{6}(a^3 + b^3) = f(a,b) = \frac{1}{2}\big(a^2(t^*a) + b^2(t^*b)\big).$$

3.0.4 Basic Exercise

In Theorem (3.0.3), put $k+1$ in place of k .

3.1 Approximate Polynomials and Remainders

In general, in the Taylor expansion of a function f,

$$T_k f(P, \mathbf{v}) := f(P) + D_{\mathbf{v}} f(P) + \cdots + \frac{1}{k!} {D_{\mathbf{v}}}^k f(P)$$

is called the kth-order **approximate value** of $f(P + \mathbf{v})$ at P. When we consider the vector $\mathbf{v}$ as a variable, the above equation is a kth-degree polynomial of $\mathbf{v} = (v_1, \ldots, v_n)$ and thus $T_k f(P, \mathbf{v})$ is called the kth-degree **approximate polynomial** or **Taylor polynomial** of f at P.

3.1.1 Example

For a function $f(x,y) = e^x \cos y + xy$,

$$
\begin{gathered}
D_1 f(x,y) = e^x \cos y + y, \qquad D_2 f(x,y) = -e^x \sin y + x \\
{D_1}^2 f(x,y) = e^x \cos y,\ D_1 D_2 f(x,y) = -e^x \sin y + 1,\ {D_2}^2 f(x,y) = -e^x \cos y \\
f(0,0) = 1, \qquad D_1 f(0,0) = 1, \qquad D_2 f(0,0) = 0, \\
{D_1}^2 f(0,0) = 1, \qquad D_1 D_2 f(0,0) = 1, \qquad {D_2}^2 f(0,0) = -1.
\end{gathered}
$$

Therefore at the origin, the linear approximation and second-order approximate polynomial of $f(x,y)$ are[19]

$$T_1 f(x,y) = 1 + x, \qquad T_2 f(x,y) = 1 + x + \frac{1}{2}(x^2 + 2xy - y^2).$$

3.1.2 Remainders

In the Taylor expansion of a function f,

$$R_k f(P, \mathbf{v}) := f(P + \mathbf{v}) - T_k f(P, \mathbf{v})$$

is called the kth-degree **remainder**. For example, the first degree remainder is

$$R_1 f(P, \mathbf{v}) = f(P + \mathbf{v}) - f(P) - \mathrm{grad} f(P) \cdot \mathbf{v}.$$

A function f is differentiable at a point P if and only if

$$R_1 f(P, \mathbf{v}) = o(\mathbf{v}).$$

3.1.3 Basic Exercise

For a polynomial

$$f(x,y) = a_0 + a_1 x + a_2 y + a_{11} x^2 + a_{12} xy + a_{22} y^2 + a_{111} x^3 + a_{112} x^2 y + a_{122} x y^2 + a_{222} y^3,$$

find the nth-degree approximate polynomial and remainder of f at the origin. (Here, $n = 1, 2, 3, 4$.)

[19]These equations can be easily obtained from Theorem 3.2.2.

Corollary 3.1.4 *For a $\mathcal{C}^2$ function f defined on an open set U in $\mathbb{R}^n$ and a line segment $\{P+t\mathbf{v} : 0 \le t \le 1\}$ contained in U, the first-degree remainder of f $R_1 f(P, \mathbf{v}) = f(P+\mathbf{v}) - f(P) - \text{grad} f(P) \cdot \mathbf{v}$ satisfies*[20]

$$|R_1 f(P, \mathbf{v})| \le \frac{M_2}{2} (|v_1| + \cdots + |v_n|)^2$$

where

$$M_2 := \max\{|D_i D_j f(P + t\mathbf{v})| : 1 \le i, j \le n, \quad 0 \le t \le 1\}.$$

Proof. From Taylor's theorem, there exists P^* on the line segment $[P, P + \mathbf{v}] = \{P + t\mathbf{v} : 0 \le t \le 1\}$ such that

$$R_1 f(P, \mathbf{v}) = \frac{1}{2} D_{\mathbf{v}}{}^2 f(P^*).$$

On the other hand,

$$\begin{aligned} |D_{\mathbf{v}}{}^2 f(P^*)| &= \left| \sum_{i,j=1}^{n} v_i v_j D_i D_j f(P^*) \right| \\ &\le \sum_{i,j=1}^{n} |v_i| \, |v_j| \, |D_i D_j f(P^*)| \le M_2 \sum_{i,j=1}^{n} |v_i| \, |v_j| \\ &= M_2 \left(\sum_{i=1}^{n} |v_i| \right) \left(\sum_{j=1}^{n} |v_j| = M_2 \right) = (|v_1| + \cdots + |v_n|)^2 . \end{aligned}$$

So we are done. □

We can generalize the above theorem to get the following theorem.

---

[20] Even if the function f is not $\mathcal{C}^2$ but merely twice differentiable, the above theorem is still true with $M_2 := \sup\{|D_i D_j f(P + t\mathbf{v})| : 1 \le i, j \le n, \ 0 \le t \le 1\}$.
For $\mathbf{v} = v_1 \mathbf{e}_1 + \cdots + v_n \mathbf{e}_n$.

$$\|\mathbf{v}\|_1 := |v_1| + \cdots + |v_n|$$

is called the **first-degree absolute value** of $\mathbf{v}$.

Corollary 3.1.5 *If the function f in Theorem 3.0.3 is $\mathcal{C}^{k+1}$, then*

$$|D_{\mathbf{v}}{}^{k+1}f(P+t^*\mathbf{v})| \le M_{k+1}(|v_1|+\cdots+|v_n|)^{k+1}$$

and

$$|R_k f(P,\mathbf{v})| \le M_{k+1}\frac{(|v_1|+\cdots+|v_n|)^{k+1}}{(k+1)!},$$

where[21]

$$\begin{aligned} M_{k+1} &:= M_{k+1}f(P,\mathbf{v}) \\ &:= \max\{|D_{i_0}\cdots D_{i_k}f(P+t\mathbf{v})| : 1 \le i_0,\ldots,i_k \le n,\ 0 \le t \le 1\}. \end{aligned}$$

Since $\lim\limits_{\mathbf{v}\to\mathbf{0}} \dfrac{(|v_1|+\cdots+|v_n|)^{k+1}}{|\mathbf{v}|^k} = 0$, we get the following theorem.

Corollary 3.1.6 *For a $\mathcal{C}^{k+1}$ function f and a fixed point P in its domain, the kth-degree remainder of f*

$$\mathbf{v} \mapsto R_k f(P,\mathbf{v})$$

is $o(|\mathbf{v}|^k)$.[22]

3.1.7 Example

For the function $f(x,y) = e^x \sin y$,

$$D_1 f = e^x \sin y, \quad D_2 f = e^x \cos y$$

$$D_1{}^2 f = e^x \sin y, \quad D_1 D_2 f = e^x \cos y, \quad D_2{}^2 f = -e^x \sin y.$$

[21]Even if the function f is not $\mathcal{C}^{k+1}$ but merely (k+1)-times differentiable, the above theorem is still true with $M_{k+1} := \sup\{|D_i D_j f(P+t\mathbf{v})| : 1 \le i,j \le n,\ 0 \le t \le 1\}$.

[22]If a function $g(\mathbf{v})$ defined near $\mathbf{0} \in \mathbb{R}^n$ satisfies that $g(\mathbf{0}) = 0$ and $\lim\limits_{\mathbf{v}\to\mathbf{0}} \dfrac{|g(\mathbf{v})|}{|\mathbf{v}|^k} = 0$, then $g(\mathbf{v})$ is called $o(|\mathbf{v}|^k)$. This theorem holds even when f is merely k-times differentiable.

So the second-degree approximate polynomial of $f(x, y) = e^x \sin y$ at the origin is

$$T_2 f = y + xy.$$

For example, the second-degree approximate value of $e^{0.01} \sin(0.02)$ is

$$e^{0.01} \sin(0.02) \approx 0.02 + (0.01)(0.02) = 0.0202.$$

On the other hand, all partial derivatives of the given function are of the form

$$\pm e^x \sin y, \qquad \pm e^x \cos y$$

and so when $x \geq 0$, we get

$$M_k(x) \leq e^x \qquad (k = 0, 1, \dots).$$

In particular, since[23]

$$M_3(0.01) \leq e^{0.01} \leq 1.02,$$

the error bound is

$$\begin{aligned}
\left|R_2 f((0,0), (0.01, 0.02))\right| &\leq M_3 \frac{(0.01 + 0.02)^3}{3!} \\
&\leq 1.02 \times \frac{(3 \times 0.01)^3}{6} \\
&= 1.02 \times \frac{3^3}{6} \times 10^{-6} \\
&= 4.59 \times 10^{-6}.
\end{aligned}$$

3.2 Uniqueness of Taylor Expansion

3.2.1 Example

Let us consider the other way to find approximate polynomial of the function $f(x, y) = e^x \sin y$ which we used as an example.

$$e^x = 1 + x + o(x)$$

and

$$\sin y = y + o(y^2)$$

[23] If we multiply $e^{0.01}$ 100 times, $e \approx 2.718 < 3$, but if we multiply 1.02 100 times, $(1 + 0.02)^{100} \geq 1 + 100 \cdot 0.02 = 3$.

and so[24]

$$\begin{aligned} e^x \sin y &= (1+x+o(x))\,(y+o(y^2)) \\ &= (1+x)y + \big(o(x)y + (1+x)o(y^2) + o(x)o(y^2)\big) \\ &= y + xy + o(x^2+y^2). \end{aligned}$$

Here if we use the "uniqueness of Taylor expansion" which will be introduced next,[25] we obtain

$$T_2 f(x,y) = y + xy.$$

Theorem 3.2.2 (Uniqueness of Taylor Expansion) *For k-times differentiable functions f and g defined near a point P in $\mathbb{R}^n$, if $f(P+\mathbf{x}) - g(P+\mathbf{x})$ as a function of $\mathbf{x}$ is $o\left(|\mathbf{x}|^k\right)$, then*

$$T_k f(P,\mathbf{x}) = T_k g(P,\mathbf{x}).$$

Proof. For

$$f(P+\mathbf{x}) = T_k f(P,\mathbf{x}) + o\left(|\mathbf{x}|^k\right), \qquad g(P+\mathbf{x}) = T_k g(P,\mathbf{x}) + o\left(|\mathbf{x}|^k\right),$$

if we examine their difference, we get

$$T_k f(P,\mathbf{x}) - T_k g(P,\mathbf{x}) = o\left(|\mathbf{x}|^k\right).$$

On the other hand, the left side of this equation is a polynomial $p(\mathbf{x})$ of degree less than or equal to k. Since $p(\mathbf{x})$ is $o\left(|\mathbf{x}|^k\right)$, $p(\mathbf{x})$ must be the **zero function.**[26] So we are done. □

[24]Since $\displaystyle\lim_{x,y\to 0} \frac{o(x)y}{x^2+y^2} = \lim_{x,y\to 0} \frac{o(x)}{x}\frac{xy}{x^2+y^2} = 0$, $o(x)y$ is $o(x^2+y^2)$, ...

[25]Since the difference between $f(x,y) = e^x \sin y$ and $g(x,y) = y + xy$ is $o(x^2+y^2)$, $T_2 f = T_2 g = g$.

[26]A function whose value is identically zero is called the zero function.

Exercise: Chapter 11 Section 3

1. Find the second-degree approximate polynomial of the following function at the origin $(0,0)$.

(1)	$\sin(xy)$	(2)	$\sin x \cos y$
(3)	$\sin(x^2+y^2)$	(4)	$\sin(e^y+x^2-2)$
(5)	$\log(1+xy)$	(6)	$x+xy+2y^2+x^2$
(7)	e^{x+y}	(8)	e^x-e^y
(9)	$\dfrac{1}{1-x^2-y^2}$	(10)	$e^x \tan y$
(11)	$(1+x)^y$	(12)	$\sin e^{xy}$

2. Find the third-degree approximate polynomial of the function $f(x,y) = \cos x \sin y$ at the origin $(0,0)$.

3. Show that

$$\log(1+x+2y) = x+2y-\frac{1}{2}(x^2+4xy+4y^2)+o(x^2+y^2).$$

4. When (x,y) approaches the origin, does the limit of $\dfrac{\sin(xy)}{x}$ exist? If so, find the limit. (We investigate it only on the region where the denominator is not zero.) How about the following function?

(1)	$(e^{xy}-1)/x$	(2)	$\dfrac{\log(1+x^2+y^2)}{x^2+y^2}$
(3)	$\dfrac{\cos xy - 1}{x}$	(4)	$\dfrac{\sin xy - xy}{xy}$

5. Find the 7th-degree term in the Taylor series[27] of the function

$$f(x,y) = x^3-2xy^4+(x-1)^2y^{10}.$$

6. When $f(x,y,z)$ is a polynomial of x,y,z, show that f is equal to the Taylor series of f itself.

7. Express the function x^2y+xy^2+1 as a power series of $(x-1)$ and $(y-1)$.

---

[27] At the origin, the **Taylor series** of $\mathcal{C}^\infty$ function $f(X)$ is

$$Tf(X) := \sum_{k=0}^{\infty} D_X{}^k f(O).$$

8. Show that if a $\mathcal{C}^2$ function $f(x, y)$ satisfies

$$f(tx, ty) = t^2 f(x, y)$$

for any (x, y) and any real number t, then

$$f(x, y) = \frac{1}{2}(xD_1 + yD_2)^2 f(0, 0).$$

9. When the value of the function is 1 at the center, find the linear approximate value of the following and its error bound.

   (1) $(1.02)^{1.01}$ (2) $(0.99)^{0.99}$

10. Show that if two functions f and g are $\mathcal{C}^\infty$ functions, then

$$T_k(fg)(P, \mathbf{v}) \equiv T_k f(P, \mathbf{v}) \cdot T_k g(P, \mathbf{v}) \qquad \mod o(|\mathbf{v}|^k)$$

for any point P in the domain, any vector $\mathbf{v}$ and any natural number k.

4 Critical Point Theorem

4.1 Critical Points

For a differentiable function f defined on an open set U in $\mathbb{R}^n$ and a point P in U, the directional derivative $D_{\mathbf{v}}f(P)$ in the direction of $\mathbf{v}$ satisfies

$$f(P+t\mathbf{v}) = f(P) + t\, D_{\mathbf{v}}f(P) + o(t).$$

Therefore if $D_{\mathbf{v}}f(P) > 0$, when the point P moves in the direction of $\mathbf{v}$ (that is, when t is a small positive number), the value of f increases locally[28]:

$$f(P+t\mathbf{v}) = f(P) + t\left(D_{\mathbf{v}}f(P) + \frac{o(t)}{t}\right) \geq f(P).$$

Similarly if $D_{\mathbf{v}}f(P) < 0$, when the point P moves in the direction of $\mathbf{v}$, the value of f decreases locally.

From these facts, let's show that if P is a maximum point of f, then

$$D_{\mathbf{v}}f(P) = 0, \qquad \forall\, \mathbf{v}.$$

From our assumption that P is a maximum point, the value of the function can not increase (locally) no matter which direction it moves in, and so

$$D_{\mathbf{v}}f(P) = \left.\frac{d}{dt}\right|_0 f(P+t\mathbf{v}) = \lim_{t\to 0+}\frac{1}{t}\big(f(P+t\mathbf{v}) - f(P)\big) \leq 0$$

for any vector $\mathbf{v}$. Similarly,

$$0 \geq D_{-\mathbf{v}}f(P) = -D_{\mathbf{v}}f(P)$$

and so

$$D_{\mathbf{v}}f(P) = 0, \qquad \forall\, \mathbf{v}.$$

Similarly, we can show that if P is a minimum point of f,

$$D_{\mathbf{v}}f(P) = 0, \qquad \forall\, \mathbf{v}.$$

Like this, a point at which every directional derivative of f is zero is called a **critical point** of f.

[28] The derivative at one point explains the local phenomenon, but cannot give information about values far away.

4.2 Extremum Points

For a given function, a point at which the function attains the maximum locally is called a **local maximum point**, and the point at which the function attains the minimum locally is called a **local minimum point**. A point is called an **local extremum point** or a **relative extremum point** if it is either a local maximum point or a local minimum point.[29]

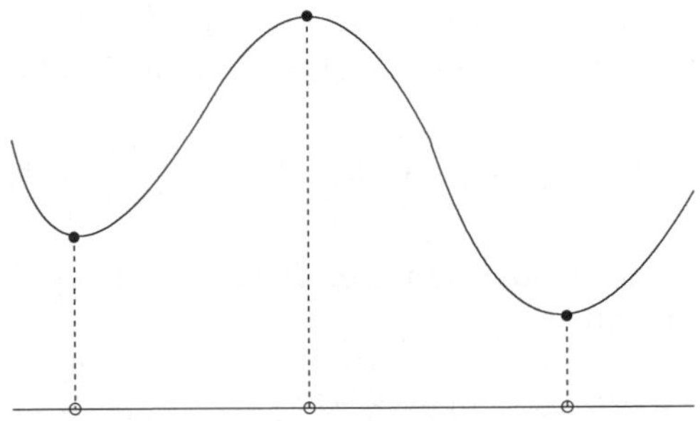

Now from the observation we made before, we get Fermat's Critical Point Theorem.

> **Theorem 4.2.1 (Critical Point Theorem)** *A local extremum point of a differentiable function defined on an open set in $\mathbb{R}^n$ is a critical point.*[30]

The critical point theorem is an example which shows that the differentiation is important to find the global extrema or local extrema. At first glance, this theorem might seem useless, because we have to investigate the directional derivatives in infinitely many directions in order to find a critical point. However, for a differentiable function f, we know that

$$D_{\mathbf{v}}f(P) = \mathbf{v} \cdot \operatorname{grad} f(P)$$

[29]The value of the function at a local maximum point is called a **local maximum (value)**, and the value at a local minimum point is called a **local minimum (value)**. Also the value of the point at a local extremum point is called an **local extreme value**.

[30]In this theorem, it is important that the domain is open.

and thus a point P where the gradient is zero

$$\operatorname{grad} f(P) = (D_1 f(P), \ldots, D_n f(P)) = O$$

is a very critical point. In other words, in order to find critical points, it suffices to examine n partial derivatives.

4.2.2 Example

The only critical point of the function $f(x, y, z) = x^2 + y^2 + z^2$ is the origin which is the minimum point.

4.2.3 Example

For given points $P_1, \ldots, P_k$ in space, let us find a point X which minimizes[31]

$$f(X) = |X - P_1|^2 + \cdots + |X - P_k|^2 .$$

If we compute the gradient vector of f,

$$\operatorname{grad} f(X) = 2\sum_{i=1}^{k}(X - P_i) = 2\big(kX - (P_1 + \cdots + P_k)\big).$$

Therefore a critical point of f— that is, a point where the gradient is zero is the very average point of $P_1, \ldots, P_k$,

$$X = \frac{1}{k}(P_1 + \cdots + P_k) .$$

This point is the minimum point of f.

---

[31]Such a problem may be thought of as the problem for building a distribution center, designing a highway, constructing a base station for mobile communication and must have the minimum value. For the given function satisfies

$$\lim_{X\to\infty} f(X) = \infty$$

and so $f(X)$ gets very big if $|X|$ is very big. Now suppose that $f(X) \geq f(O)$ if $|X| > b$. Then by the extreme value theorem, the minimum value of f is attained in a bounded, closed region $|X| \leq b$.

Exercise: Chapter 11 Section 4

1. Find critical points of the following function.

   (1) $x^2 + 4xy - y^2 - ax - by$

   (2) $x - y \sin x$

   (3) x^2y^2

   (4) $x^4 + y^2$

   (5) $e^{x^2+y^2+z^2}$

   (6) $e^{-xy/(x^2+y^2+1)}$

2. **(The Method of Least Squares)** For points $(x_1, y_1), \ldots, (x_n, y_n)$ in the plane, find the straight line $y = ax + b$ which minimizes[32]

$$\sum_{i=1}^{n} (y_i - ax_i - b)^2.$$

   (Here, we assume that $x_1, \ldots, x_n$ are all distinct.)

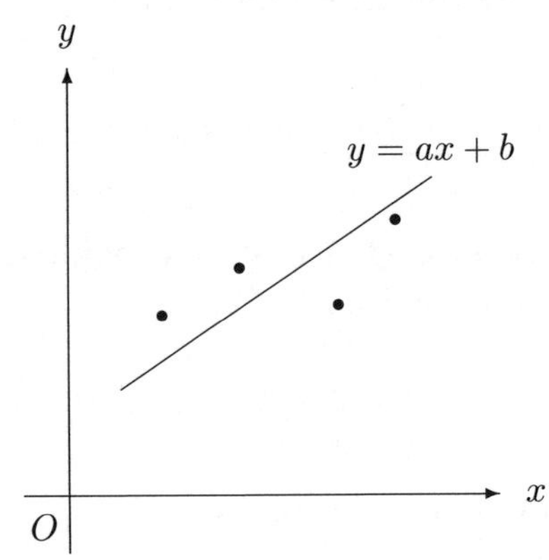

3. For the function $f(x, y) = 3x^4 - 4x^2y + y^2 = (3x^2 - y)(x^2 - y)$,

   (1) show that the origin is a critical point of f.

   (2) show that for any unit vector $\mathbf{v} = (h, k)$, the function

$$g(t) := f(t\mathbf{v}) = f(th, tk)$$

   attains a local minimum at $t = 0$.

---

[32] If a quadratic function $f(a, b)$ satisfies $f(a, b) \geq 0$ for any real numbers a, b, then f attains the minimum value.

(3) Nevertheless, show that the origin is not a local minimum point of f.[33]

(4) Sketch the curve $f(x, y) = 0$.

(5) Sketch the region where $f(x, y) > 0$ and $f(x, y) < 0$, respectively.

(6) Explain the reason why (2) doesn't contradict to (3).

4. (**Steiner's Problem**) Suppose that a triangle $\triangle ABC$ in the plane has angle less than $120°$ at each vertex. If we let P be a critical point of the function

$$f(X) := |X - A| + |X - B| + |X - C| \qquad (X \in \mathbb{R}^2 - \{A, B, C\}),$$

show that the angle between any two of line segments AP, BP, CP is $120°$.[34]

5. The following picture shows level curves of a $\mathcal{C}^1$ function. From this picture, guess the location of critical points. (Here, we assume that each level curve corresponds to a distinct value of the function.)

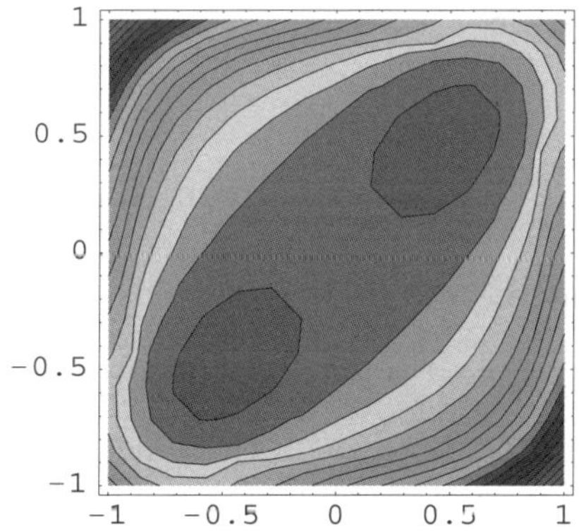

[33]Hint: Consider the curve $(x, y) = (t, 2t^2)$.

[34]Jakob Steiner (1796–1863) studied this problem. This critical point becomes a minimum point, and this point is called **Torricelli's point** or **Fermat's point**.

5 Second Derivative Test

5.1 Saddle Points

When we use differentiation to find an extremum or a local extremum, first we examine critical points, that is, points where all directional derivatives are 0. However, every critical point is not a local extremum in general.

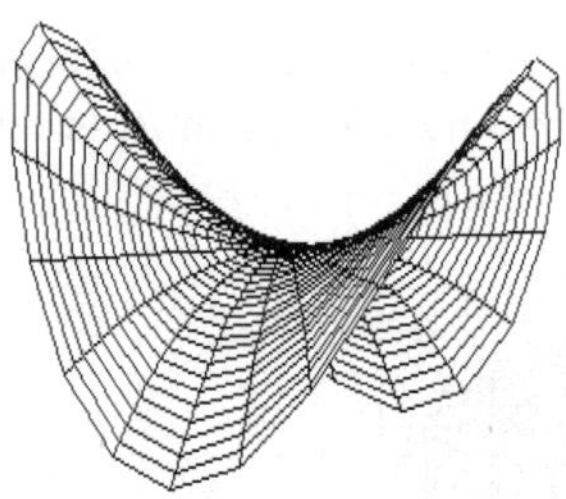

For example, consider

$$f(x, y) = xy.$$

Since

$$\frac{\partial f}{\partial x}(0,0) = 0, \qquad \frac{\partial f}{\partial y}(0,0) = 0,$$

the origin is a critical point. However, the value of the function near the origin could be positive or negative, and so it is obvious that the origin is not a local extremum. Like this, a critical point which is not a local extremum point is called a **saddle point**.

Near a saddle point, the value of the function could increase or decrease depending on the direction in which the point moves.

We may use the second partial derivatives to determine whether a critical point is a local maximum point, a local minimum point, or a saddle point.

First, recall single-variable calculus. If a point $p \in \mathbb{R}$ is a critical point of a function $f(x)$,— that is, if $f'(p) = 0$, we know that[35]

(i) $f''(p) > 0 \quad \Rightarrow \quad p$ is a local minimum point

(ii) $f''(p) < 0 \quad \Rightarrow \quad p$ is a local maximum point.

Similarly, for a function $f(x_1, \ldots, x_n)$ of several variable, we may classify critical points by examining whether "the second partial derivatives are positive or negative".

[35]For

$$f(p + x) = f(p) + \frac{1}{2}f''(p)x^2 + o(x^2)$$

near a critical point p.

5.2 Hessian

For a $\mathcal{C}^2$ function $f(x_1, \dots, x_n)$, the $n \times n$ matrix consisting of the second partial derivatives of f at a point P

$$f''(P) := \left(\frac{\partial^2 f}{\partial x_i \partial x_j}(P) \right)$$

is called the **Hessian**[36] of f at P. By the commutativity of partial derivatives, the Hessian $f''(P)$ is symmetric.

5.2.1 Example

For the function

$$f(x, y) = x^4 + y^4 - 4xy,$$

we have

$$\begin{aligned} D_1D_1f(x,y) &= 12x^2, & D_1D_2f(x,y) &= -4 \\ D_2D_1f(x,y) &= -4, & D_2D_2f(x,y) &= 12y^2. \end{aligned}$$

So the Hessian of f at $P = (1, 1)$ is

$$f''(1,1) = \begin{pmatrix} 12 & -4 \\ -4 & 12 \end{pmatrix}.$$

5.2.2 Basic Exercise

Find the Hessian of the function $f(x, y, z) = xyz$.

5.3 Positive-Definite and Negative-Definite Matrices

For a 2×2 symmetric matrix $A = \begin{pmatrix} a & b \\ b & c \end{pmatrix}$,

(i) if $a > 0$ and $\det A = ac - b^2 > 0$, then A is called **positive-definite** and denoted by $A > O$.

(ii) if $a < 0$ and $\det A = ac - b^2 > 0$, then A is called **negative-definite** and denoted by $A < O$.

---

[36]Ludwig Otto Hesse, (1811–1874) a German mathematician

5.3.1 Basic Exercise

1. Show that $\begin{pmatrix} 2 & 1 \\ 1 & 1 \end{pmatrix}$ is positive-definite.
2. For a 2×2 symmetric matrix A, show that $A > O$ if and only if $-A < O$.

Lemma 5.3.2 *Let q be a quadratic form*[37]

$$q(x, y) = ax^2 + 2bxy + cy^2 = (x\ y)\begin{pmatrix} a & b \\ b & c \end{pmatrix}\begin{pmatrix} x \\ y \end{pmatrix}$$

corresponding to a symmetric matrix $A = \begin{pmatrix} a & b \\ b & c \end{pmatrix}$.

(i) *If $A > O$, the origin $(0, 0)$ is the minimum point of $q(x, y)$.*

(ii) *If $A < O$, the origin $(0, 0)$ is the maximum point of $q(x, y)$.*

(iii) *If $\det A < 0$, the origin $(0, 0)$ is a saddle point $q(x, y)$.*

Proof. First, consider the case where $a \neq 0$. we have

$$ax^2 + 2bxy + cy^2 = a\left(\left(x + \frac{b}{a}y\right)^2 + \frac{ac - b^2}{a^2}y^2\right) \tag{11.1}$$

(i) If $A > O$, that is, $a > 0$ and $\det A > 0$, then the value of the equation (11.1) is positive for arbitrary $(x, y) \neq (0, 0)$. Therefore the origin $(0, 0)$ is the minimum point.

(ii) If $A < O$, that is, $a < 0$ and $\det A > 0$, then the value of the equation (11.1) is negative for arbitrary $(x, y) \neq (0, 0)$. Therefore the origin $(0, 0)$ is the maximum point.

---

[37] A **quadratic form** corresponding to a general $n \times n$ symmetric matrix $A = (a_{ij})$ is given by

$$q(x_1, \ldots, x_n) = (x_1\ \cdots\ x_n)A\begin{pmatrix} x_1 \\ \vdots \\ x_n \end{pmatrix} = \sum_{i,j=1}^{n} a_{ij}x_ix_j.$$

If $\det A \neq 0$, then q has a unique critical point at the origin where its value is obviously 0.

(iii) Consider the case where $\det A < 0$.

If $a \neq 0$, the equation (11.1) attains both positive and negative values. Therefore the origin is a saddle point.[38]

If $a = 0$, $0 > \det A = -b^2$ and thus $b \neq 0$. So

$$q(x, y) = 2bxy + cy^2 = (2bx + cy)y$$

attains both positive values and negative values. Therefore the origin is a saddle point. □

Theorem 5.3.3 (Second Derivative Test) *Suppose that $f(x, y)$ is a $\mathcal{C}^2$ function defined on an open set in $\mathbb{R}^2$ and P is a critical point of f.*

(i) *If $f''(P) > O$, P is a local minimum point.*

(ii) *If $f''(P) < O$, P is a local maximum point.*

(iii) *If $\det f''(P) < 0$, P is a saddle point.*

Proof. (i) If $f''(P) > O$, by the continuity of the second partial derivatives $D_i D_j f$ of f, $f''(P^*) > O$ at a point P^* near P.[39]

Since P is a critical point, if we consider the Taylor expansion of $f(P + \mathbf{v})$ for a vector $\mathbf{v} = v_1\mathbf{e}_1 + v_2\mathbf{e}_2$ of very small norm, there exists $t^* \in (0, 1)$ such that

$$f(P + \mathbf{v}) - f(P) = \frac{1}{2}\sum_{i,j=1}^{2} v_i v_j D_i D_j f(P + t^*\mathbf{v}). \tag{11.2}$$

If $\mathbf{v}$ is of very small norm, $f''(P + t^*\mathbf{v}) > O$ and thus by Lemma 5.3.2

$$f(P + \mathbf{v}) - f(P) \geq 0.$$

Therefore P is a local minimum point.

---

[38] For a quadratic form q, if $q(x_0, y_0) > 0$, then $q\big(t(x_0, y_0)\big) = t^2 q(x_0, y_0) > 0$ and thus in any neighborhood of the origin there exists a point at which $q > 0$. Similarly, if $q(x_0, y_0) < 0$, in any neighborhood of the origin there exists a point at which $q < 0$.

[39] The map

$$X \mapsto \det f''(X) = {D_1}^2 f(X) {D_2}^2 f(X) - (D_1 D_2 f(X))^2$$

is continuous.

(ii, iii) When either $f''(P) < O$ or $\det f''(P) < 0$, we can get the desired conclusion in a similar way. □

5.3.4 Example

Find local extrema of the function

$$f(x, y) = x^4 + y^4 - 4xy.$$

Since

$$\operatorname{grad} f(x, y) = 4(x^3 - y, -x + y^3),$$

There are three critical points, $(0, 0)$, $(1, 1)$, $(-1, -1)$. The Hessian of f is

$$f''(x, y) = 4 \begin{pmatrix} 3x^2 & -1 \\ -1 & 3y^2 \end{pmatrix}$$

and hence

$$f''(0, 0) = 4 \begin{pmatrix} 0 & -1 \\ -1 & 0 \end{pmatrix}, \quad f''(1, 1) = 4 \begin{pmatrix} 3 & -1 \\ -1 & 3 \end{pmatrix} = f''(-1, -1).$$

At the origin the determinant of the Hessian is negative, and so the origin is a saddle point.

On the other hand, since $f''(1, 1) = f''(-1, -1) > O$, $(1, 1)$ and $(-1, -1)$ are local minimum points.

5.3.5 Basic Exercise

Find the minimum point in the previous Example.

5.3.6 Example

Find local extrema of the function

$$f(x, y) = x^3 - 12xy + 8y^3.$$

Since

$$\operatorname{grad} f(x, y) = 3(x^2 - 4y, -4x + 8y^2),$$

its critical points are $(0, 0)$ and $(2, 1)$. Since the Hessian is

$$f''(x, y) = 3 \begin{pmatrix} 2x & -4 \\ -4 & 16y \end{pmatrix},$$

we have

$$f''(0,0) = 3\begin{pmatrix} 0 & -4 \\ -4 & 0 \end{pmatrix}, \qquad f''(2,1) = 3\begin{pmatrix} 4 & -4 \\ -4 & 16 \end{pmatrix}.$$

Since $\det f''(0,0) < 0$, the origin $(0,0)$ is a saddle point. Since $3 \cdot 4 = 12 > 0$ and $\det f''(2,1) > 0$, $(2,1)$ is a local minimum point.

5.3.7 Remark

If $\det f''(P) = 0$, we cannot use the second derivative test to determine whether a critical point P of a function $f(X)$ is a local minimum point, a local maximum point or a saddle point. When $\det f''(P) = 0$, we have to use the higher order partial derivatives or some other methods.

Exercise: Chapter 11 Section 5

1. Find the maximum and minimum values of the function

$$f(x,y) = x^4 + 2x^2y + 3xy^2 + 6xy + 3x$$

if exist.

2. Show that the function $f(x,y) = (x^2+y^2)\cos(x+2y)$ attains a local minimum value at the origin.

3. Find local maximum points, local minimum points and saddle points of the following function.

(1)	$\dfrac{1}{x} + xy - \dfrac{8}{y}$	(2)	$x^2 - y^2 - 2xy - 4x$
(3)	$x^2 + 9 + 6x\cos y$	(4)	$\log(1 + x^2 + y^2)$
(5)	$4\log x + e^y - x - y$	(6)	$x^4 + y^4 - (x+y)^2$
(7)	$x^2 + 2xy + 2y^2 + 4x$	(8)	$x^3 - y^3 + 3x^2 + 3y^2 - 9x$
(9)	$x^2 - xy + y^4$	(10)	$(x^3 - 3x)/(1 + y^2)$
(11)	$(x^2 - y^2)e^{-(x^2+y^2)/2}$		

Also find them for the following function.[40]

(12) $\sin x + \sin y + \sin(x+y)$ (13) $(x+y)^3 + (x-y)^3 - 12(x+y)$

4. Show that the following function has a saddle point at the origin.
(1) **monkey saddle**[41] $f(x,y) = x^3 - 3xy^2$
(2) **dog saddle**[42] $f(x,y) = x^3y - xy^3$

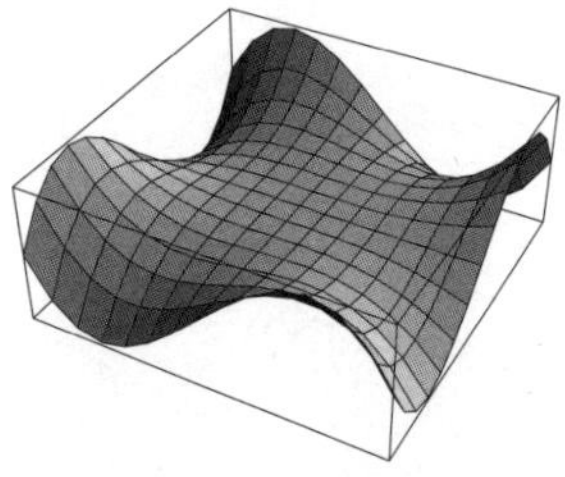

Picture: Dog saddle

[40] If you cannot use the second derivative test, use the Taylor expansion.

[41] A monkey has a tail.

[42] A dog without a tail has four legs.

5. When a level curve of a quadratic form $q(x,y) = ax^2 + 2bxy + cy^2$ doesn't pass through the origin $(0,0)$, show that this level curve is
   (1) an ellipse if $ac - b^2 > 0$,
   (2) a hyperbola if $ac - b^2 < 0$,
   (3) a straight line if $ac - b^2 = 0$.

6. For a $\mathcal{C}^2$ function f defined near a point P in $\mathbb{R}^n$ and a vector $\mathbf{v} = v_1\mathbf{e}_1 + \cdots + v_n\mathbf{e}_n$, show that

$$D_{\mathbf{v}}{}^2 f(P) = \sum_{i,j=1}^{n} v_i v_j D_i D_j f(P) = \mathbf{v} \cdot (f''(P)\mathbf{v}).$$

7. Show that for a $\mathcal{C}^2$ function $f(x,y)$

$$\frac{\partial^2 f}{\partial x^2} + \frac{\partial^2 f}{\partial y^2} \leq 0$$

at a local maximum point, and

$$\frac{\partial^2 f}{\partial x^2} + \frac{\partial^2 f}{\partial y^2} \geq 0$$

at a local minimum point.

8. Kildong's store sells two brands of apple juice. He pays 400 wons per a bottle of nationally well-known brand A, and 300 wons per a bottle of the other brand B which is less famous. He knows form his experience that when he puts the price x wons on a bottle of brand B and y wons on a bottle of brand A, $700 - 5x + 4y$ bottles of brand B and $800 + 6x - 7y$ bottles of brand A are sold. Show that in order to maximize the profit, he should put the price 530 wons on a bottle of brand B and 550 wons on a bottle of brand A.

6 Lagrange Multipliers Method

Sometimes we need to maximize or minimize a function subject to a constraint for the variables.

For instance, consider the problem to maximize

$$f(x, y) = x^2 + 2y^2$$

subject to the constraint

$$x^2 + y^2 \leq 1.$$

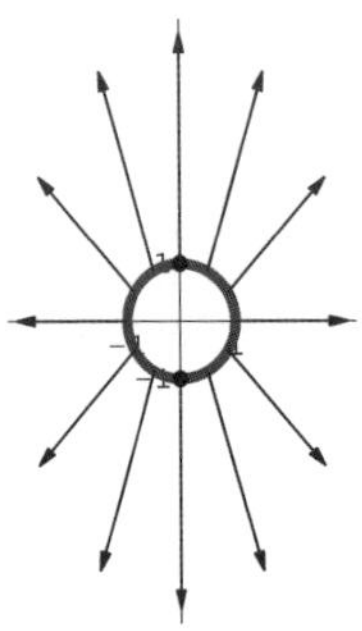

Since

$$\text{grad}\, f(x, y) = (2x, 4y),$$

the only critical point of f is the origin. Also it is obvious that the origin is the minimum point. Therefore the maximum point doesn't exists in an open region $x^2 + y^2 < 1$. On the other hand, by the extreme value theorem, a continuous function always attains the maximum value on a closed, bounded set, and thus the maximum value of f must be attained on the boundary of the domain

$$x^2 + y^2 = 1.$$

Since

$$f(x, y) = \left(x^2 + y^2\right) + y^2 = 1 + y^2$$

on the boundary, 2 is the maximum value when $y = \pm 1$. At maximum points $(0, \pm 1)$,

$$\text{grad} f(x, y) = (2x, 4y)$$

is perpendicular to the boundary. This is a general phenomenon, called the method of Lagrange multipliers.

The method of Lagrange multipliers provides the method to find local extremum points of a function when the function is constrained on a "hypersurface".[43]

[43] A level surface of a differentiable function in $\mathbb{R}^n$ is called a **hypersurface**.

Theorem 6.0.1 (Lagrange Multipliers Method) *Suppose that f and g are C^1 functions defined on an open set U in $\mathbb{R}^n$, and when f is restricted to a hypersurface*

$$S := \{X \in U \mid g(X) = c\},$$

f attains a local extreme value at $P \in S$. Then $\operatorname{grad} f(P)$ *and* $\operatorname{grad} g(P)$ *are linearly dependent. That is, there exist real numbers λ_1, λ_2, which are not both zero, such that*

$$\lambda_1 \operatorname{grad} f(P) + \lambda_2 \operatorname{grad} g(P) = \mathbf{0}.$$

In particular, if $\operatorname{grad} g(P) \neq \mathbf{0}$, *there exists a real number λ such that*[44]

$$\operatorname{grad} f(P) = \lambda \operatorname{grad} g(P).$$

Proof. When $\operatorname{grad} g(P) = \mathbf{0}$, $\operatorname{grad} f(P)$ and $\operatorname{grad} g(P)$ are obviously linearly dependent. Therefore we may assume that $\operatorname{grad} g(P) \neq \mathbf{0}$.

First, we claim that $\operatorname{grad} f(P)$ is perpendicular to (the tangent plane of) the hypersurface S at P. At P, every vector v in the tangent plane of S is given by

$$X(0) = P, \qquad X'(0) = v$$

for a curve $X(t)$ contained in S.[45] Then, since f attains a local extreme value at P, $h(t) := f(X(t))$ also attains a local extreme value when $t = 0$. Hence

$$0 = h'(0) = \operatorname{grad} f(P) \cdot X'(0)$$

and thus $\operatorname{grad} f(P)$ is perpendicular to $X'(0) = v$. Therefore $\operatorname{grad} f(P)$ is perpendicular to the level surface S at P. On the other hand, since $\operatorname{grad} g(P)$ is perpendicular to the level surface, $\operatorname{grad} f(P)$ and $\operatorname{grad} g(P)$ are parallel. It completes the proof. □

[44] This number is called a **Lagrange multiplier**.

[45] Here we need the assumption that $\operatorname{grad} g(P) \neq \mathbf{0}$. The reason for that would become clear once we study the **implicit function theorem** in Chapter 12.

6.0.2 Example

Among all rectangular boxes inscribed in a sphere of radius 1, find the one which maximizes the surface area.[46]

If we let $2x, 2y, 2z$ be the lengths of the sides, then the constraints are given by

$$g(x,y,z) := x^2 + y^2 + z^2 - 1 = 0, \quad x > 0,\ y > 0,\ z > 0,$$

and the surface area is

$$f(x,y,z) = 8(xy + yz + zx).$$

Therefore at a maximum point (x, y, z), there hold[47]

$$8(y+z, z+x, x+y) = 2\lambda(x,y,z),$$

$$x^2 + y^2 + z^2 = 1, \quad x > 0,\ y > 0,\ z > 0.$$

From the first equation, by adding all components, we get $8(x+y+z) = \lambda(x+y+z)$ and so $\lambda = 8$. From it, we get $x = y = z = 1/\sqrt{3}$. Therefore the rectangular box we want is a cube.[48]

6.0.3 Example

Find the maximum value of the function $f(x,y,z) = x^2 + 2y^2 + 3z^2$ on the sphere $x^2 + y^2 + z^2 = 1$.

The Lagrange multiplier condition, $\operatorname{grad} f = \lambda \operatorname{grad} g$, becomes

$$2(x, 2y, 3z) = 2(\lambda x, \lambda y, \lambda z)$$

and from it, we get $\lambda = 1, 2, 3$. (If $\lambda \neq 1, 2, 3$, $x = y = z = 0$ and thus we may not get a point on the sphere.)

---

[46] We need the one with large surface area to make a cooling device.

[47] In the given region, $\operatorname{grad} g \neq \mathbf{0}$. It is obvious that f attains the maximum value in the given region. If we consider f on a closed, bounded set $\{(x,y,z) \mid g(x,y,z) = 0,\ x \geq 0,\ y \geq 0,\ z \geq 0\}$, we know that the maximum exists, and we can show that a maximum point is in the interior of the set.

[48] To solve this problem, we may also use the CBS inequality

$$|xy + yz + zx| \leq |(x,y,z)| \cdot |(y,z,x)| = x^2 + y^2 + z^2.$$

When $\lambda = 1$, $y = z = 0$ and so $x = \pm 1$.
When $\lambda = 2$, $z = x = 0$ and so $y = \pm 1$.
When $\lambda = 3$, $x = y = 0$ and so $z = \pm 1$.

Now if we examine six points we found above, the maximum value of f is attained at $(0, 0, \pm 1)$[49]:

P	$(\pm 1, 0, 0)$	$(0, \pm 1, 0)$	$(0, 0, \pm 1)$
$f(P)$	1	2	3

6.0.4 Basic Exercise

Use the method of Lagrange multipliers to show that the minimum value of $f(x, y, z) = x^2 + y^2 + z^2$ subject to the constraint $yz + zx + xy = 1$ is 1.[50]

6.1 Arithmetic Mean and Geometric Mean

Consider the problem to find the maximum value of the function

$$f(x_1, \ldots, x_n) = x_1 \cdots x_n$$

on the region S consisting of points $(x_1, \ldots, x_n)$ which satisfy

$$g(x_1, \ldots, x_n) := x_1 + \cdots + x_n = 1, \qquad x_1 \geq 0, \ \ldots, \ x_n \geq 0.$$

Since a point in S satisfies

$$0 \leq x_1, \ldots, x_n \leq 1,$$

S is a closed, bounded set and hence a continuous function f has a maximum point in S, say $X = (x_1, \ldots, x_n)$. Obviously, $x_i \gneqq 0$. By the method of Lagrange multipliers,

$$\operatorname{grad} f(X) = (x_2 x_3 \cdots x_n, \ x_1 x_3 \cdots x_n, \ \ldots, \ x_1 \cdots x_{n-1})$$

is a constant multiple of

$$\operatorname{grad} g(X) = (1, 1, \ldots, 1).$$

[49]If you are interested in the theory of critical points, you may study theory of M. Morse (1892–1977).

[50]This problem can be thought of as finding the rectangular box whose diagonal has the least length among boxes with fixed surface area.

Therefore there exists a real number λ such that

$$x_2x_3\cdots x_n = \lambda, \quad x_1x_3\cdots x_n = \lambda, \quad \ldots, \quad x_1\cdots x_{n-1} = \lambda.$$

From the above equation (and the condition $g(X) = 1$), we get

$$x_1 = \cdots = x_n = \frac{1}{n},$$

and so the maximum value is $1/n^n$.

From the above equation, for any nonzero real numbers

$$a_1 \geq 0, \ \ldots, \ a_n \geq 0,$$

if we let

$$S := a_1 + \cdots + a_n,$$

then $\dfrac{a_1}{S} + \cdots + \dfrac{a_n}{S} = 1$, and so we have

$$\frac{a_1}{S}\cdots\frac{a_n}{S} \leq \frac{1}{n^n}, \qquad \text{that is,} \qquad a_1\cdots a_n \leq \left(\frac{S}{n}\right)^n.$$

From this we obtain an absolute inequality

$$(a_1\cdots a_n)^{1/n} \leq \frac{a_1 + \cdots + a_n}{n}$$

In other words, a **geometric mean** is less than or equal to an **arithmetic mean.**[51]

Here the equality holds only when $a_1 = \cdots = a_n$.

6.2 Arithmetic Mean and Quadratic Mean

Consider the problem to find the maximum value of

$$f(x_1, \ldots, x_n) = x_1 + \cdots + x_n$$

under the condition

$$g(x_1, \ldots, x_n) := x_1^2 + \cdots + x_n^2 = 1, \qquad x_1 \geq 0, \ \ldots, \ x_n \geq 0.$$

[51]There are many different ways to get this inequality. For example, it can be derived from the property of a convex function $-\log x$. For three positive numbers a, b, c, if we make the cube with the same volume as a rectangular box with sides of length a, b, c, the length of its side, $\sqrt[3]{abc}$, is the geometric mean of a, b, c.

Of course, there exists the maximum value in this problem too. If we let $X = (x_1, \ldots, x_n)$ be a maximum point, then

$$\operatorname{grad} g(X) = 2(x_1, \ldots, x_n)$$

and

$$\operatorname{grad} f(X) = (1, \ldots, 1).$$

These two vectors are parallel only when

$$x_1 = \cdots = x_n = \frac{1}{\sqrt{n}}.$$

Therefore the maximum value[52] is $n \cdot \dfrac{1}{\sqrt{n}} = \sqrt{n}$.

Now for arbitrary positive numbers

$$a_1 \geq 0, \quad \ldots, \quad a_n \geq 0,$$

if we let

$$S := \left(a_1^2 + \cdots + a_n^2\right)^{1/2},$$

then $\left(\dfrac{a_1}{S}\right)^2 + \cdots + \left(\dfrac{a_n}{S}\right)^2 = 1$, and so we get the inequality

$$\frac{a_1}{S} + \cdots + \frac{a_n}{S} \leq \sqrt{n}.$$

Therefore there holds an absolute inequality:

$$\frac{a_1 + \cdots + a_n}{n} \leq \left(\frac{a_1^2 + \cdots + a_n^2}{n}\right)^{1/2}$$

That is, the **linear mean** is less than or equal to the **quadratic mean**.[53] The equality holds only when $a_1 = \cdots = a_n$.

[52] This problem can be solved by using the CBS inequality

$$(x_1 + \cdots + x_n)^2 \leq (x_1^2 + \cdots + x_n^2)(1^2 + \cdots + 1^2).$$

[53] This inequality can be derived from the property of a convex function x^2.

6.3 Distance Between a Point and a Surface

Among the points on a curve

$$C \; : \; g(x, y) = 0$$

in the plane, let us find a point closest to the origin.[54]

Here the function that we want to **minimize** is

$$f(x, y) := \frac{1}{2}\left(x^2 + y^2\right)$$

and the domain of f is constrained to the curve

$$C := \{(x, y) \in \mathbb{R}^2 \mid g(x, y) = 0\}.$$

Obviously, f is continuous (Refer to 2.0.3 in Chapter 10).

If the function $g : \mathbb{R}^2 \to \mathbb{R}$ which defines the curve is continuous, then C is a closed set (in the plane) (Refer to Theorem 7.2.12 in Chapter 10). Now choose a very big closed ball B with the center at the origin so that this ball intersects with the curve C.[55]

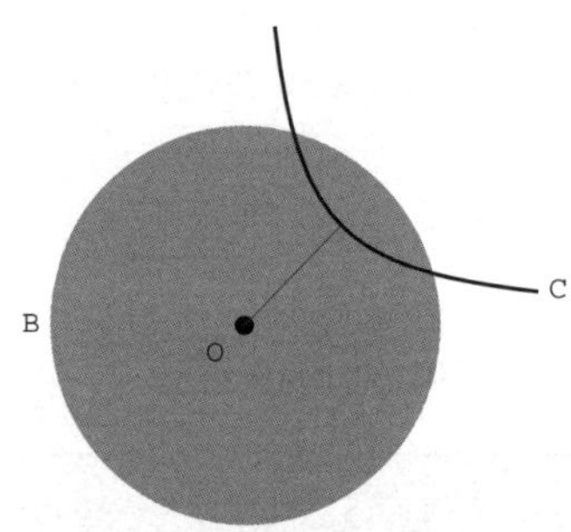

Then minimizing f is equivalent to minimizing f on $B \cap C$. On the other hand, since $B \cap C$ is closed and bounded, the extreme value theorem implies that a minimum point exists.

So the above problem must have a solution and hence it guarantees that the effort to find the solution would not be vain.

---

[54]This problem can be thought of as "the problem to build a ferry port and construct a road between some city and the ferry port".

[55]This problem implicitly assumes that the curve C is not empty.

On the other hand, since grad$f(x,y) = (x,y)$, from the method of Lagrange multipliers, we know that at a point (x,y) we try to find, grad$g(x,y) = \mathbf{0}$ or there exists a constant λ such that

$$\left(\frac{\partial g}{\partial x}(x,y), \frac{\partial g}{\partial y}(x,y)\right) = \lambda(2x, 2y).$$

In general, for a $\mathcal{C}^1$ function g defined in the whole $\mathbb{R}^n$, on the level surface $S : g = c$ there always exists a point X which minimizes the distance to a point P in n-space.[56] At a point X of the shortest distance, (grad $g(X) = 0$ or) there exists a real number λ such that[57]

$$X - P = \lambda \operatorname{grad} g(X).$$

As a concrete example, consider a hyperplane

$$S := \{(x_1, \ldots, x_n) \mid a_1x_1 + \cdots + a_nx_n = c\}.$$

If we let $A := (a_1, \ldots, a_n)$, the shortest distance between a point $P = (p_1, \ldots, p_n)$ and S is attained at a point X which satisfies

$$X - P = \lambda A, \qquad A \cdot X = c.$$

If we take the inner product of A with the first equation, we get

$$A \cdot (X - P) = \lambda |A|^2$$

and so

$$\lambda = \frac{c - A \cdot P}{|A|^2}.$$

---

[56]However, a point with maximum distance might not exist.

[57]Of course,

$$|\lambda| = \frac{|X - P|}{|\operatorname{grad} g(X)|}.$$

The given system of equations has $n+1$ unknowns $x_1, \ldots, x_n, \lambda$ and $n+1$ equations

$$x_1 - p_1 = \lambda \frac{\partial g}{\partial x_1}(x_1, \ldots, x_n), \quad \ldots, \quad x_n - p_n = \lambda \frac{\partial g}{\partial x_n}(x_1, \ldots, x_n), \quad g(x_1, \ldots, x_n) = c.$$

So it is a **well-posed** system of equations—that is, a system of which the number of equations is equal to the number of unknowns.

From this, the point X we want to find is

$$X = P + \lambda A = P + \frac{c - A \cdot P}{|A|^2} A$$

and the distance is

$$|X - P| = |\lambda A| = \frac{|c - A \cdot P|}{|A|} = \frac{|a_1 p_1 + \cdots + a_n p_n - c|}{\sqrt{a_1^2 + \cdots + a_n^2}}.$$

Exercise: Chapter 11 Section 6

1. Find the maximum and minimum values of $f(x, y) = x^2 + y^2 - x - y + 1$ on a disk $x^2 + y^2 \leq 1$.

2. Among the cylinders inscribed in a sphere of radius 1, find the surface area of the cylinder with the maximum surface area.(We assume that cylinders have neither the bottom nor the top.)

3. Find the extreme values of the function f subject to the given constraint.

(1)	$f(x,y) = xy,$	$x + y = 1$
(2)	$f(x,y) = \cos^2 x + \cos^2 y,$	$x + y = \pi/4$
(3)	$f(x,y) = x - 3y,$	$x^2 + y^2 = 1$
(4)	$f(x,y) = xy,$	$2x + y \leq 10,\ x \geq 0,\ y \geq 0$
(5)	$f(x,y) = x + y,$	$x^2 + y^2 = 1$
(6)	$f(x,y) = x - y,$	$x^2 + y^2 = 2$
(7)	$f(x,y) = 3x + 2y,$	$2x^2 + 3y^2 \leq 3$
(8)	$f(x,y,z) = x^2 + 3xy + \dfrac{1}{2}y^2 - y - 10\ ,$	$x + y = 10$
(9)	$f(x,y,z) = 3x^2 + 2y^2 + z^2 + 4x + y + 3z,$	$x + y + z = 4$

4. (**Kepler's wine barrel problem**[58]) Among the right circular cylinders with diagonal[59] of a fixed length, find the one which maximizes the volume. Also find the dimensions of the rectangular box with maximum volume that can be inscribed in a given sphere.

5. Use the method of Lagrange multipliers to prove the CBS inequality.

6. For $0 < p < q$ and $a_1 \geq 0,\ \ldots,\ a_n \geq 0$, prove the inequalities

$$(a_1^q + \cdots + a_n^q)^{1/q} \leq (a_1^p + \cdots + a_n^p)^{1/p}\,,$$

$$\left(\frac{a_1^p + \cdots + a_n^p}{n}\right)^{1/p} \leq \left(\frac{a_1^q + \cdots + a_n^q}{n}\right)^{1/q}.$$

7. Find the minimum distance from a paraboloid $z = x^2 + y^2$ to a point $(1, 1, 0)$.

8. Find all points on the surface $z = xy$ which are closest to a point $(0, 0, 5)$.

9. Find the distance from a point $(1, 2, 0)$ to a circular cone $z^2 = x^2 + y^2$.

---

[58]Kepler noted in his book "New Solid Geometry of Wine Barrel" that he was astonished by an event, occurred in the fall of 1613, that a salesman came and measured all the barrels just once to find out the volume [Tikhomirov].

[59]The longest line segment contained in the right circular cylinders

10. Find the distance from the origin to the following surface.
    (1) $z = \sqrt{x^2 - 1}$ (2) $z = 6xy + 7$
    (3) $z = 1/xy$ (4) $x - y + 2z = 3$

11. In order to make a rectangular box of volume $2000\,\text{cm}^3$, it costs 3 wons per 1cm^2 for the top, 1 won per 1cm^2 for the bottom and sides. Find the dimensions of the box with minimum total cost.

12. Find the dimensions of the open rectangular box with maximum volume whose surface area is 48.

13. Find the hottest point on the sphere $x^2 + y^2 + z^2 = 1$ where the temperature distribution is given by $T(x, y, z) = xz + yz$.

14. For an $n \times n$ symmetric matrix $Q = (q_{ij})$ and a vector $X = (x_1, \cdots, x_n)^t$, the **quadratic form** is given by

$$q(X) := X \cdot QX = \sum_{i,j=1}^{n} q_{ij}x_i x_j = \sum_{i=1}^{n} q_{ii}x_i^2 + 2\sum_{i<j} q_{ij}x_i x_j.$$

Show that the gradient vector of q is $\operatorname{grad} q(X) = 2QX$. Also when q is restricted on the unit sphere $S : x_1^2 + \cdots + x_n^2 = 1$, show that at a maximum point X, there exists a real number λ such that $QX = \lambda X$.[60]

15. When nonnegative real numbers x, y, z satisfy $x+y+z = 1$, find the maximum and minimum values of $xy + yz + zx - 2xyz$.[61]

16. Under the condition

$$g(x, y, z) := x + y - 2 \geq 0, \qquad h(x, y, z) := z^2 - 1 \geq 0,$$

find the minimum value of the function

$$f(x, y, z) := x^4 + 2y^2 + 4z^4 - 2z^2.$$

17. Find the maximum and minimum distances of the curve $x^6 + y^6 = 1$ to the origin.

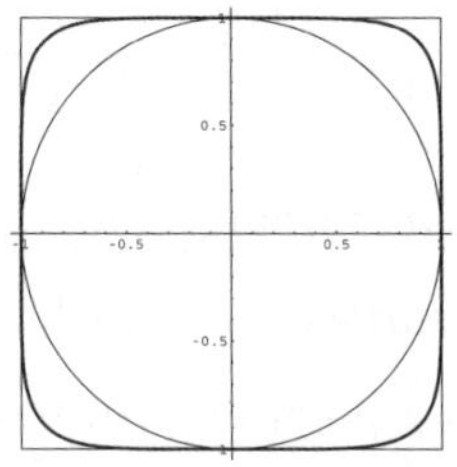

[60]Such a vector X is called an **eigenvector** of the matrix Q. Conversely, the existence of solutions of this equation tells us that "any symmetric matrix has an eigenvector".

[61]A problem of the 25th International Mathematics Olympiad problem (Prague, 1984).

7 Proof of Leibniz's Rule and Calculus of Variations

7.1 Proof of Leibniz's rule

Theorem 7.1.1 *Suppose that for a continuous function $f : [a,b]\times[c,d] \to \mathbb{R}$, its partial derivative*

$$\frac{\partial f}{\partial x} : [a,b] \times [c,d] \to \mathbb{R}$$

exists and is continuous.[62] *Then*

$$\frac{d}{dx}\int_c^d f(x,y)\,dy = \int_c^d \frac{\partial f}{\partial x}(x,y)\,dy.$$

Proof. We need to show that on the interval[63] $[x, x+\Delta x]$, for the function

$$x \mapsto F(x) := \int_c^d f(x,y)\,dy,$$

its average rate of change

$$\frac{\Delta F}{\Delta x} := \frac{1}{\Delta x}\big(F(x+\Delta x) - F(x)\big) = \frac{1}{\Delta x}\left(\int_c^d f(x+\Delta x, y)\,dy - \int_c^d f(x,y)\,dy\right)$$

is very close to $\displaystyle\int_c^d \frac{\partial f}{\partial x}(x,y)\,dy$ if $|\Delta x|$ is very small.

By the mean value theorem, there exists $x^* \in (x, x+\Delta x)$ such that

$$\frac{\Delta F}{\Delta x} = \int_c^d \frac{f(x+\Delta x, y) - f(x,y)}{\Delta x}\,dy = \int_c^d \frac{\partial f}{\partial x}(x^*, y)\,dy.$$

Therefore

$$\frac{\Delta F}{\Delta x} - \int_c^d \frac{\partial f}{\partial x}(x,y)\,dy = \int_c^d \left(\frac{\partial f}{\partial x}(x^*,y) - \frac{\partial f}{\partial x}(x,y)\right)\,dy. \tag{11.3}$$

[62]If you think of the conclusion, all assumptions are natural. A $\mathcal{C}^1$ function satisfies these assumptions. When we consider a differentiable function, usually we consider a function defined on an open interval. We say that a function defined on a closed interval $[a,b]$ is differentiable if it is a restriction of a differentiable function defined on an open interval containing the closed interval $[a,b]$.

[63]When $\Delta x < 0$, we let $[x, x+\Delta x] := [x+\Delta x, x]$.

Meanwhile, since $\dfrac{\partial f}{\partial x}$ is continuous, no matter how small a positive number ϵ is, if $|\Delta x|$ is sufficiently small,[64]

$$\left|\frac{\partial f}{\partial x}(x^*,y) - \frac{\partial f}{\partial x}(x,y)\right| \le \epsilon.$$

Therefore from the equation (11.3), we get

$$\left|\frac{\Delta F}{\Delta x} - \int_c^d \frac{\partial f}{\partial x}(x,y)\,dy\right| \le \int_c^d \left|\frac{\partial f}{\partial x}(x^*,y) - \frac{\partial f}{\partial x}(x,y)\right| dy \le \epsilon(d-c).$$

Now if we make Δx approach 0, $\displaystyle\lim_{\Delta x\to 0}\frac{\Delta F}{\Delta x} = \frac{d}{dx}\int_c^d f(x,y)\,dy$ and so

$$\left|\frac{d}{dx}\int_c^d f(x,y)\,dy - \int_c^d \frac{\partial f}{\partial x}(x,y)\,dy\right| \le \epsilon(d-c).$$

On the other hand, since ϵ is an arbitrary positive number,

$$\frac{d}{dx}\int_c^d f(x,y)\,dy - \int_c^d \frac{\partial f}{\partial x}(x,y)\,dy = 0.$$

□

7.2 Calculus of Variations

One of the greatest achievements of Lagrange is the **calculus of variations.**[65] It is believed that the calculus of variations comes from the **brachistochrone** problem[66] in modern sense.

In 1696, Johann Bernoulli in the famous family of Bernoulli in Swiss, posed the following problem in Acta Eruditorum.[67]

> Given two points O and A in a vertical plane, what is the curve traced out by a point acted on only by gravity, which starts at O and reaches A in the shortest time?

---

[64] A continuous function defined on a compact set is **uniformly continuous**.

[65] Historically, the calculus of variations can be found in the problems like "the problem of Dido, the princess of Tyre in Phoenicia, " **isoperimetric problem** [Berger, II].

[66] In Greek, *brakhus* means 'short' and *khronos* means 'time' [Schwartzman].

[67] Acta Eruditorum is the first scientific journal, published from 1682.

Unfortunately, Johann Bernoulli didn't mention the work of a pioneer of modern natural science, Galileo Galilei (1564–1642). Galileo already knew that the motion along a straight line is not the one which will minimize the time taken.

Daniel Bernoulli, who is famous for botany, anatomy, physics, is Johann's son.

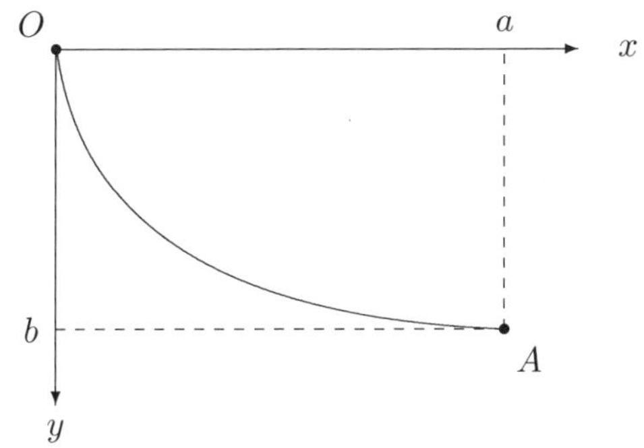

The answers for this problem were given by several mathematicians like Leibniz, L'Hôpital, Jacob (= James Bernoulli, 1654–1705) who taught mathematics to Johann, and by some anonymous one. Of course, Johann himself had the answer. When he saw the answer from the anonymous one, he said that "we know the lion by his claw."[68] They all had the right answer, the cycloid, which was the very example that Galileo gave. The cycloid is the curve traced by a fixed point on a circle as it rolls along a straight line and expressed as the following equation:

$$x(t) = r(t - \sin t), \qquad y(t) = r(1 - \cos t),$$

where r is the radius of the rolling circle and t is the angle from the center of the circle, and the curve is parameterized so that the period becomes 2π.

Let's investigate why the brachistochrone is the cycloid.

On the vertical plane, set the (rectangular) coordinate system so that O lies at the origin, the x-axis is the horizontal direction and the positive y-axis is the vertical direction downward. Let (a, b) be the coordinates of A. Here, a and b are both positive. Now if we let

$$(x, f(x))$$

be the trajectory of a moving point P,

$$f(0) = 0, \qquad f(a) = b, \qquad f'(x) \geq 0.$$

If we let g be the gravitational constant[69] and m be the mass of P, then the force acting on $P = (x, y)$ is

$$\mathbf{F}(x, y) = mg(0, 1) = -\text{grad}(-mgy).$$

Thus the **potential energy** of P is given by[70]

$$\varphi(x, y) = -mgy.$$

[68] It turned out that the anonymous one was Newton.

[69] On the ground, $g = 9.8$ m/sec^2.

[70] The potential energy can differ by some constant depending on how to define it. However the difference between two potential energies is well-defined.

On the other hand, by the **conservation of energy**, the sum of **kinetic energy** and potential energy of P

$$\frac{1}{2}mv^2 + (-mgy)$$

is constant. Since the energy of P at the origin is zero,

$$\frac{1}{2}mv^2 + (-mgy) = 0$$

and so the speed of P at any position becomes[71]

$$v = \sqrt{2gy}.$$

The time taken for P to move from $(x, f(x))$ to $(x + \Delta x, f(x + \Delta x))$ is[72]

$$\Delta T \simeq \frac{\sqrt{(\Delta x)^2 + (f(x + \Delta x) - f(x))^2}}{v} \simeq \frac{\sqrt{1 + f'(x)^2}}{\sqrt{2gf(x)}} \cdot \Delta x,$$

and so the time taken for P to move from the origin to A is

$$T = \int_0^a \frac{\sqrt{1 + f'(x)^2}}{\sqrt{2gf(x)}}\, dx.$$

Therefore the brachistochrone problem is the following:

"what is the function f which minimizes T?"

The **functional**[73] T is defined on an infinite-dimensional space

$$\mathcal{F} := \{f \in \mathcal{C}^1[0, a] \mid f(0) = 0,\ f(a) = b\}.$$

Later, the extreme value problems in an infinite-dimensional space were developed further by Euler, who was a student of Johann Bernoulli and Daniel's friend. Euler's work glittered more by young Lagrange later. The calculus in infinite-dimensional spaces became developed rigorously in the 20th century. It was not until 1934 that the method of Lagrange multipliers in finite-dimensional spaces was extended to infinite-dimensional spaces by Russian Lyusternik .

Let us consider the theory of Euler and Lagrange. Suppose that a function $f \in \mathcal{F}$ is a $\mathcal{C}^2$ function and a minimum point of a functional

$$f \mapsto \int_0^a L(f(x), f'(x))\, dx.$$

---

[71]The velocity of a body depends only on the falling distance, not on the trajectory. This principle is called **Galileo's Law (of fall)**.

[72]$dt = \dfrac{ds}{v}$

[73]A function defined on a space of functions is called a functional.

Here we assume that $L(y, z)$ is a $\mathcal{C}^2$ function. Then for any $\mathcal{C}^1$ function

$$h : [0, a] \to \mathbb{R}, \qquad h(0) = 0 = h(a),$$

if we let

$$g(t) := \int_0^a L\left(f(x) + th(x), f'(x) + th'(x)\right) dx,$$

then $g(t)$ attains its minimum at $t = 0$ and so[74]

$$0 = \left.\frac{d}{dt}\right|_{t=0} g(t) = \left.\frac{d}{dt}\right|_{t=0} \int_0^a L\left(f(x) + th(x), f'(x) + th'(x)\right) dx.$$

Therefore we get

$$\begin{aligned} 0 &= \int_0^a \left.\frac{d}{dt}\right|_0 L(f(x) + th(x), f'(x) + th'(x))\, dx \\ &= \int_0^a D_1L(f(x), f'(x)) \cdot h(x) + D_2L(f(x), f'(x)) \cdot h'(x)\, dx \\ &= \int_0^a \left(D_1L(f(x), f'(x)) - \frac{d}{dx} D_2L(f(x), f'(x))\right) h(x)\, dx. \end{aligned}$$

Here, we used the integration by parts and the boundary condition $h(0) = h(a) = 0$ to derive the last equation.

Since h is given arbitrarily,[75] we get

$$\boxed{D_1L(f(x), f'(x)) - \frac{d}{dx} D_2L(f(x), f'(x)) = 0}$$

This equation is the very **Euler-Lagrange equation**. To both sides of Euler-Lagrange equation, if we multiply $f'(x)$ and then add $f''(x) \cdot D_2L(f(x), f'(x))$, we get the equation

$$\begin{aligned} & f'(x)D_1L(f(x), f'(x)) + f''(s) \cdot D_2L(f(x), f'(x)) \\ = \; & f'(x)\frac{d}{dx} D_2L(f(x), f'(x)) + f''(x) \cdot D_2L(f(x), f'(x)) \end{aligned}$$

and thus

$$\frac{d}{dx} L(f(x), f'(x)) = \frac{d}{dx}\left(f'(x) \cdot D_2L(f(x), f'(x))\right).$$

Therefore we conclude that[76]

---

[74] This is Fermat's **critical point theorem**. The critical point theorem was discovered long before calculus was born. Fermat said that "At a critical point of a function, its value almost doesn't change." As the mountain is not steep any longer at the peak.

[75] If a continuous function $g : [a, b] \to \mathbb{R}$ satisfies $\int_a^b g(x)h(x)\, dx = 0$ for any $\mathcal{C}^\infty$ function $h : [a, b] \to \mathbb{R}$ with $h(a) = 0 = h(b)$, then $g(x) = 0$.

[76] We denote the constant of integration by c.

$$L(f(x), f'(x)) = f'(x) \cdot D_2 L(f(x), f'(x)) + c$$

If we apply this conclusion to the problem of brachistochrone

$$L(y, z) = \frac{\sqrt{1+z^2}}{\sqrt{2gy}},$$

then

$$\frac{\sqrt{1+y'(x)^2}}{\sqrt{2gy}} = y' \frac{y'}{\sqrt{2gy}\sqrt{1+y'(x)^2}} + c$$

and it can be rearranged to get the differential equation

$$y\left(1+y'(x)^2\right) = \frac{1}{2gc^2} =: 2r.$$

This can be rewritten as

$$dx = \frac{\sqrt{y}}{\sqrt{2r-y}}\, dy$$

and if we let

$$y(u) := r(1-\cos u),$$

then

$$x(u) = r(u - \sin u)$$

and the curve we want is the cycloid.

7.2.1 Basic Exercise

If $a/b > \pi/2$, show that the trajectory we've got above passes through the lowest point in the cycloid. Also show that $\dfrac{du}{dt}$ is constant.

Chapter 11 Advanced Exercise

1. Show that for a vector $\mathbf{v}$ and infinitely differentiable functions f, g in $\mathbb{R}^n$,

$$D_{\mathbf{v}}{}^k(fg) = \sum_{a+b=k} \binom{k}{a} (D_{\mathbf{v}}{}^a f)\,(D_{\mathbf{v}}{}^b g), \qquad k = 0, 1, 2, \ldots.$$

2. (**Laplacian**) In spherical coordinates, show that

$$\nabla^2 f = \frac{1}{\rho^2}\frac{\partial}{\partial\rho}\left(\rho^2\frac{\partial f}{\partial\rho}\right) + \frac{1}{\rho^2\sin\theta}\frac{\partial}{\partial\theta}\left(\sin\theta\frac{\partial f}{\partial\theta}\right) + \frac{1}{\rho^2\sin^2\theta}\frac{\partial^2 f}{\partial\varphi^2}.$$

3. Determine whether the following system of equations has a solution.

$$x^3 + 2xy - 3ye^{\sin x}\cos x = -7, \qquad y^5 + x^2 - 3e^{\sin x} = 5.$$

4. (**Hölder inequality**) For two positive numbers p, q with $\frac{1}{p} + \frac{1}{q} = 1$, show that the inequality

$$|a \cdot b| \le (|a_1|^p + \cdots + |a_n|^p)^{1/p}\,(|b_1|^q + \cdots + |b_n|^q)^{1/q}$$

for any vectors

$$a = (a_1, \ldots, a_n), \qquad b = (b_1, \ldots, b_n).$$

5. For a function $L(x, y, z)$, show that the Euler-Lagrange equation of the functional

$$\mathcal{L}(f) := \int_a^b L(u, f(u), f'(u))\,du$$

defined on the space $\mathcal{F} = \{f \in C^1[a, b] \mid f(a) = y_a,\ f(b) = y_b\}$ is

$$D_2 L(u, f(u), f'(u)) = \frac{d}{du} D_3 L(u.f(u), f'(u)).$$

6. In the brachistochrone problem, find the time taken when f is a straight line or a circular arc.

7. How should a ski slope be designed to get good record?

Chapter 12

Vector-Valued Functions

No human investigation can be called real science
if it can not be demonstrated mathematically.

— Leonardo da Vinci (1452–1519)

In this chapter, we study the differentiation of vector-valued functions of several variables. The differentiation of such a function is expressed as a Jacobian matrix. When the domain and range have the same dimension, we can consider the determinant of the Jacobian matrix and we will see that its absolute value represents the rate of change in volume. Also we study the chain rule for functions of several variables. Then we introduce the "Inverse Function Theorem," which is the extension of the "Inverse Function Theorem" for functions of a single variable, and derive the "Implicit Function Theorem".

1 Jacobian Matrices

In this chapter, we study how to differentiate a function whose domain and range are of dimension larger than 1, like the function

$$F(x, y) = (e^x \cos y, e^x \sin y).$$

A vector-valued function defined on an open set[1] in $\mathbb{R}^n$

$$F : \mathbb{R}^n \dashrightarrow \mathbb{R}^m$$

can be expressed as

$$F(x_1, \ldots, x_n) = (f_1(x_1, \ldots, x_n), \ldots, f_m(x_1, \ldots, x_n))$$

when we denote each component of F by $f_1, \ldots, f_m$.[2]

If all components f_i of F at a point P in the domain are continuous, then we call F **continuous** at P. Similarly, if all f_i's are differentiable (or $\mathcal{C}^1$, $\mathcal{C}^2$, ...), then F is called **differentiable** (or $\mathcal{C}^1$, $\mathcal{C}^2$, ...) at P.

If F is continuous (or differentiable, $\mathcal{C}^1$, $\mathcal{C}^2$, ...) at every point in the domain, then F is called **continuous** (or **differentiable**, $\mathcal{C}^1$, $\mathcal{C}^2$, ...).

If a vector-valued function $F = (f_1, \ldots, f_m)$ is differentiable at P, then we can form a matrix $F'(P)$ with the first row $f_1'(P) = \operatorname{grad} f_1(P)$, the

[1] When the domain of a function F is a subset of $\mathbb{R}^n$ but is not specified, we denotes it as $F : \mathbb{R}^n \dashrightarrow \mathbb{R}^m$. In general, a function defined on a subset of a set A which maps into a set B will be denoted as $A \dashrightarrow B$.

[2] It won't be convenient to write it by

$$F\begin{pmatrix} x_1 \\ \vdots \\ x_n \end{pmatrix} = \begin{pmatrix} f_1\begin{pmatrix} x_1 \\ \vdots \\ x_n \end{pmatrix} \\ \vdots \\ f_m\begin{pmatrix} x_1 \\ \vdots \\ x_n \end{pmatrix} \end{pmatrix},$$

as in the theory of matrices.

second row $f_2'(P) = \operatorname{grad} f_2(P)$, ..., the last row $f_m'(P) = \operatorname{grad} f_m(P)$:

$$F'(P) := \frac{\partial(f_1,\ldots,f_m)}{\partial(x_1,\ldots,x_n)}(P) := \begin{pmatrix} \frac{\partial f_1}{\partial x_1}(P) & \frac{\partial f_1}{\partial x_2}(P) & \cdots & \frac{\partial f_1}{\partial x_n}(P) \\ \frac{\partial f_2}{\partial x_1}(P) & \frac{\partial f_2}{\partial x_2}(P) & \cdots & \frac{\partial f_2}{\partial x_n}(P) \\ \vdots & & \ddots & \vdots \\ \frac{\partial f_m}{\partial x_1}(P) & \frac{\partial f_m}{\partial x_2}(P) & \cdots & \frac{\partial f_m}{\partial x_n}(P) \end{pmatrix}.$$

It is called the **Jacobian matrix**[3] of F at P.

1.0.1 Example

The Jacobian matrix of the function $F(x, y) = (e^x \cos y, e^x \sin y)$ is

$$F'(x,y) = \begin{pmatrix} e^x \cos y & -e^x \sin y \\ e^x \sin y & e^x \cos y \end{pmatrix}.$$

In particular, the Jacobian matrix at the origin is

$$F'(0,0) = \begin{pmatrix} 1 & 0 \\ 0 & 1 \end{pmatrix}.$$

1.0.2 Example

The Jacobian matrix of the function $F(x, y) = (xy, \sin x, x^2 y)$ is

$$F'(x,y) = \begin{pmatrix} y & x \\ \cos x & 0 \\ 2xy & x^2 \end{pmatrix}.$$

In particular, at $P = (\pi, 1)$,

$$F'(\pi,1) = \begin{pmatrix} 1 & \pi \\ -1 & 0 \\ 2\pi & \pi^2 \end{pmatrix}.$$

---

[3]Karl Jacobi (1804–1851), German mathematician.
The Jacobian matrix is often called just the **Jacobian**.

1.0.3 Basic Exercise

1. Find the Jacobian matrix of the function $F(x,y) = \left(x^2+y^2, e^{xy}\right)$. Also find the Jacobian matrix of F at $P=(1,1)$.
2. Find the Jacobian matrix of the identity map
$$\mathrm{id} : \mathbb{R}^n \to \mathbb{R}^n.$$
3. Find the Jacobian matrix of the linear transformation
$$\begin{aligned} u &= a_{11}x + a_{12}y + a_{13}z \\ v &= a_{21}x + a_{22}y + a_{23}z \\ w &= a_{31}x + a_{32}y + a_{33}z. \end{aligned}$$

A matrix can be thought of as a linear transformation. The following theorem tells us that at a point P, the change of the function $F : \mathbb{R}^n \dashrightarrow \mathbb{R}^m$

$$\Delta F(P, \mathbf{v}) := F(P+\mathbf{v}) - F(P)$$

is very close to the linear transformation

$$F'(P) : \mathbb{R}^n \to \mathbb{R}^m, \qquad \mathbf{v} \mapsto F'(P)\mathbf{v} \qquad (\mathbf{v} \in \mathbb{R}^n)$$

— i.e. it is the Jacobian matrix $F'(P)$ that is the **derivative** of the vector-valued function F.

Theorem 1.0.4 *If a function $F : \mathbb{R}^n \dashrightarrow \mathbb{R}^m$ defined on an open set in $\mathbb{R}^n$ is differentiable at a point P in the domain, then*[4]

$$F(P+\mathbf{v}) - F(P) - F'(P)\mathbf{v} = o(\mathbf{v}). \tag{12.1}$$

---

[4]For a vector-valued function $G = (g_1, \ldots, g_m) : \mathbb{R}^n \dashrightarrow \mathbb{R}^m$ defined near the origin in $\mathbb{R}^n$,

$$G(\mathbf{v}) = o(\mathbf{v})$$

means that $G(\mathbf{0}) = \mathbf{0}$ and

$$\lim_{\mathbf{v}\to\mathbf{0}} \frac{|G(\mathbf{v})|}{|\mathbf{v}|} = 0.$$

It is equivalent to $g_i(\mathbf{v}) = o(|\mathbf{v}|)$ for all i.

Proof. It suffices to prove that each component of the left side of the equation (12.1) is $o(\mathbf{v})$. The i-th component is

$$f_i(P+\mathbf{v}) - f_i(P) - \operatorname{grad} f_i(P) \cdot \mathbf{v}$$

and we already knew that it is $o(\mathbf{v})$. So we are done. □

1.0.5 Basic Exercise

Find the Jacobian matrix of the following map F at the origin.

$$F(x,y,z) = (x+y^2, y+z^2, z+x^2) = (x,y,z) + (y^2, z^2, x^2)$$

1.1 Jacobian Determinants

If the domain and range of a function F have the same dimension, then Jacobian matrix $F'(P)$ is a square matrix and so we can consider its determinant $\det F'(P)$. It is called the **Jacobian determinant** of F, or simply the **Jacobian** of F.

1.1.1 Example

When a function

$$F : \mathbb{R}^n \to \mathbb{R}^n$$

is a linear transformation, each component of F is expressed as

$$\begin{aligned} f_1 &= f_{11}x_1 + \cdots + f_{1n}x_n \\ &\vdots \\ f_n &= f_{n1}x_1 + \cdots + f_{nn}x_n. \end{aligned}$$

At any point $X = (x_1, \ldots, x_n)$, the Jacobian matrix of F is

$$F'(X) = (f_{ij}) = F,$$

and so the Jacobian determinant is

$$\det F'(X) = \det F.$$

For a linear transformation on $\mathbb{R}^n$, the absolute value of its determinant measures how much the unit volume is being expanded or shrunk by the linear transformation.

1.1.2 Example

For the transformation from polar coordinates to rectangular coordinates given by

$$(x, y) = F(r, \theta) = (r\cos\theta, r\sin\theta)$$

the Jacobian matrix is

$$\frac{\partial(x,y)}{\partial(r,\theta)} = \begin{pmatrix} \cos\theta & -r\sin\theta \\ \sin\theta & r\cos\theta \end{pmatrix}$$

and so the Jacobian determinant is[5]

$$\boxed{\det \frac{\partial(x,y)}{\partial(r,\theta)} = r}$$

Geometrically, the absolute value of the Jacobian determinant means the **net rate of change of area.**[6] In this example, the transformation maps an ordinary rectangle in the domain

$$r_0 \le r \le r_0 + \Delta r, \qquad \theta_0 \le \theta \le \theta_0 + \Delta\theta$$

into a polar rectangle in xy-plane. Here the rate of change of area is

$$\frac{\frac{1}{2}\left((r_0+\Delta r)^2 - r_0^2\right)\cdot\Delta\theta}{\Delta r \cdot \Delta\theta} = r_0 + \frac{1}{2}\Delta r.$$

Now if we let both Δr and $\Delta\theta$ approach 0, then the rate of change becomes

$$r_0 = \det F'(r_0, \theta_0).$$

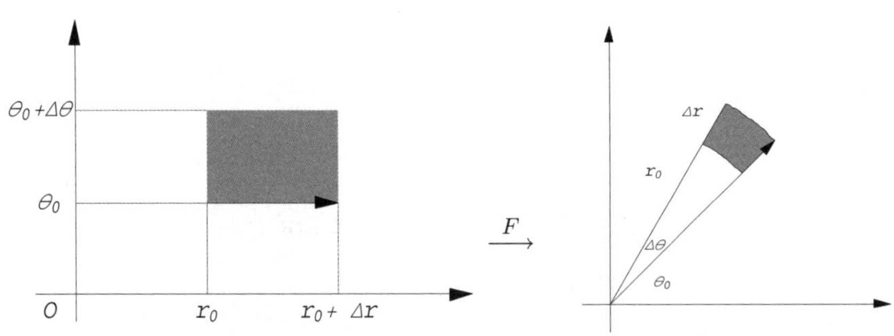

[5]It would be helpful to memorize this formula until the final exam.

[6]Here the term 'net' means 'infinitesimal'. In higher-dimension, it means the net rate of change of volume

In general, for a transformation

$$F \quad : \quad \begin{cases} y_1 = f_1(x_1, \ldots, x_n) \\ y_2 = f_2(x_1, \ldots, x_n) \\ \quad \vdots \\ y_n = f_n(x_1, \ldots, x_n), \end{cases}$$

and for a point P and a nearby point $P+\mathbf{x}$ in the domain, $F(P+\mathbf{x})$ is close to

$$F(P) + F'(P)\mathbf{x}.$$

Therefore F transforms a small parallelepiped[7] near P

$$P + \{t_1\mathbf{v}_1 + \cdots + t_n\mathbf{v}_n \mid 0 \leq t_1, \ldots, t_n \leq 1\} =: P + \varnothing(\mathbf{v}_1, \ldots, \mathbf{v}_n)$$

into approximately a parallelepiped

$$\begin{aligned} & F(P) + \{t_1 F'(P)\mathbf{v}_1 + \cdots + t_n F'(P)\mathbf{v}_n \mid 0 \leq t_1, \ldots, t_n \leq 1\} \\ = \; & F(P) + \varnothing(F'(P)\mathbf{v}_1, \ldots, F'(P)\mathbf{v}_n). \end{aligned}$$

Therefore the rate of change of volume is

$$\begin{aligned} \frac{\mathrm{vol}\,\varnothing(F'(P)\mathbf{v}_1, \ldots, F'(P)\mathbf{v}_n)}{\mathrm{vol}\,\varnothing(\mathbf{v}_1, \ldots, \mathbf{v}_n)} &= \frac{|\det(F'(P)\mathbf{v}_1, \ldots, F'(P)\mathbf{v}_n)|}{|\det(\mathbf{v}_1, \ldots, \mathbf{v}_n)|} \\ &= \frac{|\det\left(F'(P) \cdot (\mathbf{v}_1 \;\; \cdots \;\; \mathbf{v}_n)\right)|}{|\det(\mathbf{v}_1, \ldots, \mathbf{v}_n)|} \\ &= \frac{|\det F'(P) \cdot \det(\mathbf{v}_1, \ldots, \mathbf{v}_n)|}{|\det(\mathbf{v}_1, \ldots, \mathbf{v}_n)|} \\ &= \left|\det F'(P)\right|. \end{aligned}$$

1.1.3 Basic Exercise

For a point P in the domain of a $\mathcal{C}^1$ function $F : \mathbb{R}^3 \dashrightarrow \mathbb{R}^3$, if we let

$$B(r) := \{X \in \mathbb{R}^3 \mid |X - P| \leq r\},$$

explain that

$$\lim_{r \searrow 0} \frac{\mathrm{Vol}(F(B(r)))}{\frac{4}{3}\pi r^3} = |\det F'(P)|.$$

[7] Assume that $\mathbf{v}_1, \ldots, \mathbf{v}_n$ are linearly independent vectors of small norm.

1.1.4 Example

For the transformation from spherical coordinates to rectangular coordinates given by

$$\begin{aligned} x &= \rho \sin\varphi \cos\theta \\ y &= \rho \sin\varphi \sin\theta \\ z &= \rho \cos\varphi, \end{aligned}$$

the Jacobian matrix is

$$\frac{\partial(x,y,z)}{\partial(\rho,\varphi,\theta)} = \begin{pmatrix} \sin\varphi\cos\theta & \rho\cos\varphi\cos\theta & -\rho\sin\varphi\sin\theta \\ \sin\varphi\sin\theta & \rho\cos\varphi\sin\theta & \rho\sin\varphi\cos\theta \\ \cos\varphi & -\rho\sin\varphi & 0 \end{pmatrix}$$

and so its Jacobian determinant is[8]

$$\boxed{\det \frac{\partial(x,y,z)}{\partial(\rho,\varphi,\theta)} = \rho^2 \sin\varphi}$$

1.2 The Chain Rule

The chain rule is about the differentiation of composition of functions. For a function defined on an open set U in $\mathbb{R}^n$

$$F : U \to \mathbb{R}^m, \qquad y_j = f_j(x_1, \ldots, x_n) \qquad (j = 1, \ldots, m)$$

and a function defined on an open set V in $\mathbb{R}^m$

$$G : V \to \mathbb{R}^l, \qquad z_k = g_k(y_1, \ldots, y_m) \qquad (k = 1, \ldots, l)$$

consider their composition[9]

$$G \circ F : U \to \mathbb{R}^l, \qquad z_k = g_k(f_1(x_1, \ldots, x_n), \ldots, f_m(x_1, \ldots, x_n)).$$

If both F and G are differentiable, then $G \circ F$ is differentiable and we have

$$\frac{\partial z_k}{\partial x_i} = \sum_{j=1}^{m} \frac{\partial z_k}{\partial y_j} \frac{\partial y_j}{\partial x_i}.$$

[8] It would be also helpful to memorize this formula until the final exam.

[9] We assume that the image of F lies in the domain of G, that is, $F(U) \subset V$.

Therefore

$$\frac{\partial(z_1,\ldots,z_l)}{\partial(x_1,\ldots,x_n)} = \frac{\partial(z_1,\ldots,z_l)}{\partial(y_1,\ldots,y_m)}\,\frac{\partial(y_1,\ldots,y_m)}{\partial(x_1,\ldots,x_n)}.$$

In other words,

$$(G \circ F)'(P) = G'(F(P))F'(P)$$

This is called the **chain rule**.

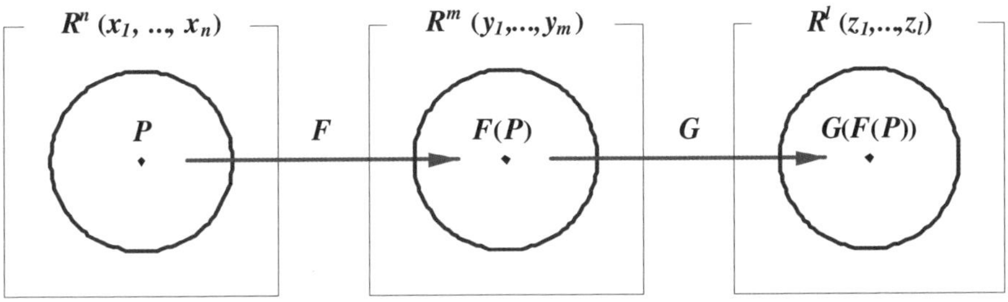

Exercise: Chapter 12 Section 1

1. Show that for the transformation from the spherical coordinates to the rectangular coordinates, the net rate of change of volume is equal to the Jacobian determinant.

2. Find the Jacobian matrix and determinant of the function[10]

$$F(x, y) = (x^2 - y^2, 2xy).$$

   Also find the Jacobian matrix and determinant of the function

$$G(x, y) = (x^3 - xy^2, x^2y - y^3).$$

3. Find the Jacobian matrix and determinant of the given function at a given point.

   (1) $F(x, y) = (x^2 + y^2, x^2 - y^2), \quad P = (1, 2)$

   (2) $F(x, y) = (x + y, x - y, xy), \quad P = (1, 2)$

   (3) $F(x, y) = (\sin x, \cos xy), \quad P = (\pi, 1)$

   (4) $F(x, y, t) = (x, y), \quad P = (1, 1, 1)$

4. Find $\partial z/\partial x$ and $\partial z/\partial y$.

   (1) $z = u^2 + 3uv - v^2, \quad u = \sin x, \quad v = -\cos x + \cos y$

   (2) $z = \sin u \cos v, \quad u = 3x^2 - 2y, \quad v = x - 3y$

5. Find $\partial(u, v)/\partial(t, s)$.

   (1) $u = x^2 + y^2, \ v = x^2 - y^2, \ x = t + s, \ y = t - s$

   (2) $u = \sin(x + y), \ v = \cos(x - y), \ x = t^2 - s^2, \ y = ts$

   (3) $u = xy, \ v = x, \ x = t^2 + s^2, \ y = t^2 - s^2$

6. Find the Jacobian determinant of the map $F(x, y) = (xy, \sin xy)$ and investigate its meaning.

[10]This function maps a complex number $z = x + y\sqrt{-1}$ to $z^2 = (x^2 - y^2) + 2xy\sqrt{-1}$.

2 Inverse Function Theorem, Implicit Function Theorem

2.1 Inverse Function Theorem

2.1.1 Continuously Differentiable Invertible Functions

If a function $f : A \to B$ gives a one-to-one correspondence between two sets A and B, then f has the **inverse function**[11]

$$g : B \to A.$$

The function $g : B \to A$ satisfies[12]

$$g \circ f = \mathrm{id}_A, \qquad f \circ g = \mathrm{id}_B.$$

If a function has the inverse function, the function is called **invertible**. However functions in which we are interested in calculus are not just functions between two sets, but usually continuous functions, linear functions, differentiable functions, $\mathcal{C}^1$ functions, $\mathcal{C}^2$ functions, ..., $\mathcal{C}^\infty$ functions, etc. Thus when we consider the inverse function, it is natural to require it to be continuous, linear, differentiable, $\mathcal{C}^1$, etc, under the given circumstance.

When we think of the inverse function theorem in multi-variable calculus, we deal with continuously differentiable functions, that is, $\mathcal{C}^1$ functions. A $\mathcal{C}^1$ function which has the inverse function and whose inverse function is also $\mathcal{C}^1$ is called a **continuously differentiable invertible function** or $\mathcal{C}^1$ **invertible function**.

2.1.2 Differentiation of Inverse Function

In the region $x > 0$ on the xy-plane, the Jacobian matrix of polar coordinates

$$r = \sqrt{x^2 + y^2}, \qquad \theta = \arctan \frac{y}{x}$$

is

$$\begin{aligned} \frac{\partial(r,\theta)}{\partial(x,y)} &= \begin{pmatrix} \dfrac{x}{\sqrt{x^2+y^2}} & \dfrac{y}{\sqrt{x^2+y^2}} \\ \dfrac{-y/x^2}{1+(y/x)^2} & \dfrac{1/x}{1+(y/x)^2} \end{pmatrix} = \begin{pmatrix} \dfrac{x}{\sqrt{x^2+y^2}} & \dfrac{y}{\sqrt{x^2+y^2}} \\ \dfrac{-y}{x^2+y^2} & \dfrac{x}{x^2+y^2} \end{pmatrix} \\ &= \frac{1}{r} \begin{pmatrix} r\cos\theta & r\sin\theta \\ -\sin\theta & \cos\theta \end{pmatrix}. \end{aligned}$$

[11] Or it is called the **inverse mapping**.

[12] Here id_A and id_B denote the **identity functions** on A and B, respectively.

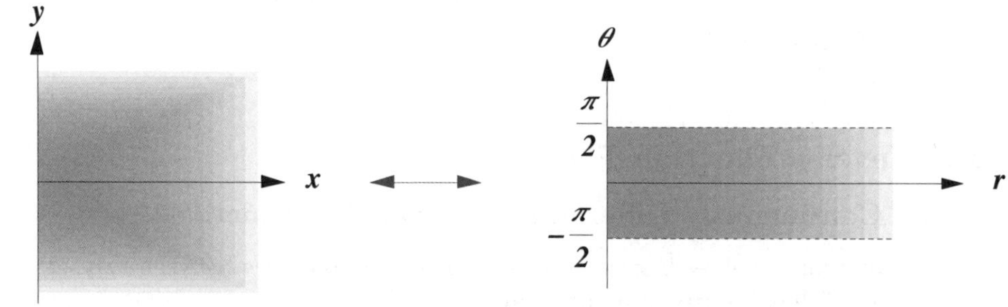

The above matrix is the inverse matrix of the Jacobian matrix for the transformation from polar coordinates to rectangular coordinates

$$x = r\cos\theta, \qquad y = r\sin\theta,$$

$$\frac{\partial(x,y)}{\partial(r,\theta)} = \begin{pmatrix} \cos\theta & -r\sin\theta \\ \sin\theta & r\cos\theta \end{pmatrix}.$$

In general, suppose that a differentiable function

$$F : \mathbb{R}^n \dashrightarrow \mathbb{R}^n, \qquad X \mapsto Y = F(X)$$

has the inverse function $X = G(Y)$:

$$(G\circ F)(X) = X, \qquad (F\circ G)(Y) = Y.$$

If G is differentiable, by the chain rule, we get

$$G'(Y)F'(X) = I_n, \qquad F'(X)G'(Y) = I_n.$$

Here I_n denotes the identity matrix of order n. Therefore $G'(Y) = (F'(X))^{-1}$, that is,[13]

$$\boxed{(F^{-1})'(Y) = (F'(X))^{-1}}$$

In particular, the Jacobian matrix $F'(X)$ is invertible. Of course, $F'(X)$ is invertible if and only if

$$\det F'(X) \neq 0.$$

Like this, *once we know that a differentiable function has the differentiable inverse function*, it is easy to find the derivative of the inverse function, that is, the Jacobian matrix. The difficult part is to know "when there exists the inverse function and why it is differentiable". It is the Inverse Function Theorem that gives the answer for the question.

[13]This principle was already seen in the inverse function theorem in single-variable calculus

$$(f^{-1})'(y) = (f'(x))^{-1}.$$

2.1.3 Basic Exercise

For a $\mathcal{C}^1$ invertible function

$$F : \mathbb{R}^n \dashrightarrow \mathbb{R}^n, \qquad Y := F(X),$$

explain[14]

$$\det \left(F^{-1}\right)'(Y) = (\det F'(X))^{-1}.$$

2.1.4 Solution of Equation and Inverse Function

If a function

$$Y = F(X)$$

has the inverse function, there always exists a solution of the equation

$$F(X) = Y \tag{12.2}$$

for any Y and it is unique. Also if F has a differentiable inverse function then for any X_0 in the domain, the linear equation

$$F'(X_0)\mathbf{x} = \mathbf{y} \tag{12.3}$$

always has a solution.

In general, it is not always easy to solve an equation. In order to solve the (nonlinear) equation (12.2), it is very helpful to solve the linear equation (12.3).[15]

The following **inverse function theorem**[16] tells us that conversely, if a $\mathcal{C}^1$ map

$$F : \mathbb{R}^n \dashrightarrow \mathbb{R}^n$$

has an invertible Jacobian matrix, then F is **locally C^1 invertible**.[17]

---

[14] The rate of change of volume for the inverse map F^{-1} is the inverse of the rate of change of volume for F.

[15] Various theories such as chaos theory, nonlinear dynamical system and fractal theory, could be understood better once linear equations were understood well.

[16] To emphasize the dimension is greater than or equal to 2, we'd like to call it the **inverse mapping theorem**.

[17] In other words, for any point X_0 in the domain, there exist an open set U near X_0 and an open set V near $F(X_0)$ such that F becomes a $\mathcal{C}^1$ invertible map between U and V. Here if F is a bijective map, then F becomes an invertible map on the whole domain.

Theorem 2.1.5 (Inverse Function Theorem) *For a $\mathcal{C}^1$ map defined on an open set in $\mathbb{R}^n$*

$$F : \mathbb{R}^n \dashrightarrow \mathbb{R}^n,$$

if the Jacobian matrix $F'(X_0)$ at a point X_0 in the domain is invertible, then F is locally $\mathcal{C}^1$ invertible near X_0. If we let G be the local inverse of F and let $Y_0 = F(X_0)$, then

$$G'(Y_0) = (F'(X_0))^{-1}.$$

To prove the inverse function theorem, it suffices to show that for any Y near $Y_0 = F(X_0)$, the equation

$$F(X) = Y$$

has a solution. This equation is equal to

$$X = X - F(X) + Y$$

and so it suffices to show that the map

$$\Phi_Y(X) := X - F(X) + Y$$

has a **fixed point**. The easiest way to prove the existence of a fixed point is to use the **contraction mapping principle**.[18] The precise proof can be found in book about analysis [Kim,Kim,Kye].

In terms of equations, the inverse function theorem states the following principle: If a linear equation

$$F'(X_0)\mathbf{x} = \mathbf{y} \in \mathbb{R}^n$$

always has a solution $\mathbf{x} \in \mathbb{R}^n$ for any $\mathbf{y}$, then the nonlinear equation

$$F(X) = Y$$

also has a solution near X_0, provided that Y is close to $Y_0 := F(X_0)$. Here the map which assigns the solution X to Y is $\mathcal{C}^1$.

---

[18]If we put a map of Korea on a table, there exists a point at which the position on the map is equal to the real position. Like this, in a complete metric space X, a contraction mapping on X which contracts the distance between two points by a certain scale has a fixed point and it is unique. If we grill a dry squid and put it back to the place where it was, there exists a unique molecule which goes back to the original position.

2.1.6 Example

For the map

$$F(x,y) = (e^x \cos y, e^x \sin y),$$

the Jacobian determinant is

$$\det \begin{pmatrix} e^x \cos y & -e^x \sin y \\ e^x \sin y & e^x \cos y \end{pmatrix} = e^{2x} \neq 0$$

and thus this function is locally invertible.

If we let G be the inverse of F near $X_0 = (0,0)$, then

$$Y_0 := F(X_0) = (1,0)$$

and[19]

$$G'(1,0) = F'(0,0)^{-1} = \begin{pmatrix} 1 & 0 \\ 0 & 1 \end{pmatrix}.$$

2.1.7 Example

For the map

$$F(x,y) = (x^3 + x - y^3, x^5 + y^7 + y),$$

the Jacobian matrix is

$$F'(x,y) = \begin{pmatrix} 3x^2+1 & -3y^2 \\ 5x^4 & 7y^6+1 \end{pmatrix}$$

and thus

$$\det F'(x,y) = (3x^2+1)(7y^6+1) + 15x^4y^2 > 0.$$

Therefore F is locally $\mathcal{C}^1$ invertible. Since $F(1,0) = (2,1)$,

$$\left(F^{-1}\right)'(2,1) = F'(1,0)^{-1} = \begin{pmatrix} 4 & 0 \\ 5 & 1 \end{pmatrix}^{-1} = \frac{1}{4}\begin{pmatrix} 1 & 0 \\ -5 & 4 \end{pmatrix}.$$

2.1.8 Basic Exercise

Show that the map

$$(u,v) := F(x,y) = (x^2 - y^2, 2xy), \qquad (x,y) \neq (0,0)$$

is locally $\mathcal{C}^1$ invertible. Also find the derivative of the inverse map $G(u,v)$.

---

[19] $(x,y) = G(u,v)$ is given by

$$x = \log\sqrt{u^2+v^2}, \qquad y = \arctan\frac{v}{u}.$$

2.2 Implicit Function Theorem

When we think of a function, usually the relation which assigns to an element in the domain an element in the codomain is given explicitly. However, when the relation between an element x in the domain and an element y in the codomain is given by the equation

$$f(x, y) = c,$$

we say that "y is defined **implicitly** as a function of x" or "y is an **implicit function** of x"[20] In some cases where y is defined implicitly as a function of x, it is possible to solve the given equation (locally) for y as an "**explicit function**" of the form

$$y = g(x).$$

For example, for an implicit equation

$$x^2 + y^2 = 1,$$

when $y > 0$, we get the function

$$y = \sqrt{1 - x^2},$$

and when $y < 0$, we get the function

$$y = -\sqrt{1 - x^2}.$$

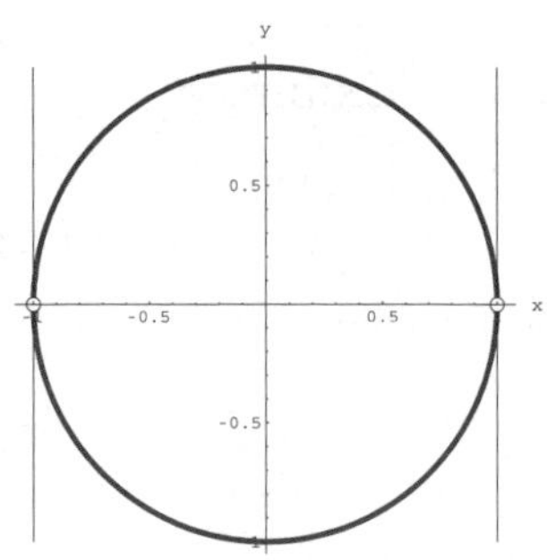

However, when $y = 0$, the tangent line of the level curve is parallel to the y-axis, and so to a point x near ± 1, there corresponds two values of y on the level curve.

The **implicit function theorem** states that the level surface

$$f(x_1, \ldots, x_n, y) = c \tag{12.4}$$

is locally a graph of some function provided that its tangent plane is not parallel to the y-axis.[21] That is, the equation (12.4) is locally given by the form of "explicit function" of y

$$y = g(x_1, \ldots, x_n).$$

---

[20] According to Boyle's(1627–1691) and Charles' law, an ideal gas satisfies the relation $PV/T =$ const for the pressure P, volume V, and absolute temperature T.

[21] Strictly speaking, the fact that we consider the tangent plane of a regular level surface and the tangent plane is perpendicular to the gradient vector results from the implicit function theorem.

Theorem 2.2.1 (Implicit Function Theorem) *Let* $f(x_1, \ldots, x_n, y)$ *be a* C^1 *function and* $P_0 := (X_0, y_0) := (x_1^0, \ldots, x_n^0, y_0)$ *be a point on the level surface* $f^{-1}(c)$. *If*

$$\frac{\partial f}{\partial y}(P_0) \neq 0,$$

then near the point P_0, $f^{-1}(c)$ *is the graph of a* C^1 *function*

$$y = g(x_1, \ldots, x_n)$$

and[22]

$$\frac{\partial g}{\partial x_j}(X_0) = -\frac{\partial f}{\partial x_j}(P_0) \Big/ \frac{\partial f}{\partial y}(P_0).$$

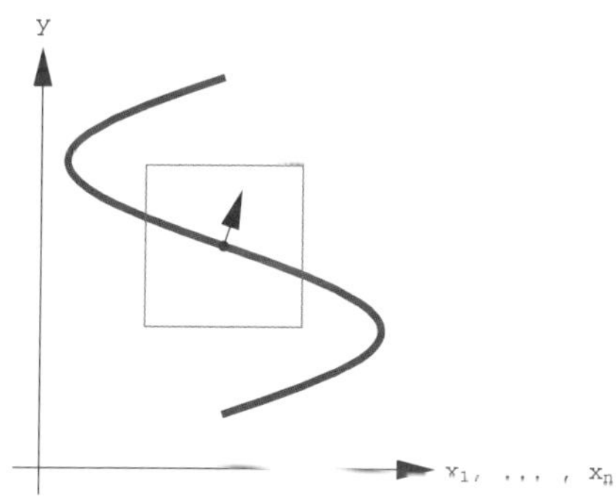

Proof. As a result of the chain rule, the last part of the implicit function theorem is obtained by differentiating the equation

$$f(X, g(X)) = c.$$

Using the inverse function theorem, we can prove the existence of the function g in the implicit function theorem.

The proof when $n \geq 2$ is similar to the proof when $n = 1$. Thus let's prove the theorem when $n = 1$.

---

[22]In the implicit function theorem, if $\frac{\partial f}{\partial y}(P_0) = 0$ then grad $f(P_0)$ has only $(x_1, \ldots, x_n)$-components and hence the tangent plane of the level surface $f^{-1}(c) \subset \mathbb{R}^n \times \mathbb{R}$ at P_0 is parallel to the y-axis. Therefore, we know that the level surface might not become a graph.

"As it can't possibly make me larger, it must make me smaller."
— In Lewis Carroll's *Alice's Adventures in Wonderland*

The map

$$F(x, y) = (x, f(x, y))$$

has the Jacobian matrix

$$F'(x_0, y_0) = \begin{pmatrix} 1 & 0 \\ \dfrac{\partial f}{\partial x}(x_0, y_0) & \dfrac{\partial f}{\partial y}(x_0, y_0) \end{pmatrix}$$

and its Jacobian determinant $\dfrac{\partial f}{\partial y}(x_0, y_0)$ is not zero. By the inverse function theorem, F has the inverse function $G = (g_1, g_2)$ near (x_0, y_0):

$$x = g_1(x, f(x, y)), \qquad y = g_2(x, f(x, y)).$$

If (x, y) is a point on the level surface $f^{-1}(c)$, then

$$y = g_2(x, c) =: g(x).$$

□

2.2.2 Example

Suppose that the amount y of crop in some farm and the amount x of fertilizer used satisfy the relation

$$10(y - x) + e^{xy} = e.$$

Also suppose that if the amount of fertilizer used is 1, i.e. $x = 1$, then the crop is $y = 1$. If the amount of fertilizer is increased to $1 + \epsilon$, then how much would this increase the crop?

Here if we let $y = y(x)$, we want to find $y'(1)$. Thus if we differentiate the given equation with respect to x, then

$$10(y' - 1) + e^{xy}(y + xy') = 0$$

and so

$$y'(1) = \left.\frac{10 - ye^{xy}}{10 + xe^{xy}}\right|_{x=1} = \frac{10 - e}{10 + e}.$$

Therefore the increase is approximately

$$y(1 + \epsilon) - y(1) \approx y'(1)\epsilon = \frac{10 - e}{10 + e}\epsilon.$$

2.2.3 Tangent Vectors of Level Surfaces

At a point $P_0 = (x_0, y_0, z_0)$ on the level surface S of a $\mathcal{C}^1$ function $f(x, y, z)$, if $\operatorname{grad} f(P_0) \neq O$, then there holds at least one of the following:

$$\frac{\partial f}{\partial x}(P_0) \neq 0, \qquad \frac{\partial f}{\partial y}(P_0) \neq 0, \qquad \frac{\partial f}{\partial z}(P_0) \neq 0.$$

If $\dfrac{\partial f}{\partial z}(P_0) \neq 0$, by the implicit function theorem, the level surface S near P_0 is the graph of a $\mathcal{C}^1$ function

$$z = g(x, y).$$

Therefore S has the parametrization

$$X(x, y) := (x, y, g(x, y)).$$

At the point P_0, the **first-coordinate curve**

$$x \mapsto (x, y_0, g(x, y_0))$$

has the velocity vector

$$X_x = \left(1, 0, \frac{\partial g}{\partial x}(x_0, y_0)\right) = \left(1, 0, -\frac{\partial f}{\partial x}(P_0) \Big/ \frac{\partial f}{\partial z}(P_0)\right),$$

and the **second-coordinate curve**

$$y \mapsto (x_0, y, g(x, y_0))$$

has the velocity vector

$$X_y = \left(0, 1, \frac{\partial g}{\partial y}(x_0, y_0)\right) = \left(0, 1, -\frac{\partial f}{\partial y}(P_0) \Big/ \frac{\partial f}{\partial z}(P_0)\right).$$

These two vectors X_x and X_y generate the tangent plane of S.

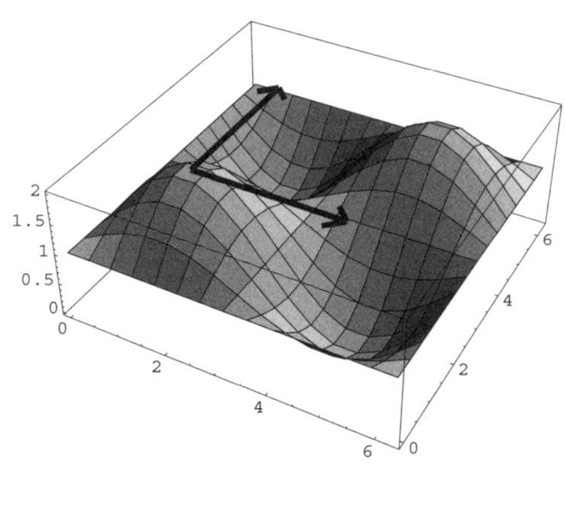

Therefore the vector

$$X_x \times X_y = \left(1, 0, \frac{\partial g}{\partial x}(x_0, y_0)\right) \times \left(0, 1, \frac{\partial g}{\partial y}(x_0, y_0)\right) = \frac{1}{\dfrac{\partial f}{\partial z}(P_0)} \operatorname{grad} f(P_0)$$

is perpendicular to tangent plane.

This principle holds similarly when $\dfrac{\partial f}{\partial x}(P_0) \neq 0$ or $\dfrac{\partial f}{\partial y}(P_0) \neq 0$.

The implicit function theorem is generalized as follows.

Theorem 2.2.4 *Let*

$$F : \mathbb{R}^n \times \mathbb{R}^m \to \mathbb{R}^m, \qquad (X, Y) \mapsto F(X, Y)$$

be a C^1 vector-valued function and (X_0, Y_0) be a point on the level surface $F^{-1}(C)$. If

$$\det\left(\frac{\partial F}{\partial Y}(X_0, Y_0)\right) \neq 0,$$

near the point (X_0, Y_0), $F^{-1}(C)$ is the graph of some C^1 function

$$Y = G(X).$$

Exercise: Chapter 12 Section 2

1. Show that the map

$$(u, v) := F(x, y) = (x^3 - 3xy^2, 3x^2y - y^3), \qquad (x, y) \neq (0, 0)$$

is locally invertible. At $(x, y) = (1, 0)$, find the derivative of the inverse map $G(u, v)$.

2. Examine whether the following function is locally C^1 invertible at a given point.

(1) $F(x, y) = (x + y, y^{1/4}), \qquad (x_0, y_0) = (1, 2)$

(2) $F(x, y) = (x, y)/(x^2 + y^2), \qquad (x_0, y_0) \neq (0, 0)$

3. For $f(x, y) := (x - 2)^2 y - xe^y$, find points (a, b) where

$$f(a, b) = 0, \qquad D_2 f(a, b) = 0.$$

4. For the following function f, explain that the level curve $f(x, y) = 0$ can be represented by the graph of a function $y = g(x)$ near the point (x_0, y_0) and find the derivative $g'(x_0)$.

(1) $f(x, y) = x^2 - xy + y^2 - 3, \qquad (x_0, y_0) = (1, 2)$

(2) $f(x, y) = x \cos xy, \qquad (x_0, y_0) = (1, \pi/2)$

(3) $f(x, y) = 2e^{x+y} - x + y, \qquad (x_0, y_0) = (1, -1)$

(4) $f(x, y) = x + y + x \sin y, \qquad (x_0, y_0) = (0, 0)$

5. For the following function f, explain that the level surface $f(x, y, z) = 0$ can be represented by the graph of a function $z = g(x, y)$ near the point (x_0, y_0, z_0) and find the derivatives $D_1 g(x_0, y_0)$ and $D_2 g(x_0, y_0)$.

(1) $f(x, y, z) = x + y + z + \cos xyz, \qquad (x_0, y_0, z_0) = (0, 0, -1)$

(2) $f(x, y, z) = x^3 + y^3 + z^3 - 3xyz, \qquad (x_0, y_0, z_0) = (0, 0, -1)$

(3) $f(x, y, z) = x + 2y + 3z - 2e^{xyz}, \qquad (x_0, y_0, z_0) = \left(0, \frac{1}{2}, \frac{1}{3}\right)$

(4) $f(x, y, z) = x^2 - 2y^2 + z^2, \qquad (x_0, y_0, z_0) = (1, 1, 1)$

6. Can the set given by the equation

$$xe^y - y + 1 = 0$$

be represented by the graph of a function near $(e^{-2}, 2)$?

7. When a function $z = f(x, y)$ defined on the plane satisfies the equation

$$x^3 + y^3 + z^3 + 3xyz = 6,$$

find $\operatorname{grad} f(1, 1)$.

8. For the equation
$$x^5 + 2x^3 + x + c = 0,$$
   (1) show that it has only one real root for any real number c.
   (2) When we let its root be $g(c)$, find $g'(0)$ and $g''(0)$.

9. The equation
$$x^3 + ax - 10 = 0$$
has only one real root when $a > 0$. When we let the root of this equation be $g(a)$, find $g'(1)$ and $g''(1)$.

10. Find the error in the following claim.

    When a function y for x is given by $y^3 = x(x-1)(x-2)$, if we differentiate this equation with respect to x, we get
$$3y^2y' = (x-1)(x-2) + x(x-2) + x(x-1).$$
    It we put $x = 0$ in both sides of the equation, we get $0 = 2$.

11. When a function z for (x, y) satisfies $x^3 - 2y^2 + z^2 = 0$, find $\dfrac{\partial z}{\partial x}(1,1)$ and $\dfrac{\partial z}{\partial y}(1,1)$.

12. When $y^5 + 3x^2y^3 - 7x^6 - 8 = 0$, find $\dfrac{dy}{dx}(0)$.

13. Explain that if a square matrix A of order n is very close to the identity matrix I then there exists a square matrix X such that
$$e^X = A.$$
Also show that for any natural number k, there exists a square matrix Y such that $Y^k = A$.

14. Show that if a $\mathcal{C}^2$ function $F : \mathbb{R}^n \to \mathbb{R}^n$ has the differentiable inverse function then the inverse function F^{-1} is also $\mathcal{C}^2$. Also show that if F is $\mathcal{C}^3$, so is F^{-1}.

Chapter 13

Vector Fields and Line Integrals

An important example of vector-valued functions of several variables is a vector field. Vector fields are used to describe velocity, acceleration, force and so on. In this chapter, we study various properties of vector fields and the way to find the work done along a path(the line integral). Also we study the **fundamental theorem of line integral** which states that if a vector field has a potential function (or potential energy) then the line integral is given by the difference of values of a potential function. In other words, we study that the line integral of a gradient vector field depends not on path but only on the initial and terminal points. The fundamental theorem of line integral is the special case of a general principle called the fundamental theorem of calculus or Stokes' theorem. In the last section, we will briefly introduce differential forms and deal with dynamical systems.

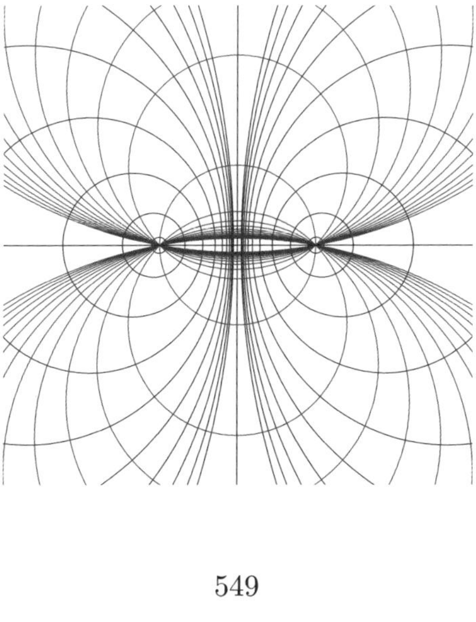

1 Vector Fields

A **vector field** on a subset U of $\mathbb{R}^n$ is a map that assigns to each point X in U a vector $\mathbf{F}(X)$ starting at X.

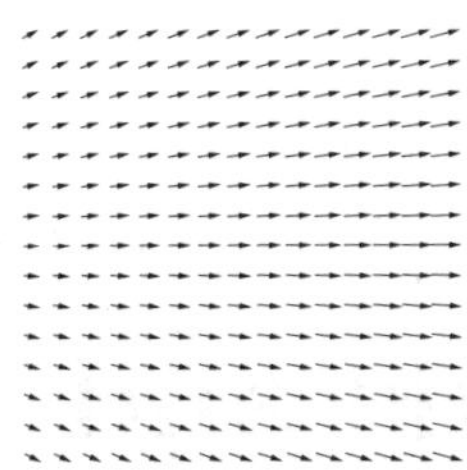

Using coordinates, we may express a vector field

$$\mathbf{F}(X) = (f_1(X), \ldots, f_n(X)) = \sum_i f_i(X)\, \mathbf{e}_i.$$

Unless stated otherwise, we will assume that its component functions

$$f_1, \ldots, f_n : U \to \mathbb{R}$$

are all continuous. If component functions are differentiable (or $\mathcal{C}^1$, $\mathcal{C}^2$, ...), then $\mathbf{F}$ is called differentiable (or $\mathcal{C}^1$, $\mathcal{C}^2$, ...).

1.1 Position Vector Field

In the plane, the **position vector field**[1]

$$\mathbf{r}(x, y) = (x, y) = x\,\mathbf{i} + y\,\mathbf{j}$$

diverges from the origin.

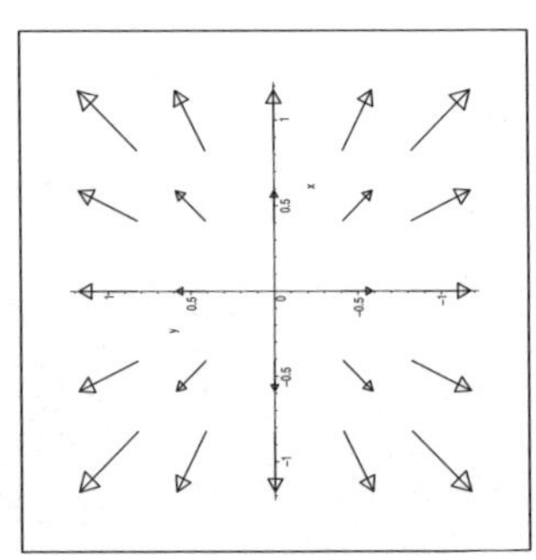

Its norm is

$$|\mathbf{r}(x, y)| = \sqrt{x^2 + y^2} =: r.$$

[1]In $\mathbb{R}^n$, the position vector field is given by

$$\mathbf{r}(x_1, \ldots, x_n) = \sum_i x_i\, \mathbf{e}_i.$$

According to Hooke's law, the amount by which the ideal spring is stretched is proportional to the force applied to the spring. Therefore the force acting on the spring is proportional to the position vector field: $\mathbf{F} = -k\mathbf{r}$. The position vector field is also called the **identity vector field** or **Euler vector field**.

1.1.1 Unit Vector Field

A vector field whose norm is identically 1 is called a **unit vector field**. A unit vector field

$$\mathbf{u}(x_1, \ldots, x_n) = \frac{\mathbf{r}}{r} = \sum_i \frac{x_i}{r} \mathbf{e}_i \qquad \left(r = \left(x_1^2 + \cdots + x_n^2\right)^{1/2}\right)$$

is not defined at the origin.

1.1.2 Example

The vector field

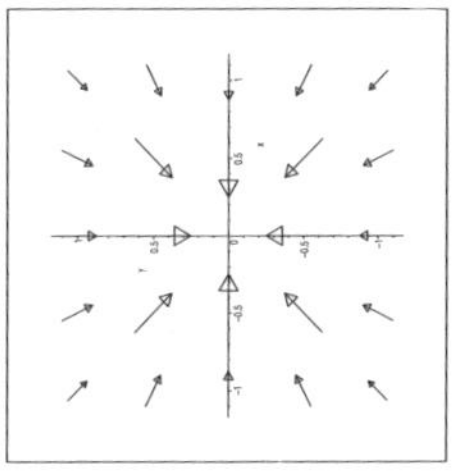

$$\mathbf{F}(x, y) = -\frac{(x, y)}{x^2 + y^2} = -\frac{\mathbf{r}}{r^2}$$

defined on the whole plane except at the origin has the norm which are proportional to the inverse of the distance to the origin:

$$|\mathbf{F}| = \frac{1}{r}.$$

1.1.3 Example

The vector field

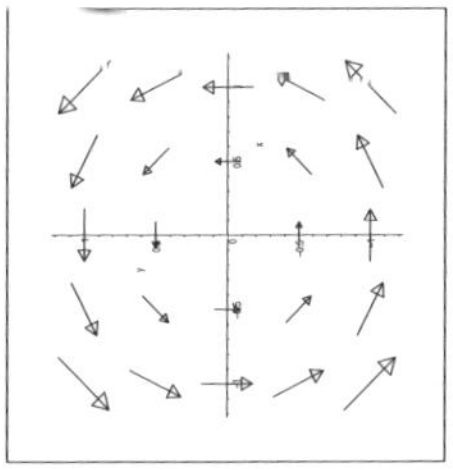

$$\mathbf{F}(x, y) = (-y, x) = -y\mathbf{i} + x\mathbf{j}$$

is obtained by rotating the position vector field $\mathbf{r}$ by 90°.

1.1.4 Example

In the plane, the rotation of points around the origin with the constant angular speed ω can be expressed as

$$x(t) = r_0 \cos \omega t, \qquad y(t) = r_0 \sin \omega t.$$

The velocity vector of this motion is given by

$$x'(t) = -\omega r_0 \sin \omega t = -\omega y(t), \qquad y'(t) = \omega r_0 \cos \omega t = \omega x(t)$$

and so the vector field

$$\mathbf{F}(x, y) := \omega(-y, x)$$

can be thought of as the vector field which describes the rotation with constant angular speed.

A vector field can represent the velocity of motion as "velocity vector field", or the acceleration of motion as "acceleration vector field".[2]

1.1.5 Poiseuille's Law

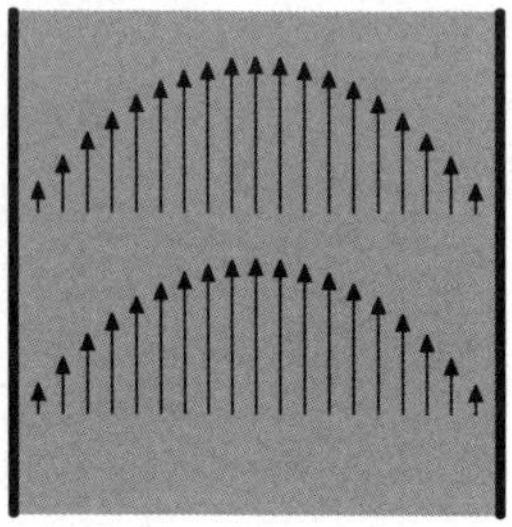

Consider a liquid flowing through a cylindrical tube of radius R with constant circular cross-section given by $x^2 + y^2 = R^2$. At a point with distance r from the central axis, the speed of the liquid is proportional to $R^2 - r^2$.[3] The vector field representing the velocity of the fluid is expressed as follows:

$$\mathbf{F}(x, y, z) := c(R^2 - r^2)\mathbf{k}.$$

1.2 Kepler's Second Law

Suppose that for each point (x, y) in the plane except the origin the vector field $\mathbf{F}(x, y)$ is proportional to the position vector $\mathbf{r}$. That is, there is a function

$$f : \mathbb{R}^2 - \{O\} \to \mathbb{R}$$

such that

$$\mathbf{F}(x, y) = f(x, y)\mathbf{r}(x, y) = f(x, y) \cdot r \cdot (\cos\theta, \sin\theta). \tag{13.1}$$

[2]However, if we increase dimension, the acceleration vector field can be thought of as a velocity vector field.

[3]Like a ray of light through Poincarè's disk in non-Euclidean geometry. A French physiologist J. Poiseuille found this law in 1830s and that if the blood pressure is uniform then the amount of blood flowing in vein per unit time is proportional to the fourth power of the radius of vein:

$$\int_0^R 2\pi k(R^2 r - r^3)\, dr = \pi k R^4/2.$$

Therefore if the radius of vein increases by 10%, then the flow of blood increases by 40%.

For the point

$$X(t) = r(t)(\cos\theta(t), \sin\theta(t))$$

which moves on the plane, the velocity vector is[4]

$$\dot{X}(t) = \dot{r}(\cos\theta, \sin\theta) + r\dot{\theta}(-\sin\theta, \cos\theta)$$

and so the acceleration vector is

$$\ddot{X}(t) = (\ddot{r} - r\dot{\theta}^2)(\cos\theta, \sin\theta) + (2\dot{r}\dot{\theta} + r\ddot{\theta})(-\sin\theta, \cos\theta). \tag{13.2}$$

If $X(t)$ satisfies Newton's law of motion

$$\mathbf{F}(X(t)) = \ddot{X}(t),$$

then by the equations (13.1) and (13.2), we get

$$f(X(t))r = \ddot{r} - r\dot{\theta}^2, \qquad 2\dot{r}\dot{\theta} + r\ddot{\theta} = 0. \tag{13.3}$$

In particular, from the second equation, we conclude that

$$\frac{d}{dt}\left(\frac{1}{2}r^2\dot{\theta}\right) = \frac{1}{2}r(2\dot{r}\dot{\theta} + r\ddot{\theta}) = 0,$$

that is,[5]

$$\boxed{\frac{1}{2}r^2\dot{\theta} = c}$$

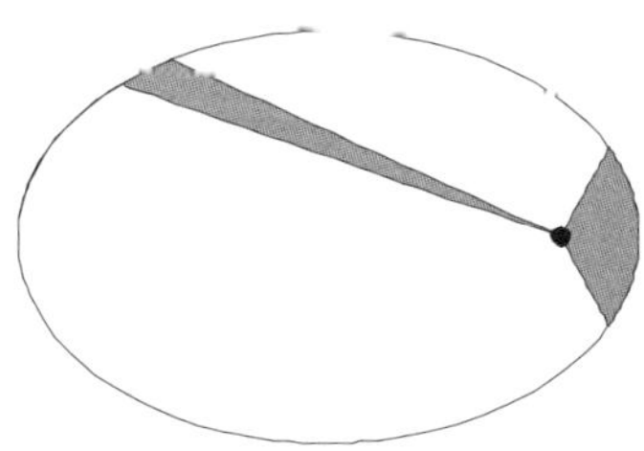

[4]To denote the derivative of a function, we may use Lagrange's notation x' or Newton's notation $\dot{x}$.

[5]A **constant function** is denoted by c or const. The above equation means the **law of conservation** of **angular momentum**:

$$\frac{1}{2}mr^2\dot{\theta} = \text{const}$$

For a particle $X(t)$ in "central force field" in the three dimensional space which follows the Newton's law of motion, there holds the "law of conservation of angular momentum"

$$X(t) \times X'(t) = \text{const}.$$

(Exercise 5).

Geometrically this equation means as follows: If we let $\theta(t)$ be the angle between the line segment OX and the x-axis at time t, then between t_0 and t_1 the line segment OX traces the region whose area is

$$\int_{\theta(t_0)}^{\theta(t_1)} \frac{1}{2} r^2 \, d\theta = \int_{t_0}^{t_1} \frac{1}{2} r^2 \dot{\theta} \, dt = c(t_1 - t_0).$$

Therefore the area of the region is proportional to time.[6]

[6]Ancient Greek scholars believed that the Earth is the center of universe and every planet goes around it. This theory explained that each planet **rotates** and the center of this rotation also rotates around earth. This theory was completed by Ptolemaeus about the second century AD. In the contrast to Ptolemaeus' theory, in 1543 Nicolaus Copernicus (1473–1543) claimed that planets **revolve** around sun. However the more precise the observation became, the more small circles Copernican system need than Ptolemaic system and so the planetary motion was described in the more complicated way. Kepler abandoned **circles** and introduced **ellipses** so that he could explain the planetary motion more precisely. Based on the tremendous observations which a Danish astronomer Tycho Brahe (1546–1601) made throughout his life, Keplar found out the following three laws of planetary motion:

(1) The orbit of every planet is an ellipse with the sun at one of the foci.

(2) The area of the region which a line joining a planet and the sun sweeps out is proportional to time during which the planet travels along its orbit.

(3) The square of the orbital period of a planet is directly proportional to the cube of the major axis (the "length" of the ellipse) of the elliptic orbit.

In *Philosophiae Naturalis Principia Mathematica*(1687), Newton was able to derive Kepler's law with basic principles of mechanics, so called, his law of universal gravitation and his three laws of motion. Like "*Ethica*" by Benedict de Spinoza (1632–1677), *Principia* was also written on the model of Euclid's *Elements*.

Exercise: Chapter 13 Section 1

1. Sketch the vector field

$$\mathbf{F}(x,y) = -2xy\mathbf{i} + (x^2 - y^2)\mathbf{j}$$

and show that this vector field is tangent to any circle through the origin with the center on the x-axis, $(x-a)^2 + y^2 = a^2$.

2. Show that the vector field $\mathbf{F}(x,y) = (-x, y)$ is tangent to the hyperbola $xy = c$.

3. Sketch the following vector field.[7]
   (1) $\mathbf{F}(x,y) = (x^2 - y^2)\,\mathbf{i} + 2xy\,\mathbf{j}$
   (2) $\mathbf{F}(x,y) = (x^3 - 3xy^2)\,\mathbf{i} + (3x^2y - y^3)\,\mathbf{j}$
   (3) $\mathbf{F}(x,y,z) = -y\,\mathbf{i} + x\,\mathbf{j}$

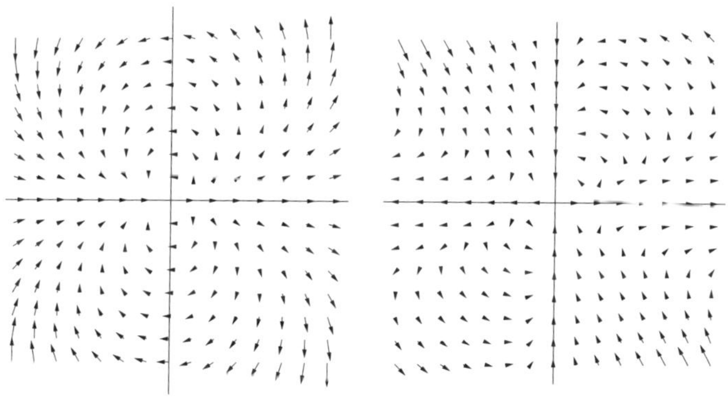

4. For the position vector field $\mathbf{r} = x\,\mathbf{i} + y\,\mathbf{j} + z\,\mathbf{k}$, sketch the set of points at which the vector field

$$\mathbf{F}(x,y,z) = \frac{-y\,\mathbf{i} + x\,\mathbf{j}}{r^3} = \mathbf{k} \times \frac{\mathbf{r}}{r^3}$$

has the same norm.[8]

[7]It would be helpful to use polar coordinates.

[8]Once we notice that the set we consider is a surface of revolution, it suffices to consider points only in the yz-plane. According to **Ampere's law** by Andre Marie Ampere (1775–1836), the **magnetic field** generated by a particle with electric charge q_0, position $\mathbf{r}_0(t)$ and velocity $\mathbf{v}_0 := \mathbf{r}_0'$ is

$$\mathbf{B}(\mathbf{r},t) := q\mathbf{v}_0 \times \frac{\mathbf{r} - \mathbf{r}_0}{|\mathbf{r} - \mathbf{r}_0|^3}.$$

The force acting on a particle with electric charge q, position $\mathbf{r}$ and velocity $\mathbf{v} := \mathbf{r}'$ in magnetic field is

$$\mathbf{F} = kq\mathbf{v} \times \mathbf{B}.$$

5. (**Law of Conservation of Angular Momentum**) In the space, the **angular momentum** of a moving point $X(t)$ with mass m is given by

$$\mathbf{L}(t) := X(t) \times (mX'(t)).$$

Show that if $X(t)$ satisfies Newton's equation of motion $\mathbf{F}(X(t)) = mX''(t)$ and the vector field $\mathbf{F}$ is proportional to the position vector field at each point, then $\mathbf{L}(t)$ is a constant vector.

6. For a given vector field $\mathbf{F}$, show that if a curve $X(t)$ satisfies Newton's equation of motion $\mathbf{F}(X(t)) = mX''(t)$, then the orientation-reversing curve $Y(s) := X(c - s)$ also satisfies the same equation.

2 Line Integrals

When we studied curves in Chapter 9, we studied **how to integrate a function** along a curve. In this section, we study **how to integrate a vector field** along a curve.

2.1 Vector Field Along Curves

For a curve $X : [a, b] \to \mathbb{R}^n$, a "vector field along a curve X" assigns to each point t in its domain $[a, b]$ a vector $\mathbf{G}(t) \in \mathbb{R}^n$ with the initial point at $X(t)$.

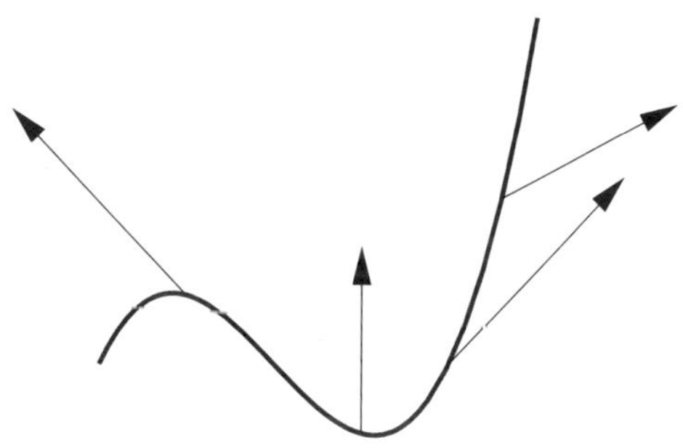

A vector field along a curve $X : [a, b] \to \mathbb{R}^n$ can be thought of as a map

$$\mathbf{G} : [a, b] \to \mathbb{R}^n.$$

We assume that every vector field is continuous, unless stated otherwise.

2.1.1 Example

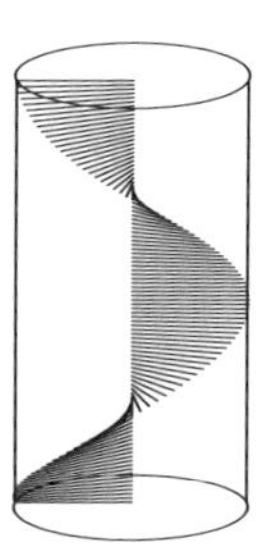

For each point t on the helix

$$X(t) := (\cos t, \sin t, t),$$

the vector field

$$\mathbf{G}(t) := (-\cos t, -\sin t, 0)$$

points the z-axis which is the central axis of the helix and gives an example of vector field along a curve X.

2.1.2

If the image of a curve $X : [a, b] \to \mathbb{R}^n$ is contained in a set $U \subset \mathbb{R}^n$ and $\mathbf{F} : U \to \mathbb{R}^n$ is a vector field defined on U, then their composition

$$\mathbf{F} \circ X : [a, b] \to \mathbb{R}^n, \qquad t \mapsto \mathbf{F}(X(t))$$

is a vector field along the curve X.

2.1.3 Unit Tangent Vector Fields

As a typical example of vector field along a curve X, we have the **velocity vector field** $X'(t)$ and **acceleration vector field** $X''(t)$. For a curve whose velocity is not zero anywhere — that is, a regular curve, we let

$$\mathbf{t}(t) = \mathbf{t}_X(t) := \frac{X'(t)}{|X'(t)|}$$

be the **unit tangent vector field**. The unit tangent vector field depends on the orientation of the curve. The standard unit tangent vector field of an orientation-reversing curve points the opposite direction of that of the original curve.

2.2 Line Integrals

When we integrate

(i) a function or
(ii) a vector field

along a curve, we get a real number.[9]

We already saw that in Chapter 9 the line integral of a function $g(t)$ defined along a $\mathcal{C}^1$ curve $X : [a, b] \to \mathbb{R}^n$ is

$$\int_X g\ ds := \int_a^b g(t)|X'(t)|\,dt.$$

For a vector field $\mathbf{G}(t) \in \mathbb{R}^n$, the integral

$$\boxed{\int_X \mathbf{G} \cdot d\mathbf{s} := \int_a^b \mathbf{G}(t) \cdot X'(t)\,dt}$$

is called the " **(line) integral**" or "**work**" of $\mathbf{G}$ along the curve X.

---

[9]People say when we integrate a vector-valued function, we get a vector …

2.2.1 Basic Exercise

For a constant f, if the force is given by $\mathbf{F}(x) = f\mathbf{i}$ at each point x on an interval $[a, b]$, show that the work is

$$\int_a^b \mathbf{F} \cdot d\mathbf{s} = f(b-a).$$

If a vector field $\mathbf{F}$ is defined on a region containing the image of a curve $X(t)$, then the line integral $\int_X \mathbf{F} \circ X \cdot d\mathbf{s}$ can be written simply as $\int_X \mathbf{F} \cdot d\mathbf{s}$. That is,

$$\int_X \mathbf{F} \cdot d\mathbf{s} := \int_a^b \mathbf{F}(X(t)) \cdot X'(t)\, dt$$

2.2.2 Example

If we integrate the vector field $\mathbf{F}(x, y) = \frac{1}{2}(-y, x)$ along the curve

$$X(t) = r(\cos t, \sin t), \qquad 0 \le t \le 2\pi$$

we get[10]

$$\begin{aligned} \int_X \mathbf{F} \cdot d\mathbf{s} &= \int_0^{2\pi} \mathbf{F}(X(t)) \cdot X'(t)\, dt \\ &= \int_0^{2\pi} \frac{1}{2}(-r\sin t, r\cos t) \cdot r(-\sin t, \cos t)\, dt \\ &= \pi r^2. \end{aligned}$$

If we integrate the position vector field $\mathbf{r}(x, y) = (x, y)$ along the above curve, we get

$$\int_X \mathbf{r} \cdot d\mathbf{s} = \int_0^{2\pi} \mathbf{r}(X(t)) \cdot X'(t)\, dt = 0.$$

[10]Once we study Green's theorem, we know that the line integral of the vector field $\mathbf{F}(x, y) = \frac{1}{2}(-y, x)$ along a **simple closed curve** is the area of the region enclosed by the curve. A curve $X : [a, b] \to \mathbb{R}^2$ is called **simple** if $X(t_0) \ne X(t_1)$ for $a \le t_0 < t_1 < b$, and X is called **closed** if the initial point $X(a)$ and terminal point $X(b)$ of X are equal.

2.2.3 Basic Exercise

1. Both curves

$$\begin{aligned} Y_+(t) &:= (\cos t, \sin t), \qquad 0 \le t \le \pi/2 \\ Y_-(t) &:= (\cos t, -\sin t), \qquad 0 \le t \le 3\pi/2 \end{aligned}$$

start at $(1,0)$ and end at $(0,1)$. Examine whether the values of line integral of the vector field $\mathbf{F}(x,y) = (-y, x)$ along two curves are equal.[11]

2. Evaluate the integral of a vector field $\mathbf{F}$ along a constant curve $X(t) \equiv X_0$.

2.2.4 Example

For a $\mathcal{C}^1$ curve $X : [a,b] \to \mathbb{R}^n$, if we integrate the velocity vector field X' along the curve X itself, we get[12]

$$\int_X X' \cdot d\mathbf{s} = \int_a^b |X'|^2 \, dt.$$

2.2.5 Basic Exercise

For a $\mathcal{C}^2$ curve $X : [a,b] \to \mathbb{R}^n$, show that if we integrate the acceleration vector field X'' along the curve X itself, we get

$$\int_X X'' \cdot d\mathbf{s} = \frac{1}{2}|X'(b)|^2 - \frac{1}{2}|X'(a)|^2.$$

For a given vector field $\mathbf{F}$, if a particle $X(t)$ satisfies the **equation of motion**[13]

$$\mathbf{F}(X(t)) = mX''(t),$$

[11] Answer: $\displaystyle\int_{Y_+} \mathbf{F} \cdot ds = \frac{\pi}{2}, \quad \int_{Y_-} \mathbf{F} \cdot ds = -\frac{3\pi}{2}$

[12] This integral is the **total kinetic energy**.

[13] This is the second among Newton's three laws of motion. It was L. Euler (1707–1783) who transformed Newton's concept for mechanics into the basic laws in the modern form which is easy to understand. [Truesdell].

then the line integral of $\mathbf{F}$ on the interval $t_0 \le t \le t_1$ is[14]

$$\begin{aligned}\int_X \mathbf{F}\cdot d\mathbf{s} &= \int_{t_0}^{t_1} \mathbf{F}(X(t))\cdot X'(t)\,dt = \int_{t_0}^{t_1} m\,X''(t)\cdot X'(t)\,dt \\ &= \frac{1}{2}m\int_{t_0}^{t_1}\frac{d}{dt}|X'(t)|^2\,dt \\ &= \frac{1}{2}m|X'(t_1)|^2 - \frac{1}{2}m|X'(t_0)|^2.\end{aligned}$$

2.2.6 Line Integrals of Vector Fields and Functions

For a regular curve X — that is, a $\mathcal{C}^1$ curve whose speed is never zero, we have the standard unit tangent vector field of X

$$\mathbf{t}(t) := \frac{X'(t)}{|X'(t)|}.$$

On the other hand, since the length element of a curve X is $ds = |X'(t)|\,dt$, the integral of a vector field $\mathbf{F}$ is

$$\int_X \mathbf{F}\cdot d\mathbf{s} = \int_a^b \mathbf{F}(t)\cdot X'(t)\,dt = \int_a^b \mathbf{F}(t)\cdot\mathbf{t}(t)\,|X'(t)|\,dt = \int_X \mathbf{F}\cdot\mathbf{t}\,ds.$$

Therefore along the regular curve X,[15]

$$\boxed{\int_X \mathbf{F}\cdot d\mathbf{s} = \int_X \mathbf{F}\cdot\mathbf{t}\,ds}$$

Thus a line integral of a vector field can be transformed into a line integral of a function.

2.2.7 Basic Exercise

For a regular curve, show that the integral of the standard unit tangent vector field is equal to the length of the curve.

---

[14]The work is the difference in **kinetic energy** $\frac{1}{2}mv^2$ where $v(t)$ denotes the velocity $|X'(t)|$.

[15]We can represent this equation as $d\mathbf{s} = \mathbf{t}\,ds$.

For a regular curve X, let's express a vector field $\mathbf{F}$ as the sum of the tangential component $\mathbf{F}_\top$ and normal component $\mathbf{F}_\perp$ to the curve X[16]

$$\mathbf{F} = \mathbf{F}_\top + \mathbf{F}_\perp.$$

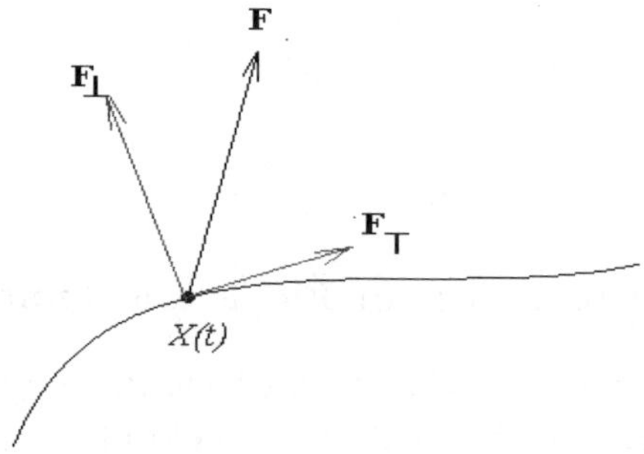

Then $\mathbf{F} \cdot \mathbf{t} = (\mathbf{F}_\top + \mathbf{F}_\perp) \cdot \mathbf{t} = \mathbf{F}_\top \cdot \mathbf{t}$ and

$$\int_X \mathbf{F} \cdot d\mathbf{s} = \int_X \mathbf{F} \cdot \mathbf{t}\, ds = \int_X \mathbf{F}_\top \cdot \mathbf{t}\, ds = \int_X \mathbf{F}_\top \cdot d\mathbf{s}.$$

That is, when we integrate a vector field, it is enough to integrate only the tangential component.

2.2.8 Basic Exercise

Show that the integral of a given vector field along a curve perpendicular to the vector field[17] is always zero.

2.3 Parametrization and Line Integrals

The line integral of a vector field doesn't change by orientation-preserving reparametrization. However by orientation-reserving reparametrization, it does change the sign. In other words, the line integral of a vector field depends on the orientation of a curve, unlike the line integral of a function.

---

[16]Here, $\mathbf{F}_\top = (\mathbf{F} \cdot \mathbf{t})\mathbf{t}$ is an orthogonal projection of $\mathbf{F}$ to $\mathbf{t}$.

[17]It means that the tangent line of the curve is perpendicular to the direction of the vector field at each point on the curve.

Theorem 2.3.1 *Let $\mathbf{F} : U \to \mathbb{R}^n$ be a vector field defined on an open subset U of $\mathbb{R}^n$ and $X : [a, b] \to U$ be a C^1 curve.*

(i) *For a C^1 function $f : [\alpha, \beta] \to [a, b]$ with $f(\alpha) = a$ and $f(\beta) = b$,*

$$\int_{X \circ f} \mathbf{F} \cdot d\mathbf{s} = \int_X \mathbf{F} \cdot d\mathbf{s}$$

(ii) *If X_- is a orientation-reversing curve of X, then*

$$\int_{X_-} \mathbf{F} \cdot d\mathbf{s} = -\int_X \mathbf{F} \cdot d\mathbf{s}$$

In particular, it follows from the theorem that if a reparametrization $\tilde{X}$ of X is

orientation-preserving, then $\displaystyle\int_{\tilde{X}} \mathbf{F} \cdot d\mathbf{s} = \int_X \mathbf{F} \cdot d\mathbf{s}$,

orientation-reserving, then $\displaystyle\int_{\tilde{X}} \mathbf{F} \cdot d\mathbf{s} = -\int_X \mathbf{F} \cdot d\mathbf{s}$.

Proof. Case (i):

$$\begin{aligned}
\int_{X \circ f} \mathbf{F} \cdot d\mathbf{s} &= \int_\alpha^\beta \mathbf{F}(X(f(u))) \cdot \frac{d}{du} X(f(u))\, du \\
&= \int_\alpha^\beta \mathbf{F}(X(f(u))) \cdot X'(f(u))\, f'(u)\, du \\
&= \int_a^b \mathbf{F}(X(t)) \cdot X'(t)\, dt \\
&= \int_X \mathbf{F} \cdot d\mathbf{s}.
\end{aligned}$$

Case (ii): Since $X_-(u) = X(a+b-u)$,

$$\begin{aligned}
\int_{X_-} \mathbf{F}\cdot d\mathbf{s} &= \int_a^b \mathbf{F}(X_-(u))\cdot\frac{d}{du}X_-(u)\,du \\
&= \int_a^b \mathbf{F}(X(a+b-u)))\cdot\big(-X'(a+b-u)\big)\,du \\
&= \int_b^a \mathbf{F}(X(t))\cdot X'(t)\,dt \qquad (\Leftarrow t := a+b-u) \\
&= -\int_a^b \mathbf{F}(X(t))\cdot X'(t)\,dt \\
&= -\int_X \mathbf{F}\cdot d\mathbf{s}.
\end{aligned}$$

□

2.3.2 Example

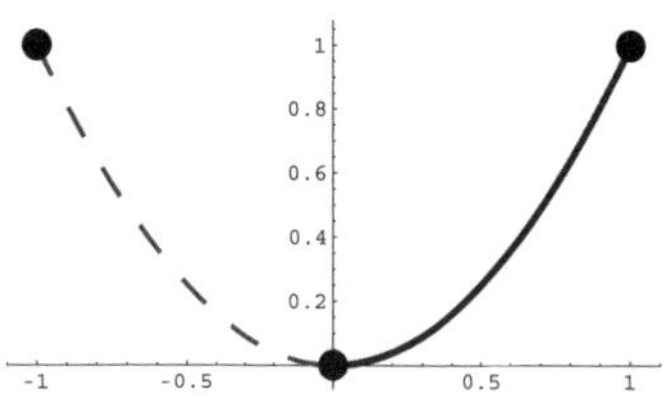

Integrate the position vector field

$$\mathbf{r}(x,y) = (x,y)$$

along the parabola $y = x^2$ from the origin to $P = (1,1)$. Also find the line integral along this parabola from P to the origin reversely.

To solve this problem, if we parameterize the given parabola as[18]

$$X(t) = (t, t^2), \qquad 0 \le t \le 1,$$

then we get[19]

$$\int_X \mathbf{r}\cdot d\mathbf{s} = \int_0^1 (t,t^2)\cdot(1,2t)\,dt = \int_0^1 (t+2t^3)\,dt = \left[\frac{1}{2}t^2 + \frac{2}{4}t^4\right]_0^1 = 1.$$

If we move backward from P to the origin, the line integral is -1.

[18]Of course, even if we parameterize the parabola in a complicated way like

$$Y(t) = \left(\frac{e^{\sin t}-1}{e-1}, \frac{\left(e^{\sin t}-1\right)^2}{(e-1)^2}\right), \qquad 0 \le t \le \pi/2$$

we will get the same conclusion.

[19]Once we study potential functions later, we can integrate it in a different way.

2.3.3 Basic Exercise

Evaluate the line integral of the position vector field along the parabola $y = x^2$ moving from $P = (1, 1)$ to $Q = (-1, 1)$ through the origin and coming back to the origin.

2.4 Angle Element Vector Field

The **angle element vector field** is defined on the plane expect at the origin by

$$\mathbf{a}(x, y) = \frac{(-y, x)}{x^2 + y^2}.$$

Its norm is[20] $|\mathbf{a}| = 1/r$.

2.4.1 Basic Exercise

Show that the angle element vector field is the velocity vector field that describes rotation with angular velocity inversely proportional to the square of distance from the origin:

$$x(t) = r\cos\frac{t}{r^2}, \qquad y(t) = r\sin\frac{t}{r^2}.$$

If we integrate the angle element vector field along a curve[21]

$$X(t) := r(t)\big(\cos\theta(t), \sin\theta(t)\big), \qquad t_0 \le t \le t_1,$$

we get

$$\begin{aligned}\int_X \mathbf{a}\cdot d\mathbf{s} &= \int_{t_0}^{t_1} \frac{r(-\sin\theta, \cos\theta)}{r^2} \cdot \left(\frac{dr}{dt}(\cos\theta, \sin\theta) + r(-\sin\theta, \cos\theta)\frac{d\theta}{dt}\right) dt \\ &= \int_{t_0}^{t_1} \frac{d\theta}{dt}\, dt = \theta(t_1) - \theta(t_0).\end{aligned}$$

If we integrate the angle element vector field, we may know the change in angle when we see it from the origin.

[20] Once we introduce the angle function $\theta(x, y) = \arctan(y/x)$ into the plane, $\mathbf{a} = \operatorname{grad}\theta$.

[21] Here we assume that r and θ are both $\mathcal{C}^1$ functions and $r(t) > 0$.

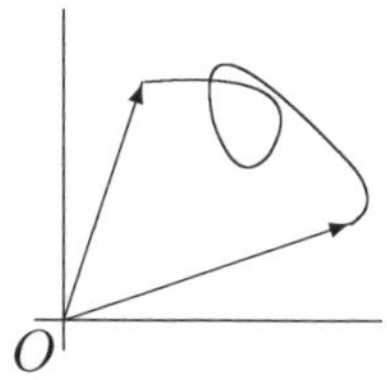

2.4.2 Basic Exercise

Evaluate the integral of the angle element vector field along the following curve.

(1) $X(t) = (\cos t, \sin t), \quad 0 \le t \le 2\pi$

(2) $Y(t) = (\cos t, \sin t), \quad 0 \le t \le 4\pi$

(3) $C_n(t) = (\cos nt, \sin nt), \quad 0 \le t \le 2\pi$ (for an integer n)

(4) $C_{n,r}(t) = r(\cos nt, \sin nt), \quad 0 \le t \le 2\pi$ (for a positive number r)

2.5 Piecewise $\mathcal{C}^1$ Curves

If a curve X is **piecewise $\mathcal{C}^1$**, in other words, if $X = X_1 * \cdots * X_k$ can be obtained by joining $\mathcal{C}^1$ curves $X_1, \ldots, X_k$, the integral of the vector field $\mathbf{F}$ along X is defined by

$$\int_X \mathbf{F} \cdot d\mathbf{s} := \int_{X_1} \mathbf{F} \cdot d\mathbf{s} + \cdots + \int_{X_k} \mathbf{F} \cdot d\mathbf{s}.$$

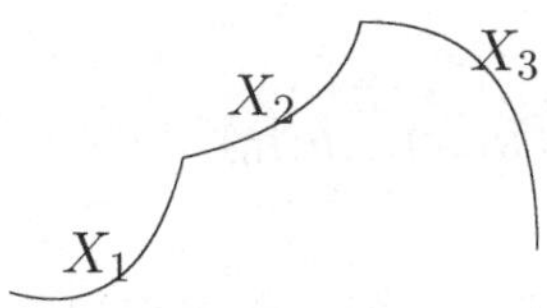

$X_1 * X_2 * X_3$

2.5.1 Example

Evaluate the integral of the angle element vector field $\mathbf{a}(x, y) = \dfrac{(-y, x)}{x^2 + y^2}$ along the square[22]

$$X: \quad \max\{|x|, |y|\} = 1.$$

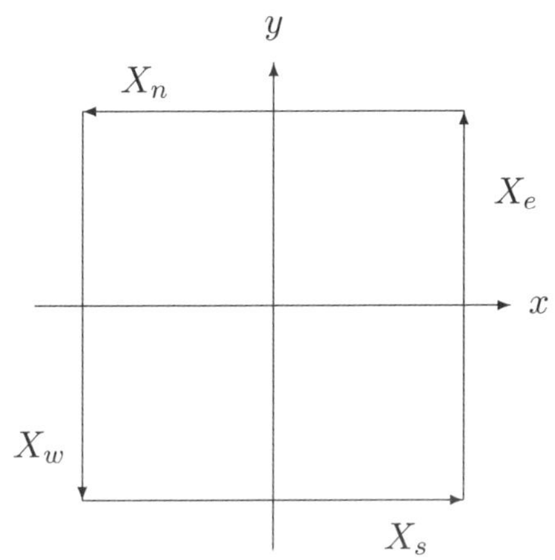

First, divide the square X into four sides "east, north, west, south" and parameterize it as follows:

$$\begin{aligned} X_e(t) &:= (1, t) && (-1 \le t \le 1) \\ X_n(t) &:= (-t, 1) && (-1 \le t \le 1) \\ X_w(t) &:= (-1, -t) && (-1 \le t \le 1) \\ X_s(t) &:= (t, -1) && (-1 \le t \le 1) \end{aligned}$$

Then

$$\begin{aligned} \int_X \mathbf{a} \cdot d\mathbf{s} &= \int_{X_e} \mathbf{a} \cdot d\mathbf{s} + \int_{X_n} \mathbf{a} \cdot d\mathbf{s} + \int_{X_w} \mathbf{a} \cdot d\mathbf{s} + \int_{X_s} \mathbf{a} \cdot d\mathbf{s} \\ &= \frac{\pi}{2} + \frac{\pi}{2} + \frac{\pi}{2} + \frac{\pi}{2} = 2\pi. \end{aligned}$$

[22]Unless stated otherwise, we always assume the curve is positively-oriented. The integral along a closed curve doesn't depend on the initial point of the curve.

2.5.2 Basic Exercise

Evaluate the angle element vector field

$$\mathbf{a} = \left(-\frac{y}{x^2+y^2}, \frac{x}{x^2+y^2} \right)$$

along the following curve.[23]

(1) (2)

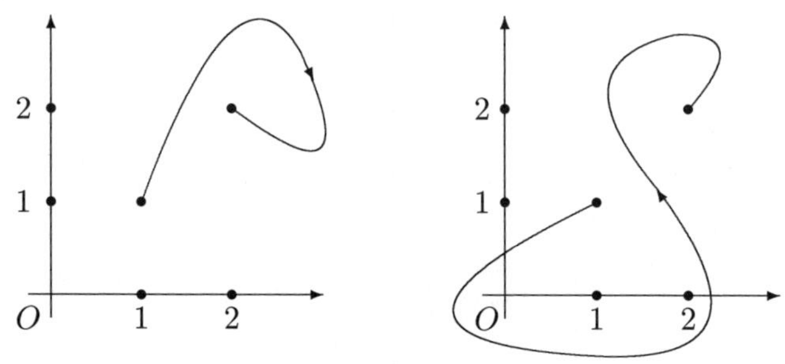

[23]Answer: (1) 0 (2) 2π

Exercise: Chapter 13 Section 2

1. Integrate the vector field $\mathbf{F}(x, y) = (-y, x)$ along the following curve.

   (1) $X(t) = (\cos t, \sin t), \quad 0 \le t \le 4\pi$

   (2) $X(t) = (\cos t, -\sin t), \quad 0 \le t \le 2\pi$

   (3) The curve X is a triangle connecting points $(0,0)$, $(3,0)$, $(0,3)$, $(0,0)$ in order.

2. Integrate the vector field

$$\mathbf{a}(x, y, z) = -\frac{y}{x^2 + y^2}\mathbf{i} + \frac{x}{x^2 + y^2}\mathbf{j}$$

   along the following curve from $P_0 = (1,0,0)$ to $P_1 = (0,1,1)$.

   (1) $y = z = 1 - x$

   (2) $x = \cos t,\ y = \sin t,\ z = 2t/\pi$

   (3) $x = \cos t,\ y = -\sin t,\ z = 2t/3\pi$

3. (**Winding Number**) For a curve which doesn't pass through the origin in the plane

$$X : [t_0, t_1] \to \mathbb{R}^2 - \{O\},$$

   we may define a continuous function $\theta(t)$ such that

$$X(t) = |X(t)|\big(\cos\theta(t), \sin\theta(t)\big), \qquad t_0 \le t \le t_1,$$

   and we define

$$\mathrm{wind}(X) := \frac{1}{2\pi}\big(\theta(t_1) - \theta(t_0)\big)$$

   to be the "**winding number** of a curve X with respect to the origin".

   (1) Find the winding number of the following curve X with respect to the origin.

   (a) $X(t) := (\cos 3t, \sin 3t), \qquad 0 \le t \le 2\pi$

   (b) $Y(t) := (\cos t, \sin 3t), \qquad 0 \le t \le 2\pi$

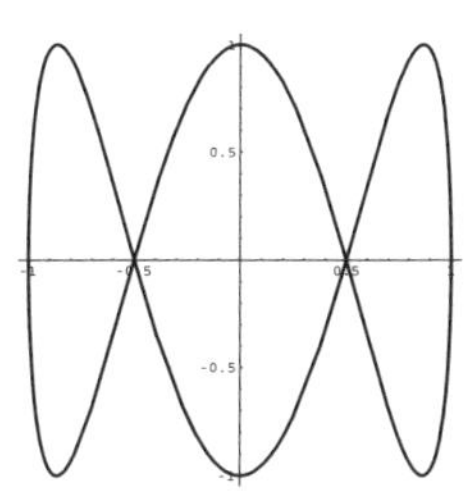

(2) Explain that the winding number is an integer if X is a closed curve.[24]

(3) If X is a differentiable curve, then show that

$$\text{wind}(X) = \frac{1}{2\pi} \int_X \mathbf{a} \cdot d\mathbf{s}.$$

(Here $\mathbf{a}$ denotes the angle element vector field.)

4. Find the winding number of the following plane curve X with respect to the origin.

(1) $X(t) = (2 + \cos 3t)(\cos 2t, \sin 2t), \qquad 0 \le t \le 2\pi$

(2) $X(t) = (2 + \cos 4t)(\cos 2t, \sin 2t), \qquad 0 \le t \le 2\pi$

(3) $X(t) = (2 + \cos 5t)(\cos 2t, \sin 2t), \qquad 0 \le t \le 2\pi$

(4) $X(t) = (2 + \cos 5t)(\cos 3t, \sin 3t), \qquad 0 \le t \le 2\pi$

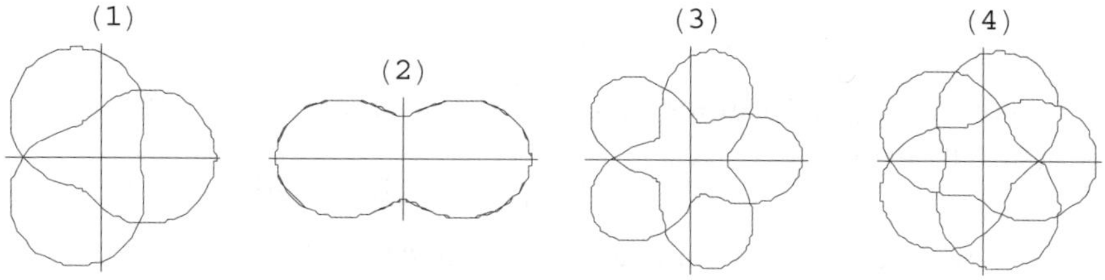

Also find the number of points where a **ray** starting from the origin meets each curve above.

[24] A curve $X(t)$, $t_0 \le t \le t_1$ is called a **closed curve**if the initial point $X(t_0)$ is the same as the terminal point $X(t_1)$.

3 Gradient Vector Fields and Potential Functions

One easy way to get a vector field is considering the gradient vector field of a $\mathcal{C}^1$ function $\varphi(x_1, \ldots, x_n)$,

$$\mathbf{F} = \operatorname{grad} \varphi = \left(\frac{\partial \varphi}{\partial x_1}, \ldots, \frac{\partial \varphi}{\partial x_n} \right).$$

Here the original function φ is called a **potential function** of the vector field $\mathbf{F}$. As we saw in Chapter 10, the gradient vector field is perpendicular to level surfaces of its potential function.

3.0.1 Example

If a vector field in the plane is given by $\mathbf{F}(x, y) = (y, x)$, the function

$$\varphi(x, y) = xy$$

is a potential function of $\mathbf{F}$. Therefore the "picture" of $\mathbf{F}$ looks perpendicular to the hyperbola $xy = c$.

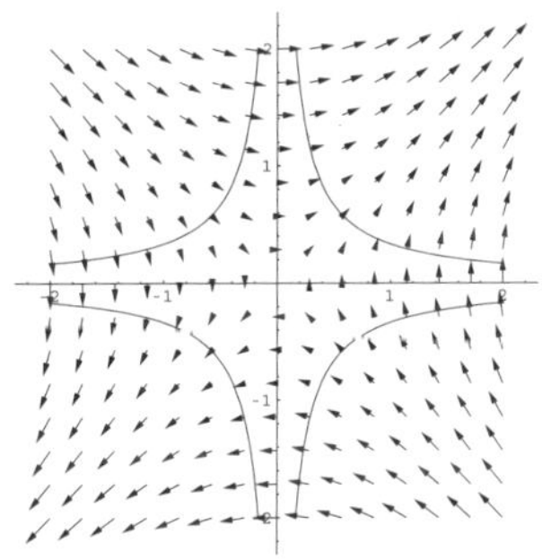

3.0.2 Example

In the first quadrant of the plane

$$x > 0, \qquad y > 0$$

the angle element vector field

$$\mathbf{a}(x, y) = \frac{(-y, x)}{x^2 + y^2}$$

has

$$\theta(x, y) := \arctan \frac{y}{x}$$

as a potential function. Of course the function

$$\varphi(x, y) := \theta(x, y) + \text{const}$$

is also a potential function of $\mathbf{a}$.

In general, if there exists a potential function φ of a vector field $\mathbf{F}$, then $\varphi + C$ is also a potential function for any constant C.

3.0.3 Basic Exercise

Find a potential function of the unit position vector field[25] $\mathbf{u} := \dfrac{\mathbf{r}}{r}$.

3.1 Fundamental Theorem of Line Integral

The following theorem says that if a vector field has a potential function, then the line integral of the vector field is given by the difference between values of the potential function at both ends of the curve.[26]

Theorem 3.1.1 (Fundamental Theorem of Line Integral) *Suppose that $\varphi : U \to \mathbb{R}$ is a C^1 function defined on an open set U in $\mathbb{R}^n$ and $X : [a,b] \to U$ is a C^1 curve. Then*[27]

$$\int_X \operatorname{grad} \varphi \cdot d\mathbf{s} = \varphi\big(X(b)\big) - \varphi\big(X(a)\big).$$

[25] Answer: $\varphi = r$ is one answer.

[26] As the amount of work done to move a unit electric charge is given by the difference of electric potential in electrostatics ....

[27] The **boundary** of a curve $X : [a,b] \to \mathbb{R}^n$ is an (ordered) pair of the **initial point** $\partial^- X := X(a)$ and **terminal point** $\partial^+ X := X(b)$,

$$\partial X := (X(a), X(b)) = (\partial^- X, \partial^+ X).$$

The "0-dimensional integral" over the boundary of a curve X is defined by

$$\int_{\partial X} \varphi := \varphi(\partial^+ X) - \varphi(\partial^- X).$$

Thus the fundamental theorem of line integral becomes

$$\int_X \operatorname{grad} \varphi \cdot d\mathbf{s} = \int_{\partial X} \varphi,$$

and the fundamental theorem of calculus $\int_a^b \varphi'(t)\,dt = \big[\varphi(t)\big]_a^b$ can be expressed by

$$\int_{[a,b]} \varphi'(t)\,dt = \int_{\partial[a,b]} \varphi.$$

In this theorem, if we replace a "straight line segment" $[a,b]$ by a general curve X, we get the fundamental theorem of line integral.

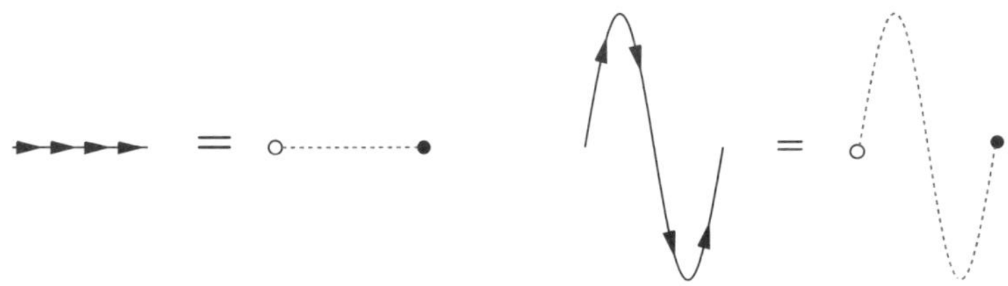

Proof.

$$\begin{aligned}\int_X \operatorname{grad}\varphi \cdot d\mathbf{s} &= \int_a^b \operatorname{grad}\varphi(X(t)) \cdot X'(t)\,dt \\ &= \int_a^b \frac{d}{dt}\varphi(X(t))\,dt \\ &= \varphi(X(b)) - \varphi(X(a))\end{aligned}$$

□

3.1.2 Example: Solid Angle Vector Field

In $\mathbb{R}^3 - \{(0,0,0)\}$, the **solid angle vector field**[28] is defined by

$$\mathbf{A}(x,y,z) := \frac{(x,y,z)}{(x^2+y^2+z^2)^{3/2}} = \frac{\mathbf{r}}{r^3}.$$

[28]It is called the "solid angle vector field" because its surface integral over a surface is the area of the shadow when the surface is projected (from the origin) onto a unit sphere. We will study the theory of surface integral soon. It would be correct to call it " solid angle element vector field" instead of "solid angle vector field". This vector field can be found in Newton's theory of gravity, theory of electromagnetic force by Charles Augustin de Coulomb (1736–1806), etc. This vector field is important in "Differential Topology" because it tells us about the **hole** of the space. In $\mathbb{R}^n$, the solid angle vector field is given by

$$\mathbf{A} = \frac{\mathbf{r}}{r^n}.$$

Using differential forms, $\mathbf{A}$ can be written as

$$\alpha := \frac{1}{r^n}\left(\sum_{i=1}^{n}(-1)^{i-1}x^i dx^1 \wedge \cdots \wedge dx^{i-1} \wedge dx^{i+1} \wedge \cdots \wedge dx^n\right)$$

and it tells us that there is one $(n-1)$-dimensional **hole** in $\mathbb{R}^n - \{O\}$ (for those who want to know the truth).

It has norm

$$|\mathbf{A}| = \frac{1}{r^2}$$

which is in inverse proportion to the square of distance, and the function

$$\varphi(x, y, z) = -1/r$$

is a potential function of $\mathbf{A}$.[29] Therefore if two points P_0, P_1 in the space have the same distance from the origin, the integral of $\mathbf{A}$ along any curve connecting these two points (which doesn't pass through the origin) is zero. If $r(P_0) = 1$ and $r(P_1) = 2$, the amount of work done to move from P_0 to P_1 is the difference of the potential energy

$$\varphi(P_1) - \varphi(P_0) = -\frac{1}{r(P_1)} + \frac{1}{r(P_0)} = -\frac{1}{2} + 1 = \frac{1}{2}.$$

3.1.3 Basic Exercise

Explain that the fundamental theorem of line integral is true also for a piecewise C^1 curve X.

3.2 Law of Conservation of Energy

For a vector field $\mathbf{F}$ defined on an open set U in $\mathbb{R}^n$, if a function $-V$ is a potential function of $\mathbf{F}$ then the **potential energy** at each point P in U is given by

$$+V(P).$$

Suppose that a particle $X(t)$ of mass m on Usatisfies Newton's law of motion

$$\mathbf{F}(X(t)) = mX''(t).$$

If we differentiate the **kinetic energy** $\frac{1}{2}m|X'(t)|^2$ of $X(t)$, then we get

$$\begin{aligned} \frac{d}{dt}\left(\frac{1}{2}m|X'(t)|^2\right) &= \frac{1}{2}m \cdot 2X''(t) \cdot X'(t) = \mathbf{F}(X(t)) \cdot X'(t) \\ &= -\operatorname{grad} V(X(t)) \cdot X'(t) \\ &= -\frac{d}{dt}V(X(t)). \end{aligned}$$

[29]The potential function with opposite sign is called the **potential energy**. In 1773, Lagrange introduced potential functions to explain the gravitational field theory [Birkhoff, p.335].

Thus we have

$$\frac{d}{dt}\left(\frac{1}{2}m|X'(t)|^2 + V(X(t))\right) = 0$$

and the sum of the kinetic energy and potential energy of $X(t)$

$$\frac{1}{2}m|X'(t)|^2 + V(X(t))$$

is constant. That is, there holds the "law of conservation of energy".[30]

We can also use line integral to show the law of conservation of energy X. For a $\mathcal{C}^1$ curve $X : [t_0, t_1] \to U$ connecting two points $X_0 := X(t_0)$ and $X_1 := X(t_1)$ in the region U, we have

$$\int_X \mathbf{F} \cdot d\mathbf{s} = \int_X \operatorname{grad}(-V) \cdot d\mathbf{s} = -V(X_1) + V(X_0). \tag{13.4}$$

On the other hand,

$$\int_X \mathbf{F} \cdot d\mathbf{s} = \int_X mX'' \cdot d\mathbf{s} = m\int_{t_0}^{t_1} X'' \cdot X'\, dt \tag{13.5}$$

$$= \frac{1}{2}m|X'(t_1)|^2 - \frac{1}{2}m|X'(t_0)|^2. \tag{13.6}$$

Therefore from (13.4) and (13.6), we get the law of conservation of energy

$$-V(X_1) + V(X_0) = \frac{1}{2}m|X'(t_1)|^2 - \frac{1}{2}m|X'(t_0)|^2.$$

3.3 Closed Curves

A curve with the same initial and terminal points is called a **closed curve**.

Corollary 3.3.1 *Suppose that a vector field* $\mathbf{F}$ *defined on an open set* $U \subset \mathbb{R}^n$ *has a potential function. If two curves* X, Y *in* U *have the same initial and terminal points, then*[31]

$$\int_X \mathbf{F} \cdot d\mathbf{s} = \int_Y \mathbf{F} \cdot d\mathbf{s}.$$

In particular, the integral of $\mathbf{F}$ *along a closed curve is zero.*

[30] It is the reason why a vector field which has a potential function— that is, a gradient vector field, is often called a **conservative field**.

Proof.
Let φ be a potential function of the vector field $\mathbf{F}$, and P and Q be the common initial and terminal points of two curves, respectively.

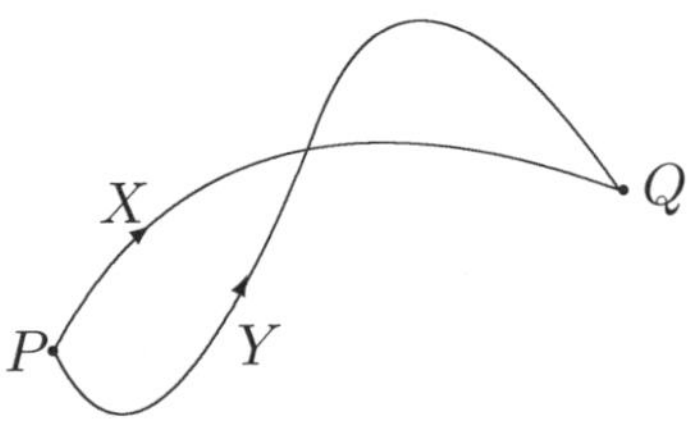

From the fundamental theorem of line integral,

$$\int_X \mathbf{F} \cdot d\mathbf{s} = \varphi(Q) - \varphi(P) = \int_Y \mathbf{F} \cdot d\mathbf{s}.$$

If X is a closed curve, then $P = Q$ and thus

$$\int_X \mathbf{F} \cdot d\mathbf{s} = \varphi(Q) - \varphi(P) = 0.$$

□

The integral of a vector field $\mathbf{F}$ along a closed curve X is written as

$$\oint_X \mathbf{F} \cdot d\mathbf{s}$$

and called the **circulation**, to emphasize that it is over a closed curve.[32] Therefore a circulation of a vector field which has a potential function is zero.

For example, consider the rotation in the three-dimensional space around the z-axis with constant angular speed ω rad/sec

$$x = r_0 \cos \omega t, \quad y = r_0 \sin \omega t, \quad z = z_0.$$

[31] Conversely, if a vector field has the line integral which is independent of paths and depends only on the initial and terminal points,—such a vector field is also called a **conservative (vector) field**— then it has a potential function. Of course, a potential function is obtained from the line integral.

[32]

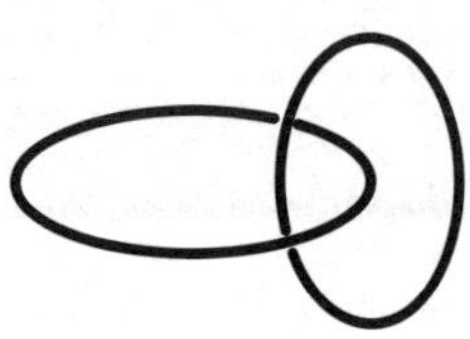

The **linking number** represents "the number of times that one **loop** winds around the other." It is related with the line integral [Flanders]. Linking numbers are used to explain the fact that electricity induces magnetism and vise versa, and the property of DNA in molecular biology [Chern, 1990]. The linking number was introduced by Gauss.

Since we have

$$\begin{aligned} x' &= -\omega r_0 \sin \omega t = -\omega y, \\ y' &= \omega r_0 \cos \omega t = \omega x, \end{aligned}$$

the **velocity vector field** is

$$\mathbf{F}(x, y, z) := \omega(-y\,\mathbf{i} + x\,\mathbf{j}),$$

and the circulation of $\mathbf{F}$ along a closed curve

$$X(t) := (r_0 \cos t, r_0 \sin t, z_0), \qquad 0 \leq t \leq 2\pi$$

is[33]

$$\oint_X \mathbf{F} \cdot d\mathbf{s} = \int_X \omega r_0 \; ds = (\omega r_0)(2\pi r_0) = 2\pi r_0^2 \omega.$$

Since the circulation is not zero, we know that the vector field $\mathbf{F}(x, y) = \omega(-y, x)$ doesn't have any potential function.

3.3.2 Basic Exercise

Show that the circulation of the angle element vector field

$$\mathbf{a}(x, y) = (-y, x)/(x^2 + y^2)$$

along a curve $x^2 + y^2 = r_0^2$ is not zero. From this, explain that the angle element vector field doesn't have any potential function.

3.4 Existence of Potential Function

We investigated several properties of vector fields which have potential functions like the solid angle vector field. When we consider a vector field, it is helpful to think of whether the vector field is induced from the more fundamental source. In other words, if we know a potential function of the

---

[33] Or

$$\begin{aligned} \oint_X \mathbf{F} \cdot d\mathbf{s} &= \int_0^{2\pi} \mathbf{F}(X(t)) \cdot X'(t)\, dt \\ &= \int_0^{2\pi} \omega(-r_0 \sin t\,\mathbf{i} + r_0 \cos t\,\mathbf{j}) \cdot (-r_0 \sin t\,\mathbf{i} + r_0 \cos t\,\mathbf{j})\, dt \\ &= 2\pi r_0^2 \omega \end{aligned}$$

vector field, we may know various properties of the vector field. However all vector fields don't have potential functions. We say that a vector field $\mathbf{F}(x_1, \ldots, x_n)$ defined on an open set $U \subset \mathbb{R}^n$ has a potential function if there exists a function $\varphi : U \to \mathbb{R}$ defined on the whole set U such that

$$\mathbf{F} = \operatorname{grad} \varphi.$$

3.5 Example

Let's find a potential function of the vector field

$$\mathbf{F}(x, y) = (y \cos xy, 1 + x \cos xy).$$

That is, let's find a function $\varphi(x, y)$ satisfying the differential equations

$$\frac{\partial \varphi}{\partial x} = y \cos xy, \qquad \frac{\partial \varphi}{\partial y} = 1 + x \cos xy. \tag{13.7}$$

If there exists a potential function, from the first equation, we know that there exists a function $g(y)$ such that

$$\varphi(x, y) = \int y \cos xy \, dx = \sin xy + g(y).$$

If we put it into the second equation in (13.7), we get

$$g'(y) = 1.$$

Therefore $g(y) = y + \text{const}$ and so

$$\varphi(x, y) = \sin xy + y + \text{const}.$$

It is obvious that this function is a potential function of $F(x, y)$.

3.5.1 Example

Previously, we saw that the vector field

$$\mathbf{F}(x, y) = (-y, x)$$

has a nonzero circulation and thus doesn't have any potential function. Let's reconsider this, using differential equations.

If φ is a potential function of $\mathbf{F}$, then there must hold the differential equations

$$\frac{\partial \varphi}{\partial x} = -y, \qquad \frac{\partial \varphi}{\partial y} = x. \tag{13.8}$$

If we integrate the first equation, we get

$$\varphi(x, y) = -xy + g(y). \tag{13.9}$$

If we differentiate (13.9) with respect to y and compare it with the second equation in (13.8), we get

$$-x + g'(y) = \frac{\partial \varphi}{\partial y} = x.$$

Therefore we have

$$g'(y) = 2x$$

which is a contradiction since the left side of this equation is independent of x but the right side depends on x. So there is no potential function of $\mathbf{F}$.

The above proof used "integration" (13.9) to show that the equations (13.8) give a contradiction, that is, they don't have a solution. The following proof use "differentiation" to show that (13.8) give a contradiction.[34] First, if we differentiate the first and second equations in (13.8) with respect to y and x respectively, we get

$$\frac{\partial^2 \varphi}{\partial y \partial x} = -1, \qquad \frac{\partial^2 \varphi}{\partial x \partial y} = 1.$$

So we get

$$\frac{\partial^2 \varphi}{\partial y \partial x} \neq \frac{\partial^2 \varphi}{\partial x \partial y},$$

which contradicts to Euler's "commutative law of partial derivatives".

In general, in order that a vector field

$$\mathbf{F}(x_1, \ldots, x_n) = \big(f_1(x_1, \ldots, x_n), \ldots, f_n(x_1, \ldots, x_n)\big)$$

has a potential function φ, there must exist a solution of the system of differential equations

$$\frac{\partial \varphi}{\partial x_i} = f_i \qquad (i = 1, \ldots, n).$$

[34] "Solving differential equations" is "integrating". In general, in order that **partial differential equations** like (13.8) have a solution, it must overcome **obstruction**.

Theorem 3.5.2 *If a $\mathcal{C}^1$ vector field $\mathbf{F} = (f_1, \ldots, f_n)$ defined on an open set in $\mathbb{R}^n$ has a potential function, then*

$$D_i f_j = D_j f_i \qquad (1 \le i, j \le n). \tag{13.10}$$

Proof. Let φ be a potential function of $\mathbf{F}$. Then, $f_j = D_j\varphi$ and so

$$D_i f_j = D_i(D_j\varphi) = D_j(D_i\varphi) = D_j f_i.$$

□

3.5.3 Example

Show that the vector field

$$\mathbf{F}(x, y) = (f(x, y), g(x, y)) := (x^2 + y^2, x^2 - y^2)$$

does not have a potential function.

Since

$$\frac{\partial f}{\partial y} = 2y, \qquad \frac{\partial g}{\partial x} = 2x$$

we know

$$\frac{\partial f}{\partial y} \neq \frac{\partial g}{\partial x}$$

and so $\mathbf{F}$ doesn't have a potential function.

3.6 Closed Vector Fields

If a $\mathcal{C}^1$ vector field $\mathbf{F} = (f_1, \ldots, f_n)$ satisfies

$$D_j f_i = D_i f_j \qquad \forall\, i, j,$$

$\mathbf{F}$ is called a **closed vector field**. Therefore the above theorem says that

A vector field is closed if it has a potential function.

Now we will see that its converse is not always true.
If we think of the **angle element vector field**

$$\mathbf{a}(x,y) = (a_1(x,y), a_2(x,y)) = \frac{(-y,x)}{x^2+y^2}$$

on $\mathbb{R}^2 - \{O\}$, we have

$$\begin{aligned} \frac{\partial}{\partial y}\left(-\frac{y}{x^2+y^2}\right) &= \frac{-x^2+y^2}{(x^2+y^2)^2}, \\ \frac{\partial}{\partial x}\left(\frac{x}{x^2+y^2}\right) &= \frac{-x^2+y^2}{(x^2+y^2)^2} \end{aligned}$$

and thus the condition

$$\frac{\partial a_1}{\partial y} = \frac{\partial a_2}{\partial x}$$

is satisfied. On the other hand, if we integrate **a** along a unit circle which is a closed curve,[35] the integral

$$\int_{\mathbf{S}^1} \mathbf{a} \cdot d\mathbf{s} = 2\pi$$

is not zero and so **a** cannot have a potential function (on the whole domain).

However, for example, if we restrict the domain of the angle element vector field to the right **half plane**

$$x > 0$$

then we can choose

$$\theta(x,y) := \arctan\frac{y}{x}$$

as a potential function of **a**.

3.6.1 Basic Exercise

Show that on the upper half plane $y > 0$,

$$\theta(x,y) = -\arctan\frac{x}{y} + \frac{\pi}{2}$$

is a potential function of the angle element vector field **a**.

[35] As we saw in section 2.4.

3.6.2 Poincaré's Lemma

Theorem 3.5.2 is a "trivial theorem" which is easy to understand and prove. To our surprise,[36] conversely, if the domain is of *good shape* then any vector field $\mathbf{F}$ satisfying the condition (13.10) has a potential function. The geometric property of the domain tells us the existence of a solution of differential equations.[37]

For your reference, A subset U in space is called **convex** if the line segment connecting any two points P and Q in U,

$$(1-t)P + tQ, \qquad 0 \le t \le 1$$

is contained in U.

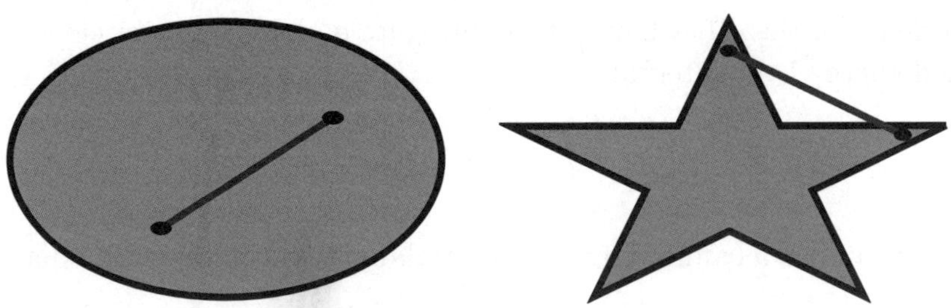

For example, an open ball $\mathbf{B}^n$ and a closed ball $\overline{\mathbf{B}}^n$ are both convex.

The following theorem is a part of **Poincaré's lemma** and its proof can be found in the last section.

Theorem 3.6.3 (Poincaré's Lemma) *A closed vector field defined on an open convex set in* $\mathbb{R}^n$ *has a potential function.*

In other word, the converse of Theorem 3.5.2 is true in a "good" domain. In particular it is true "locally", even though it is not true "globally".

[36]How could we use differentiation to know the existence of solution for problem about integration ...

[37]As grain cannot grow on a gravelly field...

3.6.4 Example

Find a potential function of the vector field $\mathbf{F}(x, y) = (2xy, x^2 + 1)$.

Solution. First, we check that

$$\frac{\partial}{\partial y}(2xy) = \frac{\partial}{\partial x}\left(x^2 + 1\right).$$

On the other hand, since the domain of $\mathbf{F}$ is the whole plane which is convex, by Poincaré's Lemma, $\mathbf{F}$ has a potential function.

For

$$\varphi(x, y) = \int 2xy\, dx = x^2 y + c(y)$$

to be a potential function of $\mathbf{F}$, it must be true that

$$x^2 + 1 = \frac{\partial \varphi}{\partial y} = x^2 + c'(y).$$

Therefore we get $1 = c'(y)$, i.e. $c(y) = y + \text{const}$ and we know that

$$\varphi(x, y) = x^2 y + y + \text{const}$$

is a potential function of $\mathbf{F}$.

3.7 Uniqueness of Potential Function

So far, we investigate the existence of potential function. Now we will consider "how many potential functions would there exist?" if there exists a potential function. Such a question is called the question of **uniqueness**.[38]

If a function φ is a potential function of a vector field $\mathbf{F}$, then obviously

$$\varphi + \text{const}$$

is also a potential function of $\mathbf{F}$.

Now, we will show conversely that if the domain U of $\mathbf{F}$ is connected then there is no other potential function. An open set U is called (path-)**connected** if for any two points P_0 and P_1 in U there exists a path in U connecting P_0 and P_1, that is, a curve $X : [0, 1] \to U$ such that $X(0) = P_0$ and $X(1) = P_1$.

[38] Many theorems in mathematics tell us about the **existence** and **uniqueness**. Even though we know the existence and uniqueness of a certain object, it is totally different and very interesting question to find it practically.

Proof. Suppose that two functions

$$\varphi,\, \psi : U \to \mathbb{R}$$

are potential functions of $\mathbf{F}$, and consider any two points P_0, P_1 in U. Then since the domain U is connected, there exists a curve

$$X : [0,1] \to U, \qquad X(0) = P_0, \quad X(1) = P_1.$$

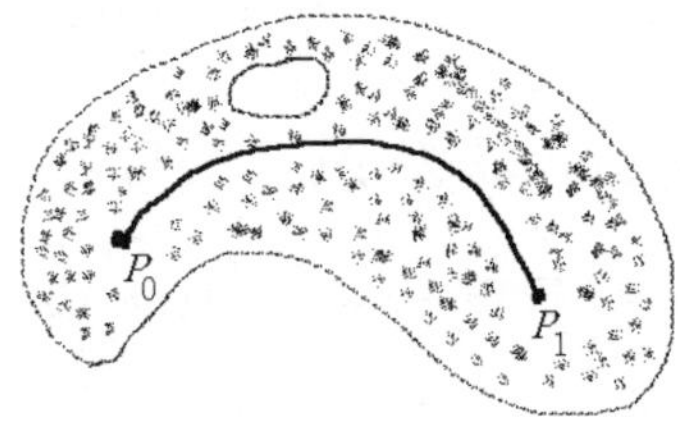

By the fundamental theorem of line integral,

$$\varphi(P_1) - \varphi(P_0) = \int_X \mathbf{F} \cdot d\mathbf{s} = \psi(P_1) - \psi(P_0)$$

and thus

$$\varphi(P_1) - \psi(P_1) = \varphi(P_0) - \psi(P_0).$$

On the other hand, since P_0 and P_1 are arbitrarily given, from the above equation we know that the function $\varphi - \psi$ is constant. □

3.7.1 Basic Exercise

When we restrict the domain of the angle element vector field

$$\mathbf{a}(x,y) = \frac{(-y,x)}{x^2+y^2}$$

to the union of the first and third quadrants

$$xy > 0,$$

find all potential functions of $\mathbf{a}$.[39]

---

[39]If we let I and III be the first and third quadrants respectively and let

$$\theta(x,y) := \begin{cases} \arctan(y/x), & (x,y) \in I \\ \pi + \arctan(y/x), & (x,y) \in III \end{cases}$$

Exercise: Chapter 13 Section 3

1. For any $\mathcal{C}^1$ function $g(r) > 0$, show that the integral of the vector field

$$\mathbf{F}(x,y) = g(r)(-y,x) \qquad \left(r = (x^2+y^2)^{1/2}\right)$$

along a unit circle is not zero, and from this explain that $\mathbf{F}$ doesn't have any potential function.

2. Find a potential function of the vector field

$$\mathbf{F}(x,y,z) = yz\mathbf{i} + (xz+y)\mathbf{j} + (xy+z^2)\mathbf{k}.$$

3. For a vector field $\mathbf{F}(x,y) = (-y,x)$ and a curve

$$X(t) := (\cos t, \sin t), \qquad 0 \le t \le \pi,$$

evaluate the line integral $\int_X \mathbf{F}\cdot d\mathbf{s}$. Also if we let

$$\tilde{X}(t) := (\cos 3t, \sin 3t), \qquad 0 \le t \le \pi,$$

show that two curves $\tilde{X},\ X\ :\ [0,\pi] \to \mathbb{R}^2$ have the same boundary but

$$\int_{\tilde{X}} \mathbf{F}\cdot ds \ne \int_X \mathbf{F}\cdot ds.$$

4. Find a potential function of the vector field

$$\mathbf{F}(x,y) = (x^2+y^2)\mathbf{i} + 2xy\mathbf{j}$$

and sketch the vector field. (You may use computer programs such as Mathematica, MatLab or Maple.)

5. Show that the vector field $\mathbf{F}(x,y,z) = (0,-z,y)$ doesn't have a potential function.

6. In $\mathbb{R}^n$, find a potential function of the following vector field.

   (1) $\mathbf{F}(X) = X/|X|$

   (2) $\mathbf{F}(X) = X/|X|^3$

$$\chi_I(x,y) := \begin{cases} 1, & (x,y)\in I \\ 0, & (x,y)\in III \end{cases}, \qquad \chi_{III}(x,y) := \begin{cases} 0, & (x,y)\in I \\ 1, & (x,y)\in III \end{cases}$$

then all potential functions of $\mathbf{a}|_{I\cup III}$ are

$$\theta(x,y) + c_1\chi_I(x,y) + c_3\chi_{III}(x,y)$$

for any constants c_1, c_3.

(3) $\mathbf{F}(X) = |X|^k X \qquad (k \neq -2)$

7. Find a potential function of the following vector field.

(1) $\mathbf{F}(x, y, z) = (yz, zx, xy)$

(2) $\mathbf{F}(x, y, z) = (2x - y - z, 2y - x, 2z - x)$

(3) $\mathbf{F}(x, y, z) = (y \cos z - yze^x, x \cos z - ze^x, -xy \sin z - ye^x)$

(4) $\mathbf{F}(x, y, z) = (yz, xz + y, xy + 1)$

8. (1) Show that Theorem 3.6.3 is still true even if U is a union of two open convex sets.

(2) What if U is a union of three convex sets?

9. Evaluate the integral of the vector field

$$\mathbf{F}(x, y, z) = (y + z, z + x, x + y)$$

along the following curve $X(t)$ from $P = (0, 0, 0)$ to $Q = (1, 1, 1)$.

(1) $X(t) = (t, t^2, t^3)$ (2) $X(t) = \left(\sin^3 t, 1 - \cos t, \dfrac{2t}{\pi}\right)$

10. For the vector field

$$\mathbf{F} = \left(\frac{x}{x^2 + y^2} + y\cos(xy + z)\right)\mathbf{i} + \left(\frac{y}{x^2 + y^2} + x\cos(xy + z)\right)\mathbf{j} + \cos(xy + z)\,\mathbf{k}$$

defined on the whole xyz-space except the z-axis, answer the following question.

(1) Find all potential functions if $\mathbf{F}$ has a potential function.

(2) For a curve

$$X(t) = (\cos t, \sin t, t), \qquad 0 \leq t \leq \frac{\pi}{2},$$

evaluate the line integral $\int_X \mathbf{F} \cdot d\mathbf{s}$.

11. For the vector field

$$\mathbf{G}(x, y, z) = -\frac{(x, y, z)}{r^3}, \qquad r = (x^2 + y^2 + z^2)^{1/2},$$

find the amount of work done along any curve (which doesn't pass through the origin) from $(1, 0, 0)$ to $(0, 3, 4)$.

12. For the vector field

$$\mathbf{a}(x, y, z) = -\frac{y}{x^2 + y^2}\,\mathbf{i} + \frac{x}{x^2 + y^2}\,\mathbf{j}$$

defined on the whole xyz-space except the z-axis, answer the following questions.

(1) Is it a closed vector field?

(2) Does it have a potential function in the whole domain?

(3) Does it have a potential function locally?

13. Show that if the difference between two vector fields **F** and **G** is a gradient vector field, then the circulation of **F** and the circulation of **G** along any closed curve are equal.

4 Differential Forms and Total Differentials

For two functions $f(x,y)$ and $g(x,y)$ defined on an open set U in the plane, an expression of the form

$$f(x,y)\,dx + g(x,y)\,dy$$

is called a **differential form**.

It corresponds to a vector field

$$f(x,y)\,\mathbf{i} + g(x,y)\,\mathbf{j}.$$

For example, the differential form which corresponds to the angle element vector field

$$\mathbf{a}(x,y) = -\frac{y}{x^2+y^2}\,\mathbf{i} + \frac{x}{x^2+y^2}\,\mathbf{j}$$

is

$$-\frac{y}{x^2+y^2}\,dx + \frac{x}{x^2+y^2}\,dy.$$

In general, for functions $f_1, \ldots, f_n : U \to \mathbb{R}$ defined on a subset U in $\mathbb{R}^n$, an expression of the form

$$f_1\,dx_1 + \cdots + f_n\,dx_n$$

is called a **differential form.**[40]

There is a one-to-one correspondence between differential forms and vector fields:

$$f_1\,dx_1 + \cdots + f_n\,dx_n \quad \Longleftrightarrow \quad f_1\,\mathbf{e}_1 + \cdots + f_n\,\mathbf{e}_n.$$

[40]For a differentiable function f defined on an open set U in $\mathbb{R}^n$, we denote the directional derivative of f at $P \in U$ in the direction of V by $D_V f(P)$. If we consider a vector $V \in \mathbb{R}^n$ as a variable, we get a linear transformation $Df(P) : \mathbb{R}^n \to \mathbb{R}$. This transformation is called the **total differential** of f at P. Now if we varies $P \in U$, we get the map Df. This map is usually denoted by df. In particular, for the projection onto the ith axis, $x_i : \mathbb{R}^n \to \mathbb{R}$, the meaning of dx_i is clear. However in calculus, such a meaning is not emphasized. To be precise differential forms which we consider here are differential 1-forms. Differential 2-forms or differential 3-forms are important too. Functions can be thought of as Differential 0-forms.

4.1 Differential Forms and Line Integrals

From now on, the integral of a differential form $f_1\,dx_1 + \cdots + f_n\,dx_n$ along an "oriented curve" X is defined by the integral of the vector field $f_1\,\mathbf{e}_1 + \cdots + f_n\,\mathbf{e}_n$:

$$\int_X f_1\,dx_1 + \cdots + f_n\,dx_n := \int_X (f_1\,\mathbf{e}_1 + \cdots + f_n\,\mathbf{e}_n)\cdot d\mathbf{s}. \tag{13.11}$$

If we let

$$X(t) = (x_1(t), \ldots, x_n(t)), \qquad a \le t \le b,$$

we have

$$\begin{aligned}
&\int_X (f_1\,\mathbf{e}_1 + \cdots + f_n\,\mathbf{e}_n)\cdot d\mathbf{s} \\
= &\int_a^b (f_1(X(t))\,\mathbf{e}_1 + \cdots + f_n(X(t))\,\mathbf{e}_n)\cdot \big(x_1'(t)\,\mathbf{e}_1 + \cdots + x_n'(t)\,\mathbf{e}_n\big)\,dt \\
= &\int_a^b \big(f_1(X(t))x_1' + \cdots + f_n(X(t))x_n'\big)\,dt.
\end{aligned}$$

Therefore from the equation (13.11), we get

$$\boxed{\int_X f_1\,dx_1 + \cdots + f_n\,dx_n = \int_a^b \big(f_1(X(t))x_1'(t) + \cdots + f_n(X(t))x_n'(t)\big)\,dt}$$

4.2 Exact Forms

For a function $f(x_1, \ldots, x_n)$ defined on an open set in the space,

$$df := \frac{\partial f}{\partial x_1}\,dx_1 + \cdots + \frac{\partial f}{\partial x_n}\,dx_n$$

is called the **total differential** of f. Thus the total differential of f is the differential form corresponding to the gradient vector field of f

$$\operatorname{grad} f = \frac{\partial f}{\partial x_1}\,\mathbf{e}_1 + \cdots + \frac{\partial f}{\partial x_n}\,\mathbf{e}_n.$$

Here f is called a potential function of the total differential df.

For example, for a function

$$\theta(x, y) = \arctan\frac{y}{x},$$

the total differential is[41]

$$d\theta = -\frac{y}{x^2+y^2}\,dx + \frac{x}{x^2+y^2}\,dy.$$

A vector field has a potential function if and only if the corresponding differential form has a potential function. From now on, a differential form which has a potential function is called an **exact form**.

Using differential forms, the fundamental theorem of line integral can be expressed as follows:

$$\boxed{\int_X df = \int_{\partial X} f}$$

4.2.1 Basic Exercise

1. For a function $r(x,y) = \sqrt{x^2+y^2}$, show that

$$dr = \frac{x}{r}\,dx + \frac{y}{r}\,dy.$$

2. A differential form is called a **closed form** if the vector field corresponding to it is closed.

   (1) Show that an exact form is closed

   (2) Examine whether a closed form is exact.

[41] In the plane except the origin, the angle function θ is not well-defined (as a continuous function), but its differential $d\theta$ is well-defined .

Exercise: Chapter 13 Section 4

1. Evaluate the following integral.

   (1) $\displaystyle\int_X xy\,dx + (x+y)\,dy,$

   where X is the curve $y = x^2$ from $(-1,1)$ to $(2,4)$.

   (2) $\displaystyle\int_X y^2\,dx + x\,dy,$ where X is the curve $x = y^3$ from $(-1,-1)$ to $(1,1)$.

   (3) $\displaystyle\int_X y\sqrt{x}\,dx + x\sqrt[3]{y}\,dy,$

   where X is the curve $y^2 = x^3$ from $(1,1)$ to $(4,8)$.

   (4) $\displaystyle\int_X x^2y\,dx + xy^3\,dy,$

   where X consists of the line segment from $(-1,1)$ to $(2,1)$ followed by the segment from $(2,1)$ to $(2,5)$.

   (5) $\displaystyle\int_X (x+2y)dx + (2x-y)dy,$

   where X consists of the line segment from $(3,2)$ to $(3,-1)$ followed by the segment from $(3,-1)$ to $(-2,-1)$.

2. For functions
$$x(r,\theta) = r\cos\theta, \qquad y(r,\theta) = r\sin\theta,$$
   show that
$$dx = \cos\theta\,dr - r\sin\theta\,d\theta, \qquad dy = \sin\theta\,dr + r\cos\theta\,d\theta.$$

3. For functions
$$\begin{aligned} x(\rho,\varphi,\theta) &:= \rho\sin\varphi\cos\theta \\ y(\rho,\varphi,\theta) &:= \rho\sin\varphi\sin\theta \\ z(\rho,\varphi,\theta) &:= \rho\cos\varphi, \end{aligned}$$
   express the total differentials
$$dx, \quad dy, \quad dz$$
   as linear combinations of $d\rho, d\varphi, d\theta$.

4. (**Leibniz's formula**) For differentiable functions f, g defined on an open set in $\mathbb{R}^n$, show that
$$d(fg) = f\,dg + g\,df.$$

5 Proof of Poincaré's Lemma, Principle of Least Action, Dynamical Systems

5.1 Proof of Poincaré's Lemma

Theorem 5.1.1 (Poincaré's Lemma) *A closed vector field defined on an open convex set in $\mathbb{R}^n$ has a potential function.*

Proof. Fix a point $P = (p_1, \ldots, p_n)$ in the given domain U which is an open convex set, and let C be the line segment connecting P and an arbitrary point $X = (x_1, \ldots, x_n)$ in U:

$$C(t) = P + t(X - P) \qquad (0 \le t \le 1).$$

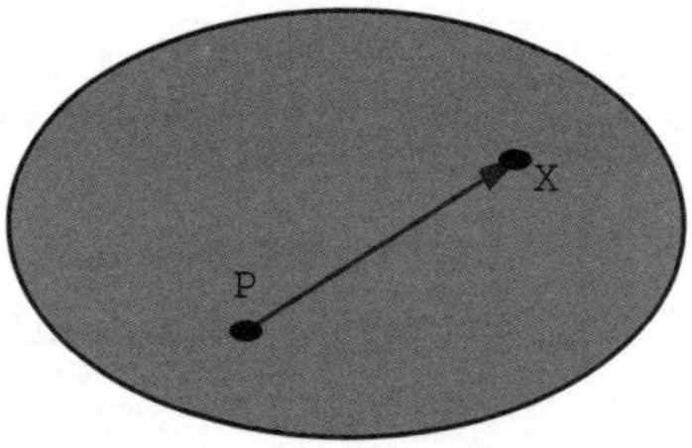

If the given closed vector field $\mathbf{F} = (f_1, \ldots, f_n)$ has a potential function φ, then $\varphi + C$ is still a potential function of $\mathbf{F}$ for any constant C. Therefore we may assume that the potential function we want to find satisfies $\varphi(P) = 0$. Then by the fundamental theorem of line integral (3.1.1),

$$\begin{aligned}\varphi(X) &= \int_C \mathbf{F} \cdot d\mathbf{s} = \int_0^1 \mathbf{F}\big(P + t(X - P)\big) \cdot (X - P)\, dt \\ &= \sum_{j=1}^n \int_0^1 f_j\big(P + t(X - P)\big) \cdot (x_j - p_j)\, dt\end{aligned}$$

Therefore if there exists a potential function φ of $\mathbf{F}$ which is zero at P, φ must be defined as above.

Now show that φ defined as in above is indeed a potential function of $\mathbf{F}$.

$$
\begin{aligned}
& D_i\varphi(X) \\
= & \sum_{j=1}^{n}\int_0^1 D_i\left(f_j(P+t(X-P))\cdot(x_j-p_j)\right)dt \\
= & \sum_{j=1}^{n}\int_0^1 \left(D_i\left(f_j(P+t(X-P))\right)\cdot(x_j-p_j)+f_j(P+t(X-P))\cdot\delta_{ij}\right)dt \\
= & \sum_{j=1}^{n}\int_0^1 \left((D_if_j)(P+t(X-P))\cdot t\right)\cdot(x_j-p_j)\,dt+\int_0^1 f_i(P+t(X-P))\,dt \\
= & \int_0^1\sum_{j=1}^{n}(D_jf_i)\left(P+t(X-P)\right)\cdot t(x_j-p_j)\,dt+\int_0^1 f_i(P+t(X-P))\,dt \\
= & \int_0^1\left(t\frac{d}{dt}\left(f_i(P+t(X-P))\right)+f_i(P+t(X-P))\right)dt \\
= & \int_0^1\frac{d}{dt}\left(tf_i(P+t(X-P))\right)dt \\
= & \left[tf_i(P+t(X-P))\right]_{t=0}^{t=1} \\
= & f_i(X).
\end{aligned}
$$

Therefore φ is a potential function of $\mathbf{F}$. □

In the above proof, we can see that even though the domain is not convex, as long as the domain is "star shape", the theorem is true. A set U is called **star shape** with respect to a point $P \in U$ if the line segment $[P, Q]$ is contained in U for any $Q \in U$.

5.2 Lagrange's Principle of Least Action

According to **Fermat's principle**, the path taken between two points by a ray of light is the path that can be traversed in the least time.[42] Lagrange generalized

---

42

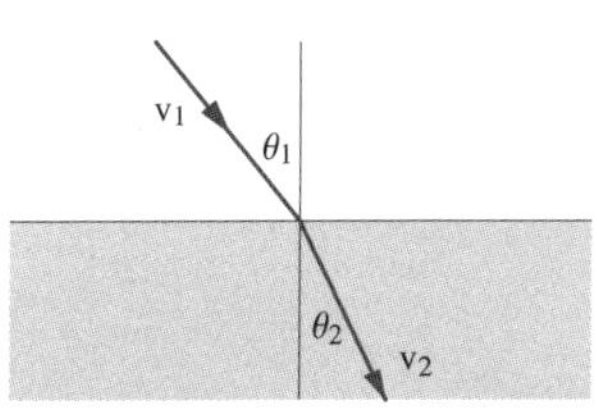

This principle explains the phenomenon that light ray is refracted through different media. For example, **Snell's principle** states that if we let v_1, v_2 be speeds of light ray in two homogeneous media respectively, and θ_1, θ_2 be angles between the light ray and the normal vectors of the boundary where two media meet respectively, then

$$\frac{\sin\theta_1}{v_1} = \frac{\sin\theta_2}{v_2}.$$

It is derived from Fermat's Principle. Snell's principle was found by a Dutch astronomer and mathematician Snellius (Willebrord van Roijen Snell, 1580–1626) in 1621. To know

the idea and thought that motions in nature must be such as to minimize a certain quantity called **action**.[43]

For example, if the potential energy function is given by φ and the force is given by

$$\mathbf{F} = -\operatorname{grad}\varphi$$

in the space U, the **action** along a path

$$X : [0,1] \to U, \qquad X(0) = P_0, \quad X(1) = P_1 \tag{13.12}$$

joining two points P_0, P_1 in U is given by the integral of the difference between kinetic energy and potential energy

$$\mathcal{L}(X) := \int_0^1 \left(\frac{1}{2}m|X'(t)|^2 - \varphi(X(t))\right)\, dt,$$

and a particle moves along the path which minimizes the value of $\mathcal{L}(X)$.

Now from Lagrange's principle, derive Newton's equation of motion.

Suppose that the path (13.12) between two points P_0, P_1 is given as to satisfy Lagrange's principle. Then for any $\mathcal{C}^1$ vector-valued function $\mathbf{w}(t)$ satisfying

$$\mathbf{w}(0) = \mathbf{0}, \quad \mathbf{w}(1) = \mathbf{0}$$

and any real number ϵ, the path

$$X_\epsilon(t) := X(t) + \epsilon\mathbf{w}(t)$$

is still through two points P_0, P_1 and thus

$$\mathcal{L}(X) \le \mathcal{L}(X_\epsilon).$$

In other words, since $\mathcal{L}(X_\epsilon)$ attains the minimum when $\epsilon = 0$, we get

$$\begin{aligned}
0 &= \left.\frac{d}{d\epsilon}\right|_0 L(X_\epsilon) \\
&= \int_0^1 \left.\frac{d}{d\epsilon}\right|_0 \left(\frac{1}{2}m|X_\epsilon'(t)|^2 - \varphi(X_\epsilon(t))\right)\, dt \\
&= \int_0^1 (mX'(t)\cdot\mathbf{w}'(t) - \operatorname{grad}\varphi(X(t))\cdot\mathbf{w}(t))\, dt \\
&= \int_0^1 \left(\frac{d}{dt}(mX'\cdot\mathbf{w}(t)) - (mX''(t) + \operatorname{grad}\varphi(X(t)))\cdot\mathbf{w}(t)\right)\, dt \\
&= \big[mX'(t)\cdot\mathbf{w}(t)\big]_0^1 - \int_0^1 (mX''(t) + \operatorname{grad}\varphi(X(t)))\cdot\mathbf{w}(t)\, dt \\
&= -\int_0^1 (mX''(t) - \mathbf{F}(X(t)))\cdot\mathbf{w}(t)\, dt.
\end{aligned}$$

---

that how Euclid or Heron described the properties of light and how Ptolemy misunderstood the properties of light, refer to [Schiffer, Bowden].

[43]It would be more appropriate to call Lagrange's principle the **critical action principle**.

On the other hand, since $\mathbf{w}(t)$ is arbitrary, we get

$$mX''(t) - \mathbf{F}(X(t)) = 0.$$

This equation is nothing but Newton's equation of motion $\mathbf{F} = m\mathbf{a}$.

5.3 Vector Fields and Dynamical Systems

A vector field $\mathbf{F}$ defined on an open set U in $\mathbb{R}^n$ can describe the movement of points in U. If $X(t)$ denotes the position of a point after time t which is located at X_0 initially, the **equation of motion** is given by

$$X'(t) = \mathbf{F}(X(t)), \qquad X(0) = X_0.$$

From now on, $X(t)$ is called the "**integral curve** of $\mathbf{F}$" (with the initial position X_0).[44]

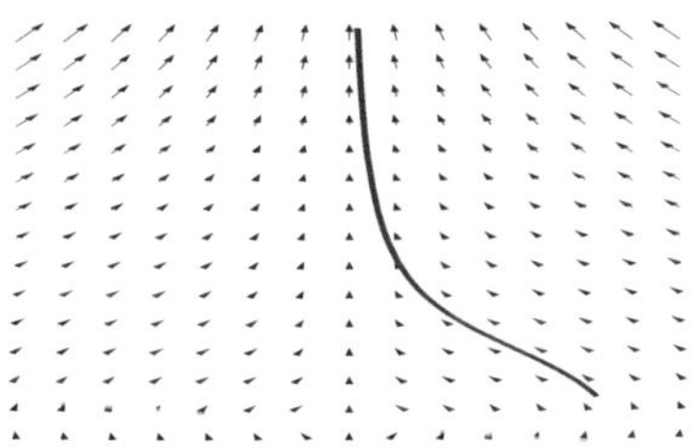

5.3.1 Basic Exercise

For a constant vector field

$$\mathbf{i}(x, y) - (1, 0)$$

show that the integral curve starting at $X_0 \in \mathbb{R}^2$ is given by

$$X(t) = X_0 + t(1, 0).$$

---

[44]If a vector field $\mathbf{F}$ is differentiable, the integral curve through a given point is unique. This theorem is called the **fundamental theorem of ordinary differential equations**. Newton's equation of motion

$$\mathbf{f}(Y(t)) = Y''(t)$$

can be written as

$$\big(Y(t), Y'(t)\big)' = \big(Y'(t), \mathbf{f}(Y(t))\big).$$

Here, if we let $X(t) = \big(Y(t), Y'(t)\big)$ and $\mathbf{F}(Y, Z) := \big(Z, \mathbf{f}(Y)\big)$, then the above equation becomes

$$X'(t) = \mathbf{F}(X(t))$$

and thus Newton's equation of motion is expressed as the form which we use for the equation of motion.

5.3.2 Example

For a plane vector field $\mathbf{F}(x, y) = -y\,\mathbf{i} + x\,\mathbf{j}$, the integral curve starting at $(1, 0)$ is given by

$$X(t) = (\cos t, \sin t).$$

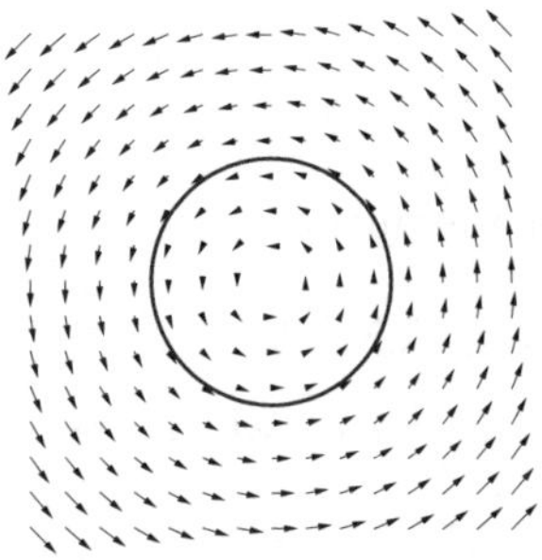

5.3.3 Linear Vector Fields

The previous example is an example of **linear vector field**. In general, the linear vector field $\mathbf{F}_A$ in the plane determined by a matrix $A = \begin{pmatrix} a & b \\ c & d \end{pmatrix}$ is defined by

$$\mathbf{F}_A \begin{pmatrix} x \\ y \end{pmatrix} := \begin{pmatrix} ax + by \\ cx + dy \end{pmatrix} = A \begin{pmatrix} x \\ y \end{pmatrix}.$$

Then the integral curve starting at $X_0 := \begin{pmatrix} x_0 \\ y_0 \end{pmatrix}$ is

$$X(t) = e^{tA} X_0.$$

Therefore the map

$$e^{tA} : \mathbb{R}^2 \to \mathbb{R}^2$$

tells us where a point flows to after time t. In other words, in the **dynamical system** defined by a vector field $\mathbf{F}_A$, a point X_0 is located at $e^{tA} X_0$ after time t.

For $\mathbf{F}(x, y) = (-y, x)$, the corresponding matrix is $\begin{pmatrix} 0 & -1 \\ 1 & 0 \end{pmatrix}$ and

$$\exp\left(t \begin{pmatrix} 0 & -1 \\ 1 & 0 \end{pmatrix} \right) = \begin{pmatrix} \cos t & -\sin t \\ \sin t & \cos t \end{pmatrix}.$$

This **rotation** describes the **dynamical system** of the vector field

$$\mathbf{F}(x, y) = (-y, x).$$

5.3.4 Basic Exercise

Show that in the plane the integral curve $\big(x(t), y(t)\big)$ of the vector field

$$\mathbf{A} = \frac{x}{x^2+y^2}\,\mathbf{i} + \frac{y}{x^2+y^2}\,\mathbf{j} = \frac{\mathbf{r}}{r^2}$$

satisfies

$$x(t)x'(t) + y(t)y'(t) = 1.$$

Use this to show that if the integral curve starts at $(x_0, y_0) \neq (0,0)$ then

$$x(t)^2 + y(t)^2 = 2t + r_0^2 \qquad (r_0^2 := x_0^2 + y_0^2).$$

Finally, show that

$$\big(x(t), y(t)\big) = \sqrt{\frac{2t}{r_0^2} + 1}\,(x_0, y_0).$$

Chapter 13 Advanced Exercise

1. For a vector field $\mathbf{F}$ defined on an open set U in $\mathbb{R}^n$, show that if

$$\int_X \mathbf{F} \cdot d\mathbf{s} = \int_Y \mathbf{F} \cdot d\mathbf{s}$$

for arbitrary curves X, Y in U which satisfy

$$\partial X = \partial Y,$$

then $\mathbf{F}$ has a potential function.

2. (**Integrating factor**) Show that for any differential form

$$\omega = a(x,y)\,dx + b(x,y)\,dy$$

defined on an open set in the plane, there exists a function $f(x,y) > 0$ such that $f\,\omega$ is locally an exact form.

3. In $\mathbb{R}^n$, the **solid angle vector field** with respect to the point P is defined by

$$\mathbf{A}_P(X) := \frac{\overrightarrow{PX}}{\left|\overrightarrow{PX}\right|^n}.$$

For points $P_1, \ldots, P_k$ in $\mathbb{R}^n$, show that the vector field

$$\mathbf{F}(X) := \mathbf{A}_{P_1}(X) + \cdots + \mathbf{A}_{P_k}(X)$$

is closed.

4. For $P = (1,0,0)$, $Q = (-1,0,0)$, find a potential function of the vector field

$$\mathbf{F}(X) := \mathbf{A}_P(X) + \mathbf{A}_Q(X)$$

defined on the **upper half space**

$$z > 0.$$

Part V

Multiple Integrals and Green's Theorem

Chapter 14

Multiple Integrals

In this chapter, we study integration and change of variables for functions of several variables. The elementary calculus deals with **Riemann integral**. In the theory by Riemann (1826–1866), the "space of all integrable functions" doesn't have the "completeness" and hence in modern mathematics we prefer **Lebesgue integral**, which extends Rie-mann integral. For Riemann integral, the given domain is subdivided into "subintervals" (or "boxes") and the upper and lower sums are defined, whereas for the integral defined by a French mathematician Lebesgue (1875–1941), the domain is divided by "measurable subsets" to define the upper and lower sums [Boas, p.208]. However, for (piecewise) continuous functions, these two integrals coincide and hence it suffices to consider Riemann integral. We can find the theory for integral in the work of Archimedes, who is one of the greatest ancient mathematicians. In his tombstone, how to get the surface area and volume of a sphere was described. The theory of integral gives the way to measure the length in the 1-dimensional space, the area in the 2-dimensional space, the volume in the 3-dimensional space. In the 0-dimensional space, the integral is nothing but "counting". The theory of integrals can be considered as "connection between spaces and functions" and in this point of view it is the fundamental theorem of calculus that tells us the **duality** of spaces and functions.

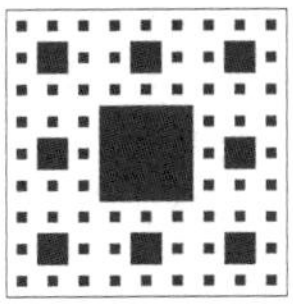

Picture. Carpet of W. Sierpinski (1882–1969)

1 Area and Volume

1.1 Banach's Question about Measure

When we try to define the "volume" $\mu(A)$ of a subset A of the Euclidean space $\mathbb{R}^n$ as a real number, Banach asked the following natural question:

"Is there a function μ which satisfies the following four properties?":

(i) For any bounded set A in the space, a real number $\mu(A)$ is defined and $\mu(A) \geq 0$.

(ii) If A and B are disjoint, then $\mu(A \cup B) = \mu(A) + \mu(B)$.

(iii) For any length-preserving transformation $T : \mathbb{R}^n \to \mathbb{R}^n$, $\mu(T(A)) = \mu(A)$.

(iv) For a regular box $[0,1]^n := [0,1] \times \cdots \times [0,1]$, $\mu([0,1]^n) = 1$.

Banach showed the existence of such a function μ for $n = 1, 2$, but Felix Hausdorff (1868–1942) proved that if $n \geq 3$, such a function μ doesn't exist.[1]

In mathematics, to avoid such a paradox(?), we consider and find the volume of only <u>measurable sets</u> instead of <u>all</u> bounded sets.

---

[1]Stephan Banach (1892–1945) and Alfred Tarski (1902–1983) found out the following phenomenon in 1924: Consider a partition $U_1, \ldots, U_k$ of a ball of radius 1 and a partition $V_1, \ldots, V_k$ of a ball of radius 2 in 3-dimensional space. If we take proper partitions, then for all $i \in \{1, \ldots, k\}$, U_i can be mapped onto V_i by a **Euclidean transformation** (that is, the length-preserving transformation such as rotations, parallel transformations or reflections). Such a phenomenon is called the **Banach-Tarski Paradox** [Chae, p.142].

In the second ICM (International Congress of Mathematicians) held in Paris in 1900, Hilbert provided 23 most important unsolved problems in those days. Among them, the third one, based on earlier writings by Gauss, is the following [Hopf], [Klee, Wagon]:

"Is a regular tetrahedron scissors-congruent to a regular hexahedron?"

We call two sets A and B are **scissors-congruent** if A is a union of polyhedrons $A_1, \ldots, A_k$ and B is also a union of polyhedrons $B_1, \ldots, B_k$, the volumes of $A_i \cap A_j$ and $B_i \cap B_j$ are 0 for $i \neq j$ and there is a Euclidean transformation T_i such that $T_i(A_i) = B_i$.

If the answer for this third question is positive, then we can deduce "without using calculus" that the volume of a tetrahedron is "(bottom area) $\times$ height/3".

Among Hilbert's 23 questions, this was the first one solved. Max Dehn(1878–1952) gave the negative answer to this question in the very same year.

1.2 Measurable Sets in Two-Dimensional Plane

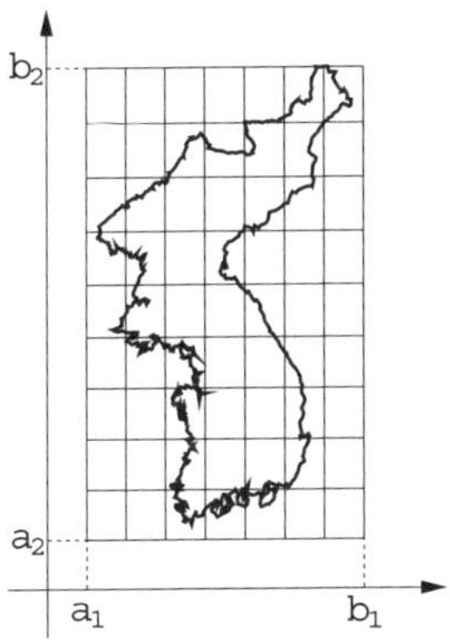

Let A be a bounded subset in $\mathbb{R}^2$. Then A is contained in a rectangle $[a_1, b_1] \times [a_2, b_2]$. Now for a partition of a rectangle $[a_1, b_1] \times [a_2, b_2]$

$$P \ : \ \begin{array}{l} a_1 = x_0 < x_1 < \cdots < x_l = b_1 \\ a_2 = y_0 < y_1 < \cdots < y_m = b_2, \end{array}$$

consider a small **boxes**[2]

$$P_{ij} := [x_{i-1}, x_i] \times [y_{j-1}, y_j]$$

for $1 \le i \le l,\ 1 \le j \le m$. Among these boxes, let $L(P, A)$ be the sum of areas of boxes contained in A and $U(P, A)$ be the sum of areas of boxes which intersects A. Then

$$L(P, A) \le U(P, A)$$

and as the partition P gets finer, $L(P, A)$ increases and $U(P, A)$ decreases. Therefore both[3]

$$\sup_P L(P, A), \qquad \inf_P U(P, A)$$

exist, and those values are called the **inner area** and **outer area** of A, respectively.

In general, the inner area of a bounded set is less than or equal to the outer area. If the inner area is equal to the outer area, then the common value is called the **area** and the set is called **measurable**.

1.2.1 Basic Exercise

Show that a closed box $[a, b] \times [c, d]$ and an open box $(a, b) \times (c, d)$ both have the area $(b - a) \times (d - c)$.

1.2.2 Example

If we consider the set of all **rational points** in the square $[0, 1] \times [0, 1]$

$$A = \{(x, y) \mid 0 \le x, y \le 1,\ x, y \in \mathbb{Q}\}$$

[2]When sides of a rectangle are parallel to the axis, we call it a box. This term is used in the three-dimensional space too.

[3]Refer to Chapter 1 Section 8 for sup and inf.

there is no box contained in A and thus the inner area of A is 0. However boxes which cover A also cover the square $[0,1] \times [0,1]$ and thus the outer area of A is 1. Therefore A is not a measurable set.

1.3 Measurable Sets in Three-Dimensional Space

As in the plane, for a bounded subset A in the three-dimensional space $\mathbb{R}^3$, we may define the **inner volume** and **outer volume** of A.

That is, pick a box containing A

$$[a_1, b_1] \times [a_2, b_2] \times [a_3, b_3]$$

and for a partition of this box

$$P \ : \quad \begin{aligned} a_1 &= x_0 < x_1 < \cdots < x_l = b_1 \\ a_2 &= y_0 < y_1 < \cdots < y_m = b_2 \\ a_3 &= z_0 < z_1 < \cdots < z_n = b_3, \end{aligned}$$

consider small boxes

$$P_{ijk} = [x_{i-1}, x_i] \times [y_{j-1}, y_j] \times [z_{k-1}, z_k].$$

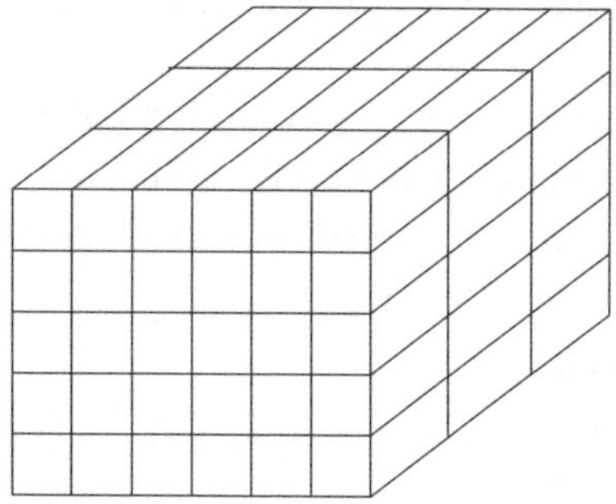

We let $L(P, A)$ be the sum of volumes of boxes contained in A and $U(P, A)$ be the sum of areas of boxes which intersects A. Then as the partition P gets finer their limits exist and are called the **inner volume** and **outer volume** of A. If the inner volume is equal to the outer volume, the common value is called the **volume** of A and denoted by

$$\mathrm{Vol}(A).$$

The set whose inner volume and outer volume are equal is called **measurable**.

In general, in the n-dimensional space, we can define a bounded measurable set A and its n-dimensional **volume** is denoted by

$$\mathrm{Vol}_n(A) \quad \text{or} \quad \mathrm{Vol}(A).$$

In particular, if A is a subset of the plane, its area is denoted by $\mathrm{Vol}_2(A)$.

2 Multiple Integrals

There are several ways to integrate a function over a given domain, but we will consider the simplest one, the definite integral of a "bounded function defined on a bounded domain".

Let's recall how to integrate a function of one variable. The (definite) integral of a continuous function f defined on an interval $[a, b]$ is denoted as

$$\int_a^b f \qquad \text{or} \qquad \int_a^b f(x)\,dx$$

and is computed as follows. First, for a **partition** of the interval $[a, b]$

$$P: \quad a = x_0 < x_1 < \cdots < x_n = b,$$

let[4]

$$\Delta x_i := x_i - x_{i-1}, \qquad |P| = \max\{x_i - x_{i-1} \mid 1 \le i \le n\}.$$

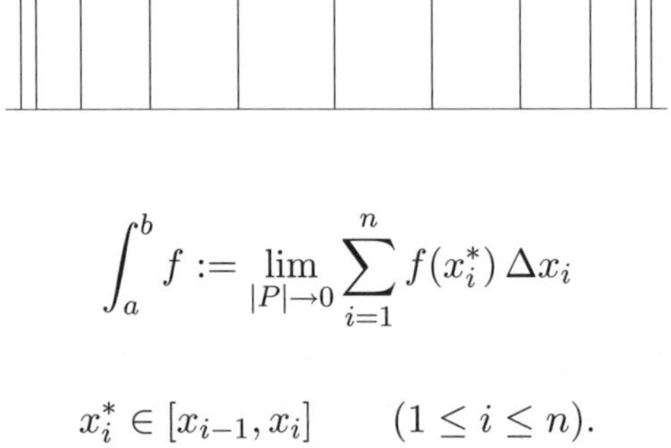

Then

$$\int_a^b f := \lim_{|P| \to 0} \sum_{i=1}^{n} f(x_i^*)\,\Delta x_i \tag{14.1}$$

for

$$x_i^* \in [x_{i-1}, x_i] \qquad (1 \le i \le n).$$

For a continuous function or a piecewise continuous function, there exists the integral defined as above, but the integral for general functions doesn't always exist.

[4]The **minimum** and **maximum** of a real set A are denoted by $\min A$ and $\max A$ respectively.

For a bounded function $f : [a, b] \to \mathbb{R}$, if the right side of the equation (14.1) converges, then f is called **integrable**.

To explain in detail, the integrability of a bounded function $f(x)$ is determined as follows. For a partition $P = \{a = x_1 < x_2 < \cdots < x_n = b\}$ of an interval $[a, b]$, we can define the **lower sum** and **upper sum** by

$$L(P, f) := \sum_{i=1}^{n} \big(\inf f[x_{i-1}, x_i]\big)\Delta x_i, \quad U(P, f) := \sum_{i=1}^{n} \big(\sup f[x_{i-1}, x_i]\big)\Delta x_i,$$

respectively. And then we define the **lower integral** and **upper integral** of f by

$$L(f) := \lim_{|P| \to 0} L(P, f), \qquad U(f) := \lim_{|P| \to 0} U(P, f),$$

respectively. If these two integrals have the same value, then $f(x)$ is integrable and this common value is denoted by

$$\int_a^b f \qquad \text{or} \qquad \int_a^b f(x)\, dx.$$

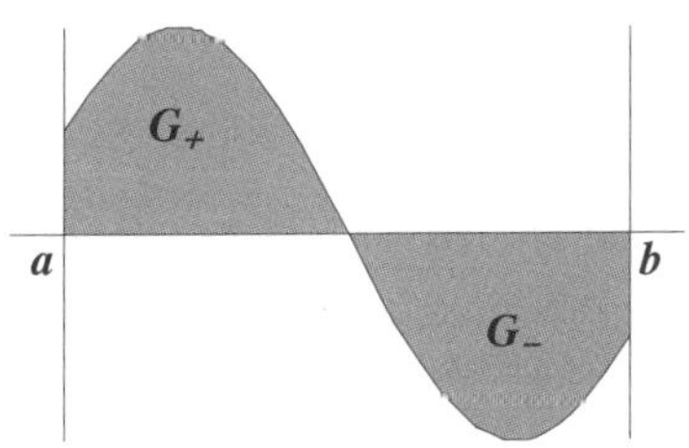

The following theorem states that if an integrable function f is positive, then the integral means the "area" between the graph of the function and the x-axis, and that without the condition that f is positive, the integral of f is the difference between the area of the positive part and that of the negative part.

Theorem 2.0.1 *A bounded function $f : [a, b] \to \mathbb{R}$ is integrable if and only if the sets*

$$\begin{aligned} G_+(f) &:= \{(x, y) \in \mathbb{R}^2 \mid x \in [a, b], \quad 0 \le y \le f(x)\}, \\ G_-(f) &:= \{(x, y) \in \mathbb{R}^2 \mid x \in [a, b], \quad f(x) \le y \le 0\} \end{aligned}$$

are both measurable. Moreover,

$$\int_a^b f(x)\, dx = \mathrm{Vol}_2\,(G_+(f)) - \mathrm{Vol}_2\,(G_-(f)).$$

Like this, integrating a function is equivalent to finding the area of some region. Each viewpoint has its own merit and both can be extended to the n-dimensional space.

2.1 Double Integrals

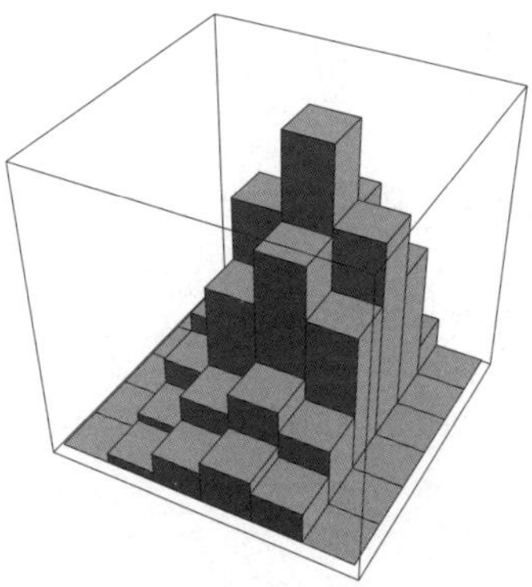

As for a function of a single variable, we can define the integrability of a bounded function $f(x, y)$ on a domain[5] A in the plane as follows.

First, pick a rectangle $[a, b] \times [c, d]$ containing A. Then for a partition of the rectangle

$$P : \begin{cases} a = x_0 < x_1 < \cdots < x_n = b \\ c = y_0 < y_1 < \cdots < y_m = d, \end{cases}$$

let

$$P_{ij} := [x_{i-1}, x_i] \times [y_{j-1}, y_j],$$

$$\mathrm{Vol}_2 P_{ij} := (x_i - x_{i-1})(y_j - y_{j-1}) = \Delta x_i \, \Delta y_j,$$

and let $f(x, y) = 0$ for $(x, y) \notin A$. Also let[6]

$$L(P, f) := \sum_{i,j} \big(\inf f(P_{ij})\big) \mathrm{Vol}_2 P_{ij}, \quad U(P, f) := \sum_{i,j} \big(\sup f(P_{ij})\big) \mathrm{Vol}_2 P_{ij}.$$

Now we define the **lower integral** and **upper integral** of $f(x, y)$ by

$$L(f) := \lim_{|P| \to 0} L(P, f), \qquad U(f) := \lim_{|P| \to 0} U(P, f).$$

If the value of two integrals are equal, then we say that $f(x, y)$ is **integrable** and this common value is denoted as

$$\iint_A f \, dV_2, \quad \int_A f, \quad \iint_A f, \quad \iint_A f(x, y) \, dx \, dy, \quad \iint_A f(x, y) \, dV_2(x, y).$$

[5]For convenience, we assume that it is a measurable bounded subset. If you don't want to restrict the domain, you can only consider a function defined on the whole $\mathbb{R}^2$ for which

$$\{X \mid f(X) \neq 0\}$$

is bounded.

[6]They are called the **lower sum** and **upper sum** of f with respect to the partition P respectively.

Like this, determining the integrability of a function $f(x, y)$ is equivalent to determining whether the region enclosed by the graph of the function is measurable. Two solids

$$\begin{aligned} G_+(f) &:= \{(x,y,z) \mid (x,y) \in A, \quad 0 \le z \le f(x,y)\} \\ G_-(f) &:= \{(x,y,z) \mid (x,y) \in A, \quad f(x,y) \le z \le 0\} \end{aligned}$$

are measurable if and only if the function f is integrable in the region A. Here the "integral of a function f over the region A" is

$$\iint_A f\, dV_2 = \mathrm{Vol}\big(G_+(f)\big) - \mathrm{Vol}\big(G_-(f)\big).$$

2.2 Triple Integrals

We can define integrals of functions defined in the higher dimensional space just as we defined double integrals.

For a function $f : A \to \mathbb{R}$ defined on a subset A of the three-dimensional space $\mathbb{R}^3$, the (definite) integral of f is denoted by

$$\int_A f, \qquad \iiint_A f, \qquad \iiint_A f(x,y,z)\, dx\, dy\, dz, \qquad \iiint_A f\, dV_3.$$

If f represents a density function of A, then $\int_A f$ represents the mass of A.

2.3 Multiple Integrals

For a function $f : A \to \mathbb{R}$ defined on a set A in the n-dimensional space $\mathbb{R}^n$, the (definite) integral of f is denoted by

$$\int_A f, \qquad \int_A f\, dV_n \qquad \int_A f(x_1, \ldots, x_n)\, dV(x_1, \ldots, x_n).$$

2.4 Characteristic Functions

For a set $A \subset \mathbb{R}^n$, the **characteristic function** $\chi_A : \mathbb{R}^n \to \mathbb{R}$ is defined by

$$\chi_A(X) := \begin{cases} 1 & (X \in A) \\ 0 & (X \notin A). \end{cases}$$

If A is bounded, then the characteristic function χ_A is integrable if and only if A is measurable (in the n-dimensional space). The (n-dimensional) volume of A is

$$\mathrm{Vol}_n(A) = \int_{\mathbb{R}^n} \chi_A = \int_A 1.$$

2.5 Properties of Integrable Function

From now on, we will call a measurable set a **region**.

Theorem 2.5.1 *For integrable functions f, g defined on a region A in $\mathbb{R}^n$ and a real number c,*

(i) $\int_A (f+g) = \int_A f + \int_A g, \qquad \int_A cf = c\int_A f$

(ii) *If* $f \le g$, $\int_A f \le \int_A g$

(iii) *If the region A is a union of two regions A_1 and A_2, then*[7]

$$\int_{A_1 \cup A_2} f + \int_{A_1 \cap A_2} f = \int_{A_1} f + \int_{A_2} f.$$

Also a bounded continuous function defined on A is integrable on A.

Corollary 2.5.2 *For an integrable function f defined on a region A in $\mathbb{R}^n$,*

$$\left|\int_A f\right| \le \int_A |f|.$$

Proof. From the inequalities

$$f \le |f|, \qquad -f \le |f|,$$

we get

$$\int_A f \le \int_A |f|, \qquad -\int_A f \le \int_A |f|$$

which imply the conclusion we want. □

[7]This equation implies that "f is integrable on $A_1, A_2, A_1 \cap A_2$".

2.5.3 Basic Exercise

Let M, m be the maximum and minimum of an integrable function f defined on a region A in $\mathbb{R}^n$ respectively.[8] Show that the **average** of f

$$\bar{f} := \frac{1}{\mathrm{Vol}\,(A)} \int_A f$$

satisfies the inequality

$$m \leq \bar{f} \leq M.$$

2.6 Trivial Extensions

For a function $f : A \to \mathbb{R}$ defined on $A \subset \mathbb{R}^n$ and a set $\tilde{A}$ ($\subset \mathbb{R}^n$) containing A, the function $\tilde{f} : \tilde{A} \to \mathbb{R}$ defined by

$$\tilde{f}(X) = \begin{cases} f(X), & X \in A \\ 0, & X \in \tilde{A} - A \end{cases}$$

is called "function which trivially extends f to $\tilde{A}$" or simply the "**extension** of f". If A and $\tilde{A}$ are both measurable and $f : A \to \mathbb{R}$ is integrable, then the extension $\tilde{f} : \tilde{A} \to \mathbb{R}$ is also integrable and two integrals are equal:

$$\int_A f \, dV_n = \int_{\tilde{A}} \tilde{f} \, dV_n.$$

[8]For convenience, assume that there exist the maximum and minimum of f. Otherwise, we may consider the supremum and infimum of f instead.

Exercise: Chapter 14 Section 2

1. Show that a triangle is measurable. Also show that any polygon or disk are measurable.

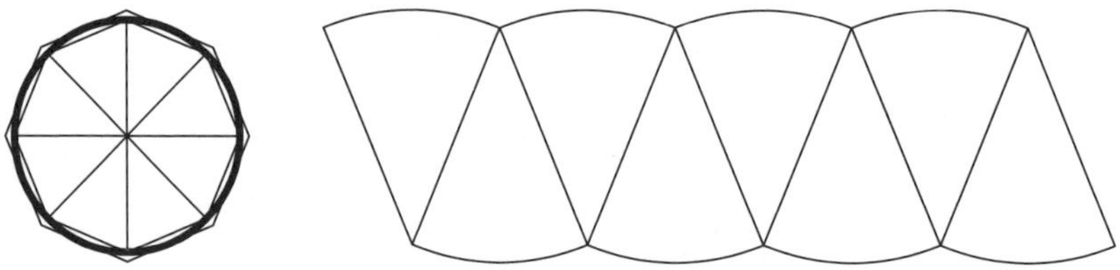

Picture: Archimedes and Kepler

2. (**Mean Value Theorem for Integrals**) Suppose that in a region $A \subset \mathbb{R}^n$, a continuous function f attains the maximum and minimum.[9] Show that if A is path-connected,there exists a point $P \in A$ such that

$$\frac{1}{\mathrm{Vol}(A)} \int_A f = f(P).$$

3. For continuous functions $f, g : A \to \mathbb{R}$ defined in a connected region $A \subset \mathbb{R}^n$, if we let $\bar{f}$ and $\bar{g}$ be the averages of f and g respectively, show that there exists a point $P \in A$ such that

$$\bar{f} : \bar{g} = f(P) : g(P).$$

4. Suppose that the amount of carbon monoxide in the air of some region K is given by $f(x, y)$ in appropriate coordinates. Express the average amount of carbon monoxide in this region K.

5. When the amount of rainfall last month at each point (x, y) in Jeju is given by $f(x, y)$, express the average amount of of rainfall last month in Jeju.

6. When the density of radon detected at each point (x, y, z) in a subway station A is given by $h(x, y, z)$, express the average density of radon in this subway station.

7. A point (x, y) in the plane whose x- and y-coordinates are both integers is called an **integral point**. If we let $n(r)$ be the number of integral points contained in a disk of radius r with the center at the origin in the plane, explain that $\displaystyle\lim_{r \to \infty} \frac{n(r)}{r^2} = \pi$.

[9]In fact, it is enough for f to be bounded and integrable.

3 Fubini's Theorem

In the previous section, we investigated several concepts and properties of multiple integral, but we didn't provide any practical and efficient way to calculate integrals. Now, we will introduce Fubini's Theorem,[10] which gives a practical way to compute multiple integrals by expressing them as **iterated integrals**. You may find its proof in the last section.

Theorem 3.0.1 (Fubini's Theorem) *Let $f(x,y)$ be an integrable function in a rectangle $R=[a,b]\times[c,d]$. If the function*

$$f_x:[c,d]\to\mathbb{R},\qquad y\mapsto f(x,y)$$

is integrable on $[c,d]$ for any $x\in[a,b]$, then the function

$$x\mapsto\int_c^d f(x,y)\,dy$$

is integrable on $[a,b]$, and

$$\iint_R f=\int_a^b\left(\int_c^d f(x,y)\,dy\right)dx.$$

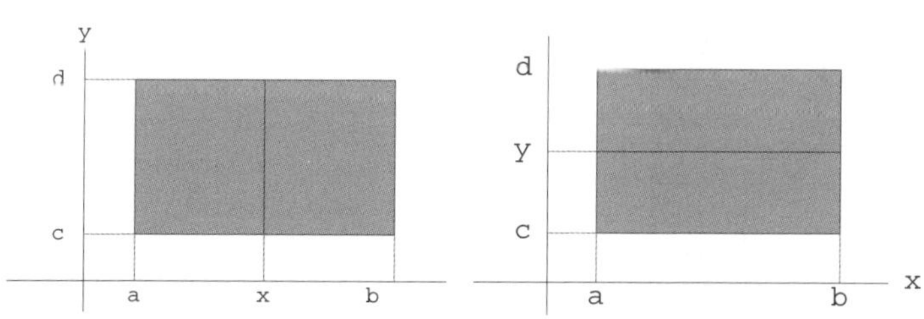

Similarly, if the function $x\mapsto f(x,y)$ is integrable on $[a,b]$ for any $y\in[c,d]$, then the function $y\mapsto\displaystyle\int_a^b f(x,y)\,dx$ is integrable on $[c,d]$ and

$$\iint_R f=\int_c^d\left(\int_a^b f(x,y)\,dx\right)dy.$$

In particular, if f is continuous on the rectangle $R=[a,b]\times[c,d]$, then

[10] G. Fubini (1879–1943). Italian mathematician

$$\iint_R f = \int_a^b \left(\int_c^d f(x,y)\,dy \right) dx = \int_c^d \left(\int_a^b f(x,y)\,dx \right) dy$$

and so we may change the order of iterated integrals.

3.0.2 Example

If we calculate the integral of the function $f(x,y) = x^2 y$ on the rectangle $R = [1,2] \times [3,4] \subset \mathbb{R}^2$, then

$$\begin{aligned} \iint_R f &= \int_1^2 \left(\int_3^4 x^2 y\,dy \right) dx \\ &= \int_1^2 \frac{x^2}{2} \left(4^2 - 3^2\right) dx \\ &= \frac{1}{6}\left(2^3 - 1^3\right) \cdot \left(4^2 - 3^2\right) \\ &= \frac{49}{6}. \end{aligned}$$

This value is equal to $\displaystyle\int_3^4 \int_1^2 x^2 y\,dx\,dy.$

3.0.3 Example

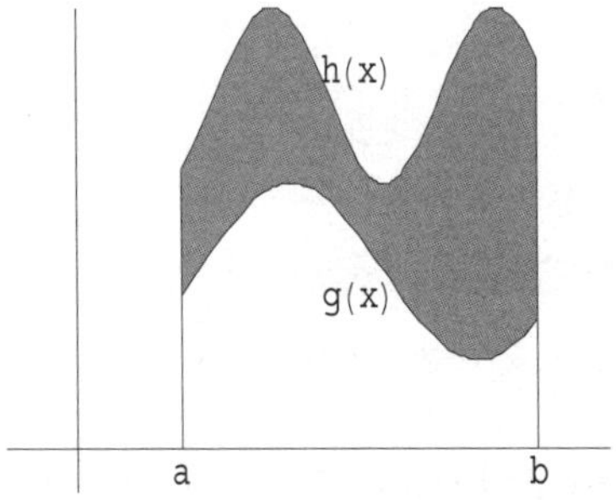

If we integrate a function $f(x,y)$ over the region

$$A \;:\; a \le x \le b, \qquad g(x) \le y \le h(x)$$

given by two continuous functions

$$g, h : [a,b] \to \mathbb{R}, \qquad g(x) \le h(x),$$

then we have

$$\iint_A f(x,y)\,dV_2 = \int_a^b \int_{g(x)}^{h(x)} f(x,y)\,dy\,dx.$$

Similarly, if we integrate a function $f(x,y)$ over the region

$$\tilde{A} \,:\, \tilde{g}(y) \leq x \leq \tilde{h}(y), \qquad c \leq y \leq d$$

given by two continuous functions

$$\tilde{g}, \tilde{h} : [c,d] \to \mathbb{R}, \qquad \tilde{g}(y) \leq \tilde{h}(y)$$

then we have

$$\iint_{\tilde{A}} f(x,y)\, dV_2 = \int_c^d \int_{\tilde{g}(y)}^{\tilde{h}(y)} f(x,y)\, dx\, dy.$$

3.0.4 Example

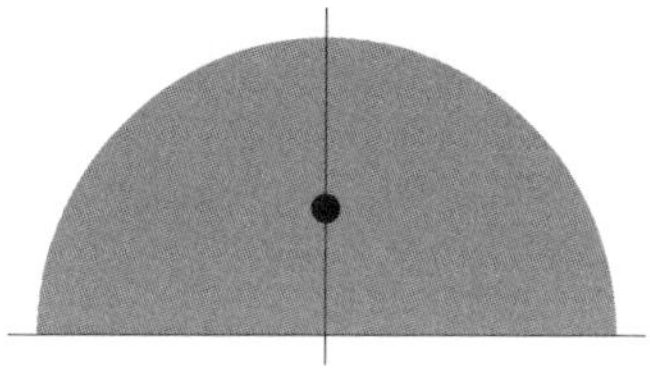

Find the (geometric) center[11] $(\bar{x}, \bar{y})$ of the half disk

$$A = \{(x,y) \mid x^2 + y^2 \leq 1,\ y \geq 0\}.$$

From the symmetry of the half disk, we know that $\bar{x} = 0$.[12] On the other hand,

$$\begin{aligned}
\bar{y} &= \frac{1}{\pi/2} \iint_A y\, dV_2 = \frac{2}{\pi} \int_{-1}^{1} \int_0^{\sqrt{1-x^2}} y\, dy\, dx \\
&= \frac{2}{\pi} \int_{-1}^{1} \left[\frac{1}{2} y^2\right]_0^{\sqrt{1-x^2}} dx = \frac{2}{\pi} \int_0^1 (1 - x^2)\, dx \\
&= \frac{2}{\pi} \left[x - \frac{1}{3} x^3\right]_0^1 \\
&= \frac{2}{\pi} \cdot \frac{2}{3} \approx 0.424413
\end{aligned}$$

and thus

$$(\bar{x}, \bar{y}) = \left(0, \frac{4}{3\pi}\right).$$

[11]The center $(\bar{x}, \bar{y})$ of a region A is given by

$$\bar{x} := \frac{1}{\mathrm{area}(A)} \iint_A x\, dV_2, \qquad \bar{y} := \frac{1}{\mathrm{area}(A)} \iint_A y\, dV_2.$$

[12]On the region which remains unchanged when we replace x by $-x$, $\bar{x}$ is equal to $-\bar{x}$, and thus $\bar{x} = 0$. This can be shown by change of variables.

3.0.5 Example

For a region A in the plane given by the equation

$$x^2 \le y \le x,$$

answer the following question.

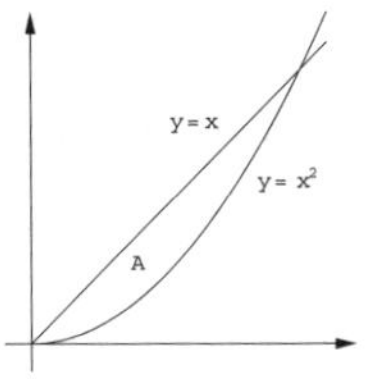

(i) Find the area of A.
(ii) Find the center of A.
(iii) When $f(x, y) = x^2 + y^2$ is the "density function" of A, find the "mass" of A.
(iv) Find the average density of A.

The area of A is

$$V = \iint_A dV_2 = \int_0^1 \int_{x^2}^x dy\, dx = \int_0^1 (x - x^2)\, dx = \frac{1}{2} - \frac{1}{3} = \frac{1}{6}.$$

If we let $(\bar{x}, \bar{y})$ be the center of A, then

$$\begin{aligned} \bar{x} &= \frac{1}{V} \iint_A x\, dV_2 = 6 \int_0^1 \int_{x^2}^x x\, dy\, dx \\ &= 6 \int_0^1 (x^2 - x^3)\, dx = 6 \left(\frac{1}{3} - \frac{1}{4} \right) = \frac{1}{2} \end{aligned}$$

$$\begin{aligned} \bar{y} &= \frac{1}{V} \iint_A y\, dV_2 = 6 \int_0^1 \int_{x^2}^x y\, dy\, dx \\ &= \frac{6}{2} \int_0^1 (x^2 - x^4)\, dx = 3 \left(\frac{1}{3} - \frac{1}{5} \right) = \frac{2}{5}. \end{aligned}$$

Thus, the center of A is $\left(\frac{1}{2}, \frac{2}{5} \right)$. On the other hand,

$$\begin{aligned} \iint_A f &= \int_0^1 \int_{x^2}^x (x^2 + y^2)\, dy\, dx \\ &= \int_0^1 \left(x^2(x - x^2) + \frac{1}{3} (x^3 - x^6) \right) dx \\ &= \int_0^1 \left(\frac{4}{3} x^3 - x^4 - \frac{1}{3} x^6 \right) dx \\ &= \frac{1}{3} - \frac{1}{5} - \frac{1}{3 \cdot 7} = \frac{3}{35}, \end{aligned}$$

and so the mass of A is $\frac{3}{35}$ and the average density is $\frac{3}{35} \Big/ \frac{1}{6} = \frac{18}{35}$.

3.0.6 Basic Exercise

Let A be the triangle enclosed by straight lines

$$y = 0, \qquad y = x, \qquad x + y = 4$$

in the plane. Find the average of the function $f(x, y) = 2xy$ on A.

3.0.7 Example

Find the iterated integral

$$\int_0^1 \int_{\sqrt{x}}^1 \sqrt{1 + y^3}\, dy\, dx.$$

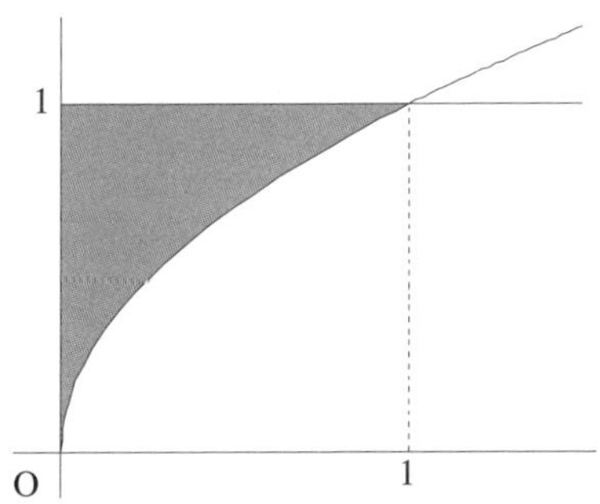

Since it is not easy to find the antiderivative of the function $\sqrt{1 + y^3}$, let's change the order of integrals. The given region

$$\sqrt{x} \le y \le 1, \qquad 0 \le x \le 1$$

can be expressed as

$$0 \le x \le y^2, \qquad 0 \le y \le 1.$$

So the integral becomes

$$\begin{aligned} \int_0^1 \int_0^{y^2} \sqrt{1 + y^3}\, dx\, dy &= \int_0^1 y^2 \sqrt{1 + y^3}\, dy = \left[\frac{2}{3} \cdot \frac{1}{3} \cdot \left(1 + y^3\right)^{3/2} \right]_0^1 \\ &= \frac{2}{9}\left(2\sqrt{2} - 1\right). \end{aligned}$$

3.0.8 Basic Exercise

When R is the triangle enclosed by the x-axis, $y = x$ and $x = 1$ in the plane, find[13]

$$\iint_R \frac{\sin x}{x}\, dx\, dy.$$

[13] Answer: $1 - \cos 1$

3.0.9 Fubini's Theorem in Higher Dimension

Fubini's theorem also holds in the general dimension. For example, for a continuous function $f(x,y,z)$ defined on a three-dimensional region $[a_1,b_1]\times[a_2,b_2]\times[a_3,b_3]$,

$$\begin{aligned}&\iiint_{[a_1,b_1]\times[a_2,b_2]\times[a_3,b_3]} f(x,y,z)\,dV(x,y,z)\\ =&\int_{a_3}^{b_3}\int_{a_2}^{b_2}\int_{a_1}^{b_1} f(x,y,z)\,dx\,dy\,dz\\ =&\quad\cdots\\ =&\int_{a_1}^{b_1}\int_{a_2}^{b_2}\int_{a_3}^{b_3} f(x,y,z)\,dz\,dy\,dx.\end{aligned}$$

3.0.10 Basic Exercise

For a rectangular cube

$$R\,:\,0\le x\le\pi,\ 2\le y\le 3,\ -1\le z\le 1$$

and a function $f(x,y,z)=\sin x$, show that

$$\iiint_R f = 4.$$

3.0.11 Example

We can express a tetrahedron with four vertices $(0,0,0)$, $(1,0,0)$, $(0,1,0)$, $(0,0,1)$ as

$$0\le x\le 1,\quad 0\le y\le 1-x,\quad 0\le z\le 1-x-y.$$

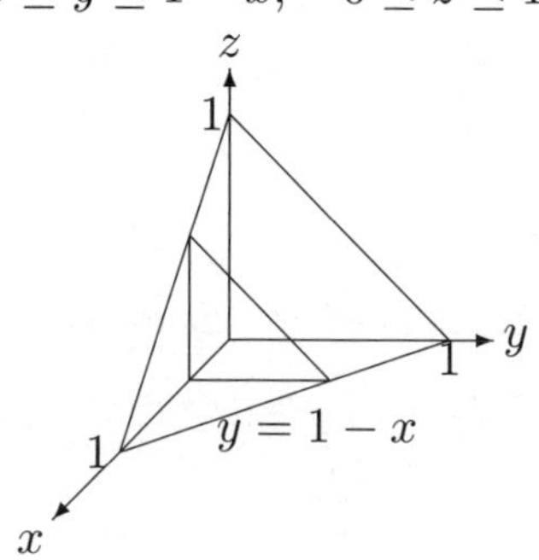

When the density function of T is given by $f(x,y,z) = x^2+y^2$, the mass of T is[14]

$$\begin{aligned}
\iiint_T f &= \int_0^1 \int_0^{1-x} \int_0^{1-x-y} (x^2+y^2)\, dz\, dy\, dx \\
&= \int_0^1 \int_0^{1-x} (x^2+y^2)(1-x-y)\, dy\, dx \\
&= \int_0^1 \left[x^2(1-x)y - \frac{x^2}{2}y^2 + \frac{1}{3}(1-x)y^3 - \frac{1}{4}y^4 \right]_{y=0}^{y=1-x} dx \\
&= \int_0^1 \left(x^2(1-x)^2 - \frac{x^2}{2}(1-x)^2 + \frac{1}{3}(1-x)^4 - \frac{1}{4}(1-x)^4 \right) dx \\
&= \frac{1}{2}\int_0^1 x^2(1-x)^2\, dx + \frac{1}{12}\int_0^1 (1-x)^4\, dx \\
&= \frac{1}{2}\cdot\frac{2\cdot 2}{5!} + \frac{1}{12}\cdot\frac{1}{5} = \frac{1}{30.}
\end{aligned}$$

Since the volume of T is $\dfrac{1}{6}$ and thus the average density of T is

$$\frac{1}{30}\Big/\frac{1}{6} = \frac{1}{5}.$$

3.1 Cavalieri's Principle

A student of Galileo, Bonaventura Cavalieri (1598–1647) found out that two solids have the same volume if the areas of their corresponding cross-sections made by arbitrary planes parallel to the xy-plane are always equal. Such a principle is immediately followed from Fubini's theorem.[15]

14

$$\int_0^1 x^n(1-x)^m\, dx = \frac{n!m!}{(n+m+1)!} \qquad \text{for nonnegative integers } n, m.$$

This can be obtained by comparing coefficients on both sides of a polynomial for t

$$\begin{aligned}
\sum_{n+m=l} \binom{l}{n} t^n \int_0^1 x^n(1-x)^m\, dx &= \int_0^1 \sum_{n+m=l} \binom{l}{n} (tx)^n(1-x)^m\, dx \\
&= \int_0^1 (tx+1-x)^l\, dx = \frac{t^{l+1}-1}{(l+1)(t-1)} \\
&= \frac{1}{l+1}\left(t^l + t^{l-1} + \cdots + t + 1\right).
\end{aligned}$$

15 A Chinese mathematician Zu Geng found this principle in the 3rd century [Hsiang, 1994].

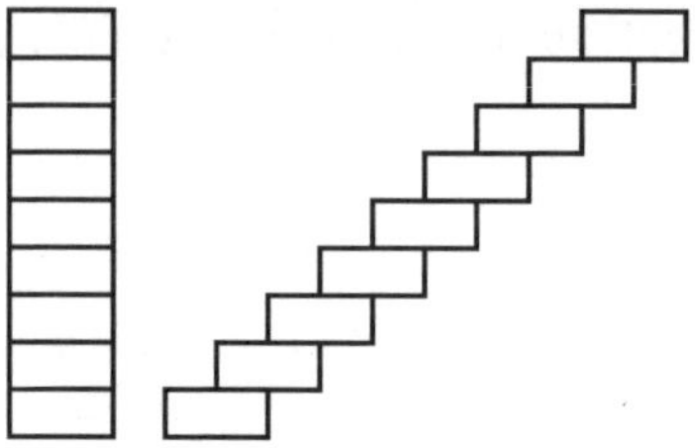

3.1.1 Basic Exercise

Find the volumes of an oblique cylinder and an oblique cone.

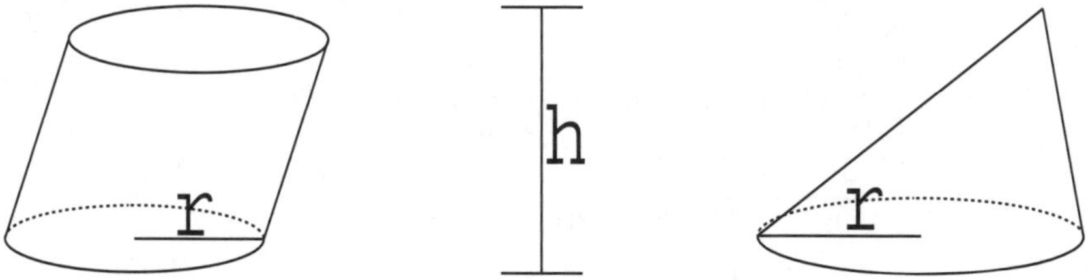

3.1.2 Basic Exercise

1. Consider the cylinder circumscribing a given ball and the circular cone inscribed the cylinder. Show that the volume of the cone and the volume of the ball add up to the volume of the cylinder.[16] Use this to show that the volume of the ball is $\frac{2}{3}$ times the volume of the cylinder.

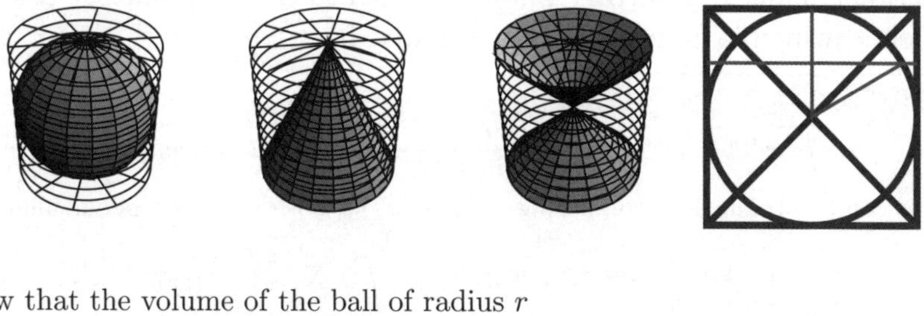

2. Show that the volume of the ball of radius r

$$B(r) := \{(x, y, z) \mid x^2 + y^2 + z^2 \leq r^2\}$$

is $\frac{4}{3}\pi r^3 \approx 4.18879\, r^3$.

[16]It suffices to show that in the last picture, the cross-sectional area of the ball and that of the cone add up to the cross-sectional area of the cylinder.

Exercise: Chapter 14 Section 3

1. Calculate the integral of the following function on the given region.

   (1) $f(x, y) = 2x^2y + xy^2 - 6xy,$ $\quad -1 \le x \le 3,\ -1 \le y \le 3$

   (2) $f(x, y) = 3\sqrt{x+y+4} - 2\sqrt{x} - \sqrt{y},$ $\quad 0 \le x \le 6,\ 0 \le y \le 6$

   (3) $f(x, y) = \sin x + \sin y + \sin(x+y),$ $\quad 0 \le x \le 2\pi,\ 0 \le y \le 2\pi$

   (4) $f(x, y) = xe^{-(x^2+y^2)/8},$ $\quad -6 \le x \le 6,\ -6 \le y \le 6$

2. When we let A be the region enclosed by the tetrahedron with four vertices $(0, 0, 0)$, $(1, 0, 0)$, $(0, 1, 0)$, $(0, 0, 1)$ in the three-dimensional space, find the average of the function $f(x, y, z) = e^x$ on A.

3. Suppose that a region is of the triangular shape given by three vertices $(0, 0)$, $(1, 0)$, $(1, 1)$ and the price of the land per unit area is given by $f(x, y) = 400\, xe^{-y}$ at each point. Find the average price of the land in this region.

4. When we put and fix a ball on the xy-plane and then cut it by the plane parallel to the xy-plane whose distance from the xy-plane is $1/4$ of the diameter of the ball, find the ratio of the volumes of two parts.

5. For a region R in the xy-plane and a point P which is not on the xy-plane, the **cone** with base R and vertex P is defined by

   $$C := \text{Cone}(R, P) := \{(1-t)X + tP \mid X \in R,\ 0 \le t \le 1\}.$$

   If we let h be the height of the cone and A be the area of the base, show that the volume of the cone is $\frac{1}{3}Ah$. Also show that the center of the cone is the point which divides the vertex P and the center of the region R internally with ratio 3:1.

6. Show that in a triangle, the point which divides a vertex and the corresponding side internally with the ratio 2:1 is the center of the triangle.

7. Show that the center of a tetrahedron is equal to the average of four vertices of the tetrahedron.

8. Change the order of the following integration. (Here, $0 < a < b$.)

   (1) $\displaystyle\int_a^b \int_a^x f(x, y)\, dy\, dx$ $\qquad$ (2) $\displaystyle\int_0^a \int_0^{\sqrt{x}} f(x, y)\, dy\, dx$

   (3) $\displaystyle\int_0^a \int_y^{2a-y} f(x, y)\, dx\, dy$ $\qquad$ (4) $\displaystyle\int_0^a \int_{\sqrt{a^2-x^2}}^{x+2a} f(x, y)\, dy\, dx$

9. Evaluate the following iterated integral.

(1) $\int_1^2 \int_1^3 (x+y)\,dx\,dy$

(2) $\int_0^2 \int_1^{x^2} y\,dy\,dx$

(3) $\int_0^1 \int_0^y \sqrt{x}\,dx\,dy$

(4) $\int_0^\pi \int_0^\pi x \sin y\,dy\,dx$

(5) $\int_1^2 \int_y^{y^2} dx\,dy$

(6) $\int_0^\pi \int_0^{\sin x} y\,dy\,dx$

(7) $\int_1^5 \int_0^y \frac{3\,dx\,dy}{x^2+y^2}$

(8) $\int_{\pi/4}^{\pi/2} \int_{\pi/2}^r \frac{1}{\sin^2\theta}\,d\theta\,dr$

(9) $\int_0^1 \int_y^1 \sqrt{1-x^2}\,dx\,dy$

(10) $\int_0^1 \int_{\sqrt[3]{x}}^1 \sqrt{1+y^4}\,dy\,dx$

(11) $\int_0^1 \int_{\sqrt[4]{y}}^1 \sqrt{1+x^5}\,dx\,dy$

(12) $\int_0^1 \int_{\sqrt{y}}^1 y\sqrt{1+x^5}\,dx\,dy$

(13) $\int_0^1 \int_{x^2}^{\sqrt{x}} 160\,xy^3\,dy\,dx$

(14) $\int_0^1 \int_0^y y^2 e^{xy}\,dx\,dy$

(15) $\int_0^1 \int_{1-y}^1 e^y\,dx\,dy$

(16) $\int_0^1 \int_0^{\pi/2} (\sin 2x)\, e^{y\sin x}\,dx\,dy$

(17) $\int_0^2 \int_0^{x^2/2} \frac{x}{\sqrt{x^2+y^2+1}}\,dy\,dx$

10. Evaluate the following integral.

(1) $\int_0^3 \int_0^{2\pi} \int_1^{xz} xyz\,dy\,dx\,dz$

(2) $\int_0^{\pi/2} \int_{\sin 2z}^0 \int_0^{2yz} \sin\frac{x}{y}\,dx\,dy\,dz$

(3) $\int_0^1 \int_0^y \int_0^{\sqrt{9y^2-3z^2}} 10xy\,dx\,dz\,dy$

(4) $\int_0^{\pi/2} \int_0^z \int_0^y \sin(x+y+z)\,dx\,dy\,dz$

11. Find the average of the following function over the given region.

(1) $x\cos(x+y)$ over the triangle with vertices $(0,0)$, $(\pi,0)$, (π,π)

(2) e^{x+y} over the region given by $|x|+|y|\le 1$

(3) x^2-y^2 over the region enclosed by $y=\sin x$, $0\le x\le\pi$ and the x-axis

(4) x^2+y over the triangle with vertices $\left(-\frac{1}{2},\frac{1}{2}\right)$, $(1,2)$, $(1,-1)$

(5) x^2 over the region determined by $x\ge 0$, $x^2+y^2\le 2$, $x^2+y^2\ge 1$

(6) $\frac{1}{x+y}$ over the region determined by four lines $y=x$, $x=1$, $x=2$, $y=0$

(7) $x\sin(xy)$ over the region determined by $0\le x\le\pi$ and $0\le y\le 1$

(8) $\dfrac{1}{x+y+1}$ over the square $0 \le x \le 1$, $0 \le y \le 1$

12. Sketch the following solid and find its center.

    (1) The tetrahedron enclosed by the plane $20x + 12y + 15z = 60$ and three coordinate planes

    (2) The solid enclosed by two surfaces $b^2x^2 + a^2y^2 - a^2b^2z = 0$, $b^2x^2 + a^2y^2 = k^2$ and the plane $z = 0$

13. Find the center of the following region.

    (1) $0 \le x \le \dfrac{2}{3}, \quad x^2 \le y \le 1 - x^4$

    (2) $\dfrac{\pi}{3} \le x \le \dfrac{\pi}{2}, \quad \cos x \le y \le x^3$

    (3) $1 \le x \le 2, \quad \log x \le y \le e^x$

    (4) $\dfrac{\sqrt[4]{y}}{2} \le x \le \dfrac{1}{y^2+1}, \quad 0 \le y \le \dfrac{1}{\sqrt{3}}$

14. (**Delesse's rule**) Achille Ernest Delesse was a mid-nineteenth century mineralogist interested in determining the composition of rocks. To find out how much of a particular mineral is contained in a rock, he cut it through, polished an exposed face, and covered the face with transparent waxed paper, trimmed to size. He then traced on the paper the exposed portions of the mineral that interested him. After weighing the paper, he cut out the mineral traces and weighed them. The ratio of the weights gave not only the proportion of the surface occupied by the mineral but also the proportion of the entire rock occupied by the mineral. Explain this law.[17]

15. In a container containing liquid with density[18] δ kg/m^3, the pressure at the point of depth h m is δh kg/m^2. Explain that the total pressure acting on the dam of length L m and with liquid of depth W m is

$$F = \int_0^W L\delta h \, dh = L\delta W^2/2.$$

[17]Refer Exercise 3 in Section 2. Delesse's Rule is still used by petroleum geologists today, It is also used for examining a hole of ceramic filter to measure organic molecule in chemical laboratory, or for investigating the time passed since an apple is picked. (Once an apple is picked, sugar contained in the apple start to change into dextrin.) If it is hard to choose a cross section properly, we can do the sampling test-examine several cross sections and then take the average-.

[18]mass per unit volume

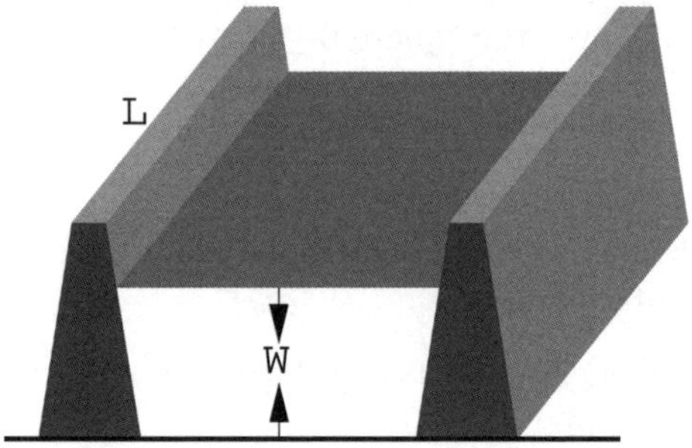

16. Suppose that when an iron plate of area A cm^2 is immersed in a swimming pool perpendicularly to the surface, the center of the plate is h cm apart from the surface. Show that the pressure of water acting on one side of the plate is

$$F = hA\delta.$$

(Here δ denotes the density of water.)

17. Suppose that two kettles with the same base area hold water of the same height. Find the difference between pressures of water acting on bases of two kettles.

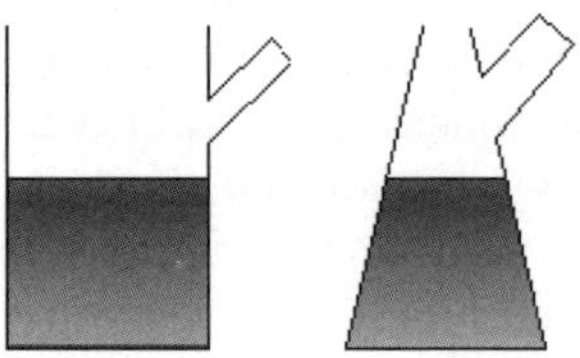

18. Let $A(c)$ be the area of the region enclosed by a curve $y = x^4 - 5x^2 + 4$ and a line $y = c$ in the plane. Find $A'(c)$ when $c = 3$.[19]

[19]From Special Entrance Examination in 1998

4 Change of Variables

When we integrate a function $f(y)$ of a single variable over the interval $[a, b]$, if we replace the variable y by a bijective function

$$y = g(x), \qquad \alpha \leq x \leq \beta, \qquad g([\alpha, \beta]) = [a, b]$$

we can get the following change of variables formula:[20]

$$\boxed{\int_{[a,b]} f(y)\, dy = \int_{[\alpha,\beta]} f(g(x)) \left|g'(x)\right| dx.}$$

This formula can be extended to functions of several variables.

4.1 $\mathcal{C}^1$ Invertible Functions

In the plane, the differentiable function

$$G \quad : \quad u = e^x \cos y, \quad v = e^x \sin y$$

maps a rectangle

$$1 < x < 2, \qquad -\frac{\pi}{2} < y < \frac{\pi}{2}$$

into a sector

$$e < \left(u^2 + v^2\right)^{1/2} < e^2, \qquad u > 0.$$

[20]In this equation, $a < b$ and

$$\int_{[a,b]} f(y)\, dy := \int_a^b f(y)\, dy$$

If the substituting function $y = g(x)$ is increasing then

$$g'(x) \geq 0, \qquad g(\alpha) = a, \quad g(\beta) = b,$$

and thus

$$\int_a^b f(y)\, dy = \int_\alpha^\beta f(g(x)) g'(x)\, dx = \int_{[\alpha,\beta]} f(g(x)) \left|g'(x)\right| dx.$$

If the substituting function $y = g(x)$ is decreasing then

$$g'(x) \leq 0, \qquad g(\alpha) = b, \quad g(\beta) = a,$$

and thus

$$\int_a^b f(y)\, dy = \int_\beta^\alpha f(g(x)) g'(x)\, dx = -\int_\alpha^\beta f(g(x)) g'(x)\, dx = \int_{[\alpha,\beta]} f(g(x)) \left|g'(x)\right| dx.$$

Its inverse function is given by

$$G^{-1} \quad : \quad x = \frac{1}{2}\log(u^2+v^2), \qquad y = \arctan\frac{v}{u}$$

and it is also differentiable.

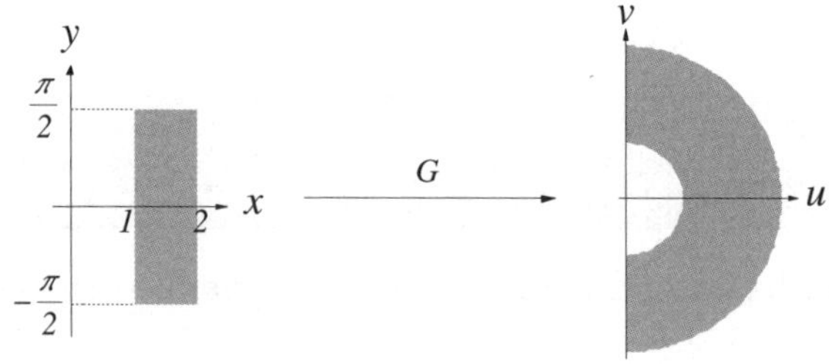

In general, if a $\mathcal{C}^1$ function G between two open sets U, W in $\mathbb{R}^n$ has the inverse function which is also $\mathcal{C}^1$, G is called a $\mathcal{C}^1$ **invertible function.**[21]

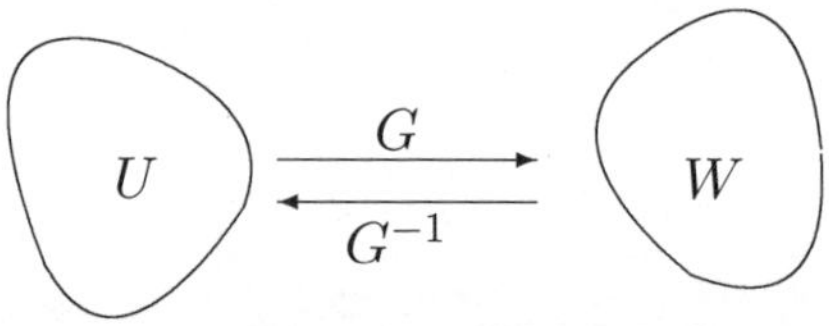

Theorem 4.1.1 (Change of Variables) *Let U and W be open sets in $\mathbb{R}^n$ and let*

$$G : U \to W, \qquad Y = G(X)$$

be a $\mathcal{C}^1$ invertible function. Then for an integrable function f on W,

$$\int_W f(Y)\,dV(Y) = \int_U f(G(X))\,|\det G'(X)|\,dV(X).$$

Let's sketch the idea of the proof for change of variables. To calculate the integral, we may divide the domain into small pieces and add up the values over the small regions. In a small region the map G is approximately equal to a linear map $G'(X)$ and the integrand f is approximately equal

[21] In mathematics, we consider appropriate functions depending on situation. When we consider the inverse function in calculus, it would be natural to consider not just the set-theoretic inverse function but the inverse function which is still differentiable.

to a constant function.[22] Therefore to prove Theorem 4.1.1 it suffices to prove it when $f = \chi_W = 1$ and G is a linear map L. In that case, we have $\int_U |\det L|\, dV = \int_W \chi_W \, dV$, that is,

$$|\det L| \operatorname{Vol}(U) = \operatorname{Vol}(L(U))$$

which we already dealt with.

4.2 Volume of Ellipsoid

The **ellipse**

$$\frac{x^2}{a^2} + \frac{y^2}{b^2} \leq 1$$

is the image of the unit disk

$$x^2 + y^2 \leq 1$$

by the linear transformation

$$\begin{pmatrix} x \\ y \end{pmatrix} \mapsto \begin{pmatrix} a & 0 \\ 0 & b \end{pmatrix} \begin{pmatrix} x \\ y \end{pmatrix}$$

which stretches the unit disk by the factor a in the direction of x-axis and by the factor b in the direction of y-axis. The determinant of this transformation is ab, and thus the area of the ellipse is ab times the area of the unit disk π.[23]

Similarly, the volume of the **ellipsoid**

$$\frac{x^2}{a^2} + \frac{y^2}{b^2} + \frac{z^2}{c^2} \leq 1$$

is[24]

$$\frac{4\pi}{3} abc.$$

[22] By Lebesgue's theorem, an integrable function f is continuous **almost everywhere**.

[23] Here, a and b are both positive.

[24] We assume that a, b, c are all positive. The volume of the ball with radius 1 is $\frac{4}{3}\pi$.

4.3 Integration in Polar Coordinates

Let G be the transformation from polar coordinates (r, θ) into rectangular coordinates (x, y),

$$G: \qquad x = r\cos\theta, \quad y = r\sin\theta.$$

We already saw in Chapter 12 that its Jacobian determinant is

$$\det G'(r, \theta) = r.$$

Therefore if we use polar coordinates to integrate a function $f(x, y)$ defined on a domain W in the plane, we have

$$\iint_W f(x, y)\, dx\, dy = \iint_{G^{-1}(W)} f(r\cos\theta, r\sin\theta)\, r\, dr\, d\theta$$

We will denote it shortly by

$$\boxed{dx\, dy = r\, dr\, d\theta}$$

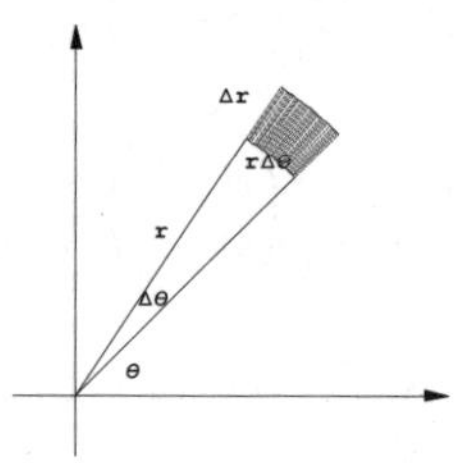

4.3.1 Gaussian Integral

We will show that[25]

$$\boxed{\int_{-\infty}^{\infty} e^{-x^2}\, dx = \sqrt{\pi}}$$

Let's denote the disk $x^2+y^2 \le a^2$ by $D(a)$ and the square $[-a, a]\times[-a, a]$ by $R(a)$. Then

$$D(a) \subset R(a) \subset D(\sqrt{2}a).$$

[25]This equation was found by Abraham de Moivre (1667–1754). The equation

$$n! \sim \sqrt{2\pi n}e^{-n}n^n,$$

is usually called **Stirling's formula**, but it was also found by de Moivre.

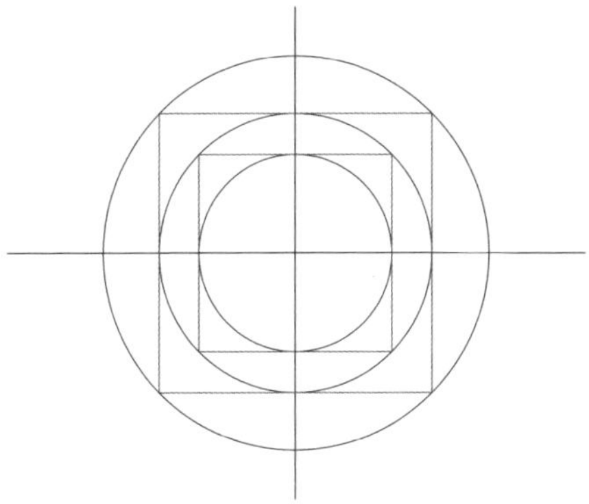

$$\iint_{D(a)} e^{-x^2-y^2}\,dV = \int_0^{2\pi}\int_0^a e^{-r^2} r\,drd\theta = \pi(1-e^{-a^2})$$

and

$$\iint_{D(a)} e^{-x^2-y^2}\,dV \le \iint_{R(a)} e^{-x^2-y^2}\,dV \le \iint_{D(\sqrt{2}a)} e^{-x^2-y^2}\,dV. \qquad (14.2)$$

Since the middle term in (14.2) becomes

$$\iint_{R(a)} e^{-x^2-y^2}\,dV = \left(\int_{-a}^{a} e^{-x^2}\,dx\right)\cdot\left(\int_{-a}^{a} e^{-y^2}\,dy\right) = \left(\int_{-a}^{a} e^{-x^2}\,dx\right)^2,$$

the inequalities (14.2) imply

$$\pi(1-e^{-a^2}) \le \left(\int_{-a}^{a} e^{-x^2}\,dx\right)^2 \le \pi(1-e^{-2a^2}).$$

If we let $a \to \infty$ here, we get

$$\int_{-\infty}^{\infty} e^{-x^2}\,dx = \sqrt{\pi}.$$

4.3.2 Example

In $\mathbb{R}^3$, when the force by which a point $(x, y, 0)$ in the xy-plane attracts a point $P = (0, 0, z_0)$ in the z-axis[26] is given inversely proportional to the square of the distance by

$$\mathbf{F}(x,y) = \frac{1}{x^2+y^2+z_0^2}\cdot\frac{(x,y,0)-(0,0,z_0)}{\sqrt{x^2+y^2+z_0^2}},$$

[26] Assume that $z_0 > 0$.

the **k**-component of the force by which the whole xy-plane pulls the point P is

$$\begin{aligned}
\iint_{\mathbb{R}^2} \mathbf{F}(x,y)\cdot\mathbf{k}\,dx\,dy &= \int_{\mathbb{R}^2} \frac{-z_0}{\left(x^2+y^2+z_0^2\right)^{3/2}}\,dx\,dy \\
&= -z_0 \int_0^{2\pi}\int_0^{\infty} \frac{1}{\left(r^2+z_0^2\right)^{3/2}}\,r\,dr\,d\theta \\
&= -2\pi z_0 \int_0^{\infty} \frac{r\,dr}{\left(r^2+z_0^2\right)^{3/2}} \\
&= -2\pi z_0 \int_{z_0}^{\infty} \frac{t\,dt}{t^3} \qquad \left(\Leftarrow\ (r^2+z_0^2)^{1/2} =: t\right) \\
&= -2\pi z_0 \left[-\frac{1}{t}\right]_{z_0}^{\infty} \\
&= -2\pi.
\end{aligned}$$

4.3.3 Basic Exercise

1. Explain that an object falling to an infinite homogeneous plane has a constant acceleration.

2. Explain that if an object lies between two parallel planes, no force acts on this object from two planes.

4.3.4 Basic Exercise

1. Let A be a region give by

   $$0 \neq x^2+y^2 \leq 1, \qquad x \geq 0, \qquad y \geq 0,$$

   and

   $$(u,v) = G(x,y) = (x^2-y^2, 2xy).$$

   Evaluate the integral of the function

   $$f(u,v) = 1/(u^2+v^2)^{1/2}$$

   over the region $G(A)$.

2. For the function G given above, find the area of $G([0,2]\times[0,2])$.

3. For a function
$$f(z) = z^2$$
defined on the unit disk
$$D := \{z \in \mathbb{C} : |z| \leq 1\},$$
find the Jacobian determinant of f, and explain the reason why the change of variables formula doesn't hold. Also find the Jacobian determinant of the function $g(z) = z^n$.

4.4 Integration in Cylindrical Coordinates

If we let G be the transformation from cylindrical coordinates (r, θ, z) to rectangular coordinates (x, y, z),

$$G \ : \ x = r\cos\theta, \qquad y = r\sin\theta, \qquad z = z$$

its Jacobian determinant is

$$\det\begin{pmatrix} \cos\theta & -r\sin\theta & 0 \\ \sin\theta & r\cos\theta & 0 \\ 0 & 0 & 1 \end{pmatrix} = r.$$

Therefore the integral of a function $f(x, y, z)$ defined on $W \subset \mathbb{R}^3$ is

$$\iiint_W f(x,y,z)\,dx\,dy\,dz = \iiint_{G^{-1}(W)} f(r\cos\theta, r\sin\theta, z)\,r\,dr\,d\theta\,dz.$$

We denote this equation simply by

$$\boxed{dx\,dy\,dz = r\,dr\,d\theta\,dz}$$

4.4.1 Example

The mass of the cylindrical region

$$x^2 + y^2 \leq 1, \quad 2 \leq z \leq 3$$

with the density function $f(x, y, z) = 4 - z$ is

$$\begin{aligned} \iiint f(x,y,z)\,dx\,dy\,dz &= \int_2^3 \int_0^{2\pi} \int_0^1 (4-z)\,r\,dr\,d\theta\,dz \\ &= \left(\int_2^3 (4-z)\,dz\right)\left(\int_0^{2\pi} d\theta\right)\left(\int_0^1 r\,dr\right) \\ &= \left(4 - \frac{1}{2}(3^2 - 2^2)\right) \cdot 2\pi \cdot \frac{1}{2} = \frac{3}{2}\pi. \end{aligned}$$

4.5 Integration in Spherical Coordinates

As we studied in Chapter 12, for the transformation G from spherical coordinates (ρ, φ, θ) into rectangular coordinates (x, y, z)

$$G : \quad x = \rho \sin\varphi \cos\theta, \quad y = \rho \sin\varphi \sin\theta, \quad z = \rho \cos\varphi,$$

its Jacobian determinant is

$$\det G'(\rho, \varphi, \theta) = \rho^2 \sin\varphi.$$

Therefore the integral of a function $f(x, y, z)$ over a region W is

$$\begin{aligned} &\iiint_W f(x, y, z)\, dx\, dy\, dz \\ = &\iiint_{G^{-1}(W)} f(\rho \sin\varphi \cos\theta, \rho \sin\varphi \sin\theta, \rho \cos\varphi)\, \rho^2 \sin\varphi\, d\rho\, d\varphi\, d\theta. \end{aligned}$$

This equation is denoted by[27]

$$\boxed{dx\, dy\, dz = \rho^2 \sin\varphi\, d\rho\, d\varphi\, d\theta}$$

For example, for a ball of radius r

$$B : x^2 + y^2 + z^2 \leq r^2$$

its volume is[28]

$$\iiint_B dx\, dy\, dz = \int_0^{2\pi} \int_0^{\pi} \int_0^{r} \rho^2 \sin\varphi\, d\rho\, d\varphi\, d\theta = 2\pi \cdot 2 \cdot \frac{r^3}{3} = \frac{4}{3}\pi r^3.$$

---

[27] In this equation, if we let

$$d\Theta := \sin\varphi\, d\varphi\, d\theta,$$

we will see in Chapter 16 that it is the **area element** of the unit sphere. And

$$dx\, dy\, dz = d\rho \cdot \rho^2\, d\Theta$$

is the polar coordinates formula in the three-dimensional space. In the general n-dimensional space, the **change of variable formula to polar coordinates** is

$$dV_n = d\rho \cdot \rho^{n-1}\, d\Theta_n.$$

[28] The volume of a ball is 2/3 of the volume of the circumscribed cylinder.

4.5.1 Example

Find the volume of the region given by inequalities

$$x^2 + y^2 \le z^2, \qquad x^2 + y^2 + z^2 \le z.$$

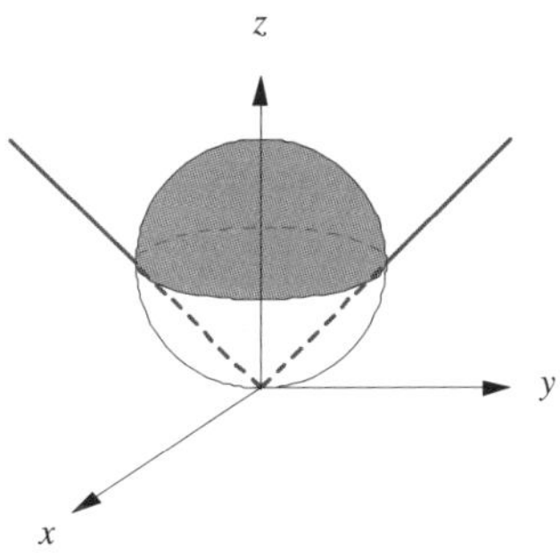

Solution. If we express the region in spherical coordinates, we get

$$\sin^2\varphi \le \cos^2\varphi, \qquad \rho^2 \le \rho\cos\varphi.$$

We can simplify it as

$$0 \le \varphi \le \frac{\pi}{4}, \qquad 0 \le \rho \le \cos\varphi.$$

Therefore the volume of the solid is[29]

$$\begin{aligned}
\int_0^{2\pi}\int_0^{\pi/4}\int_0^{\cos\varphi} \rho^2 \sin\varphi \, d\rho \, d\varphi \, d\theta &= \int_0^{2\pi} d\theta \cdot \int_0^{\pi/4} \sin\varphi \cdot \frac{1}{3} \cdot \cos^3\varphi \, d\varphi \\
&= (2\pi) \cdot \frac{1}{3} \cdot \left(-\frac{1}{4}\right) \left[\cos^4\varphi\right]_0^{\pi/4} \\
&= (2\pi) \cdot \frac{1}{3} \cdot \left(-\frac{1}{4}\right) \left(\frac{1}{4} - 1\right) \\
&= \frac{\pi}{8}.
\end{aligned}$$

4.5.2 Basic Exercise

Find the mass and average density of the region given above when the density function is $\rho = (x^2 + y^2 + z^2)^{1/2}$.

[29]The sum of the volume of half ball $\pi/12$ and volume of circular cone $\pi/24$

Exercise: Chapter 14 Section 4

1. For the **region enclosed by the ellipse** $A : \dfrac{x^2}{a^2} + \dfrac{y^2}{b^2} \leq 1$ and a 'linear map' $G := \begin{pmatrix} 3 & 0 \\ 1 & 5 \end{pmatrix}$, find the area of $G(A)$.

2. When a region A is given by $0 \leq x \leq 1,\ 0 \leq y \leq \pi$ in the plane and the map $G : \mathbb{R}^2 \to \mathbb{R}^2$ is given by $G(x, y) := (e^x \cos y, e^x \sin y)$, find the area of $G(A)$.

3. Find the area of the region enclosed by a quadratic curve $(a_1x + b_1y)^2 + (a_2x + b_2y)^2 = k^2$. (Here we assume that a_1, a_2, b_1, b_2, k are all constants and $a_1b_2 \neq a_2b_1$.)

4. Find the area of the following region given in polar coordinates.

   (1) The intersection of the region enclosed by $r = 4\cos\theta$ and the region given by $r \geq 2$

   (2) The region enclosed by one arc of $r = a \sin 2\theta$

   (3) The region enclosed by $r = a(1 + \cos\theta)$

5. Evaluate the following integral.

$$\int_{-\infty}^{\infty}\int_{-\infty}^{\infty} (x^2 + y^2 + 1)^{-3/2}\, dx\, dy.$$

6. Show that in $\mathbb{R}^3$ the volume of the intersection of three cylinders

$$x^2 + y^2 \leq 1, \qquad y^2 + z^2 \leq 1, \qquad z^2 + x^2 \leq 1$$

   is $16 - 8\sqrt{2} \approx 4.6863$.

7. Find the volume of the solid bounded by three surfaces

$$z = x^2 + y^2 + 9, \qquad z = -x^2 - y^2 - 9, \qquad x^2 + y^2 = 1.$$

8. Find the volume of the following solid.

   (1) $x^2 + y^2 + z^2 \leq 1, \qquad \left(x - \dfrac{1}{2}\right)^2 + y^2 \leq \dfrac{1}{4}$

   (2) $x^2 + y^2 \leq z \leq \sqrt{1 - x^2 - y^2}$

9. Evaluate the following integral.

   (1) $\displaystyle\iint_D \sqrt{\frac{1 - x^2 - y^2}{1 + x^2 + y^2}}\, dx\, dy, \quad D : x \geq 0,\ y \geq 0,\ x^2 + y^2 \leq 1$

   (2) $\displaystyle\iint_D (x \sin\alpha - y\cos\alpha)^2\, dx\, dy, \quad D : \frac{x^2}{a^2} + \frac{y^2}{b^2} \leq 1$

(3) $\iint_D (x^2+y^2+1)^{-3/2}\,dx\,dy, \quad D : x^2+y^2 \le a^2$

(4) $\iint_D 40\,x^2 y\,dV,$ D is the region enclosed by $y=\sqrt{x}$ and $y=x$

(5) $\iint_D y\,dV,$ D is the region enclosed by $y=\log x,\ y=1,\ x=0,\ y=0$

(6) $\iint_D e^{y^2}\,dV,$ D is the region enclosed by $y=x/2,\ y=1,\ x=0$

(7) $\iint_D (2x^2+xy-y^2)\,dx\,dy, \quad D : 1 \le 2x-y \le 3,\ -2 \le x+y \le 0$

(8) $\iiint_R \sqrt{x^2+y^2}\,dx\,dy\,dz, \quad R : 0 \le z \le 4-\sqrt{x^2+y^2}$

(9) $\iiint_R \left(x^2+\frac{y^2}{4}+\frac{z^2}{9}\right)^{-2} dx\,dy\,dz, \quad R : 1 \le x^2+\frac{y^2}{4}+\frac{z^2}{9} \le 9$

(10) $\iiint_R \left(x^2+y^2+z^2\right) dx\,dy\,dz,$

$R : 0 \le x \le \sqrt{y},\ 0 \le y \le 4,\ 0 \le z \le \frac{3}{2}x$

(11) $\iiint_R \left(\sqrt{x}+\sqrt{y}+\sqrt{z}\right) dx\,dy\,dz,$

$R : 0 \le x \le y^2,\ 0 \le y \le \sqrt{z},\ 0 \le z \le 1$

(12) $\iiint_R \frac{y-2x}{x}\,dx\,dy\,dz,$ R is the solid bounded by a cylinder $y^2+z^2=1$ and the planes $x=1$, and $x=4$ in the first octant

(13) $\iiint_R z\,dx\,dy\,dz,$ R is the solid bounded by $z=\sqrt{x^2+y^2},\ z=0,$ $x^2+y^2=1,\ x=0$ and $y=0$

10. We define the "n-dimensional geometric **center**" $\bar{X} := (\bar{x}_1, \ldots, \bar{x}_n)$ of a region A in $\mathbb{R}^n$ by[30]

$$\bar{x}_j := \frac{1}{\mathrm{Vol}(A)} \int_A x_j\,dV_n. \qquad (j=1,\ldots,n)$$

For an $n \times n$ invertible matrix L and an n-vector $\mathbf{v}$, we define

$$f : \mathbb{R}^n \to \mathbb{R}^n, \qquad f(X) := LX + \mathbf{v}.$$

If we let $\bar{Y}$ be the center of the region $f(A)$, then show that $f(\bar{X}) = \bar{Y}$.

[30] The above equation can be expressed as a "vector-valued integral"

$$\bar{X} = \frac{1}{\mathrm{Vol}(A)} \int_A X\,dV_n.$$

11. Find the center of the following solid.

    (1) the region bounded by the surface $y = \cos x$ and the planes $z = y$, $x = 0$, $x = \frac{\pi}{2}$, $z = 0$

    (2) the region bounded above by the sphere $x^2 + y^2 + z^2 = 1$ and below by the paraboloid $z = x^2 + y^2$

    (3) the region that lies inside the sphere $x^2 + y^2 + z^2 = 1$ and outside the cylinder $r = \sin\theta$

    (4) the region that lies above the cone $z^2 = x^2 + y^2$ and inside the sphere $\rho = 2a\cos\varphi$

    (5) $x^2 + y^2 + z^2 \leq 1, \quad z \geq 0$

    (6) $\dfrac{x^2}{a^2} + \dfrac{y^2}{b^2} + \dfrac{z^2}{c^2} \leq 1, \quad z \geq 0$

12. (**Pappus' Theorem**[31] **I**) Let d be the distance between the x-axis and the center of a region D in the upper half plane. Show that the volume of the body of revolution obtained from rotating D about the x-axis is $2\pi d\cdot \mathrm{Vol}_2(D)$.

13. (**Gaussian Function**) Show that for positive numbers A and B,

$$\int_{-\infty}^{\infty} Ae^{-B^2x^2}\, dx = \frac{A}{B}\sqrt{\pi}$$

and

$$\int_{-\infty}^{\infty} x^2 Ae^{-B^2x^2}\, dx = \frac{A\sqrt{\pi}}{2B^3}.$$

From these, show that the Gaussian function

$$g(x) = \frac{1}{\sqrt{2\pi}}e^{-x^2/2}$$

is the probability density function with expectation 0 and divergence 1. This function is called Gauss' **standard normal distribution function.**

14. (**Error Function**) For the function defined by

$$\mathrm{erf}(x) := \frac{2}{\sqrt{\pi}}\int_0^x e^{-t^2}\, dt,$$

show that

$$\lim_{x\to\infty} \mathrm{erf}(x) = 1.$$

15. Find the volume of the region bounded by $0 \leq r \leq \cos 2\theta$ in cylindrical coordinates (r, θ, z) and inside the ball $x^2 + y^2 + z^2 \leq 1$.

---

[31] Pappus is a Greek mathematician in the third century.

16. For a real number p and the unit disk

$$D := \{(x,y) \mid x^2 + y^2 \le 1\},$$

show that

$$\iint_D \frac{dx\,dy}{(x^2+y^2)^{p/2}} < \infty$$

if and only if $p < 2$.[32]

17. For a real number p and

$$D_+ := \{(x,y) \mid x^2 + y^2 \le 1, \quad x \ge 0,\ y \ge 0\},$$

find all values of p for which

$$\iint_{D_+} \frac{dx\,dy}{(x^2+y^2)^{p/2}} < \infty.$$

18. Integrate the following function over the standard unit ball

$$B := \{(x,y,z) \mid x^2 + y^2 + z^2 \le 1\}.$$

$$x, \quad y, \quad z, \quad x^2, \quad y^2, \quad z^2, \quad xy, \quad yz, \quad zx, \quad x^2y, \quad xyz$$

19. For the standard unit ball $B \subset \mathbb{R}^3$ and a real number p, show that

$$\iiint_B \frac{dx\,dy\,dz}{(x^2+y^2+z^2)^{p/2}} < \infty$$

if and only if $p < 3$.

20. For a real number p and

$$B_+ := \{(x,y,z) \mid x^2 + y^2 + z^2 \le 1, \quad x \ge 0, \quad y \ge 0, \quad z \ge 0\},$$

find all values of p for which

$$\iiint_{B_+} \frac{dx\,dy\,dz}{(x^2+y^2+z^2)^{p/2}} < \infty.$$

---

[32]Such a improper integral is thought of as

$$\iint_D = \lim_{\epsilon \to 0} \iint_{D - D_\epsilon}.$$

Here D_ϵ is a disk of radius ϵ with center at the origin.

21. For a density function[33] μ defined on a region A in the plane, let $\bar{X}$ be the center of mass of A. For a plane H through $\bar{X}$ perpendicular to a unit vector $\mathbf{v}$, we define

$$h(X) := (X - \bar{X}) \cdot \mathbf{v}.$$

This function is the "signed" distance function between a point X and the plane H. Show that

$$\iiint_A h(X)\mu(X)\,dV(X) = 0.$$

[33] An integrable function is called a **density function**.

5 Proof of Fubini's Theorem, Gravity, Moments of Inertia

5.1 Proof of Fubini's Theorem

Theorem 5.1.1 (Fubini's Theorem) *Let $f(x, y)$ be an integrable function in a rectangle $R = [a, b] \times [c, d]$. If the function*

$$f_x : [c, d] \to \mathbb{R}, \qquad y \mapsto f(x, y)$$

is integrable on $[c, d]$ for any $x \in [a, b]$, then the function

$$x \mapsto \int_c^d f(x, y)\, dy$$

is integrable on $[a, b]$, and

$$\iint_R f = \int_a^b \left(\int_c^d f(x, y)\, dy \right) dx.$$

Proof. For $x \in [a, b]$, if we let

$$F(x) := \int_c^d f(x, y)\, dy,$$

we want to prove the equation

$$\iint_R f\, dV_2 = \int_a^b F(x)\, dx.$$

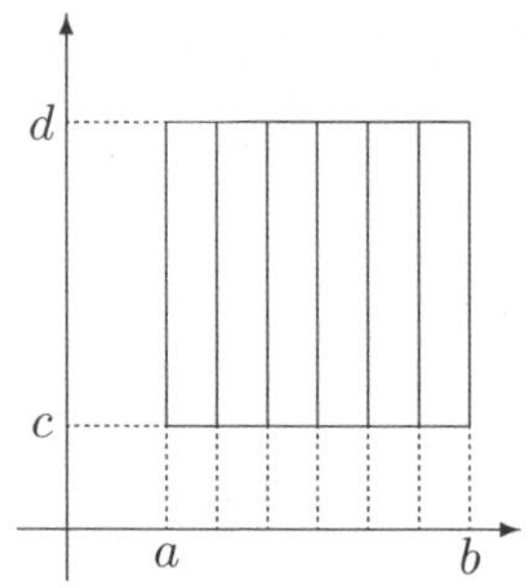

A partition P of the rectangle R consists of a partition of the interval $[a,b]$

$$P_1 \ : \ a = x_0 < \cdots < x_n = b$$

and a partition of the interval $[c,d]$

$$P_2 \ : \ c = y_0 < \cdots < y_m = d.$$

Suppose that we prove the inequality

$$L(P,f) \le L(P_1,F) \le U(P_1,F) \le U(P,f). \tag{14.3}$$

Then from the integrability of f, we may take a partition P so that the difference $|L(P,f)-U(P,f)|$ is arbitrarily small and thus the difference between $L(P_1,F)$ and $U(P_1,F)$ becomes arbitrarily small. So F is integrable and the value of its integral is equal to $\iint_R f$. Therefore $\iint_R f = \int_a^b F(x)\,dx$, which completes the proof.

Now it suffices to prove the inequality (14.3). Since

$$\begin{aligned}
L(P,f) &= \sum_{i=1}^{n}\sum_{j=1}^{m} (\inf f(P_{ij}))\,(x_i - x_{i-1})(y_j - y_{j-1}) \\
&= \sum_{i=1}^{n}\left(\sum_{j=1}^{m} \inf f(P_{ij})(y_j - y_{j-1})\right)(x_i - x_{i-1}) \\
&\le \sum_{i=1}^{n} \inf_{x_{i-1}\le x\le x_i}\left(\sum_{j=1}^{m} \inf_{y_{j-1}\le y\le y_j} f(x,y)(y_j - y_{j-1})\right)(x_i - x_{i-1}) \\
&\le \sum_{i=1}^{n} \inf_{x_{i-1}\le x\le x_i} L(P_2, f_x)(x_i - x_{i-1}) \qquad (\Leftarrow\ f_x(y) := f(x,y)) \\
&\le \sum_{i=1}^{n} \inf_{x_{i-1}\le x\le x_i} F(x)(x_i - x_{i-1}) = L(P_1, F),
\end{aligned}$$

we get the first inequality in (14.3). The inequality in the middle of (14.3) is trivial, and we can get the last inequality just as we got the first inequality. □

5.2 Newton's Law of Gravitation

Newton made a great discovery about gravity between planets, and provided the way to get the gravitational force of attraction between two planets. Since the size of sun or earth is relatively very small compared with their distance,[34] we may think of the force of attraction between them as the force acting on two particles, and the magnitude of the force is inversely proportional to the square of the distance r between two particles[35] and proportional to the product of their masses m, M[36]:

$$f = G\frac{m \cdot M}{r^2}.$$

Moreover Newton discovered that such a principle is universal. On the other hand, since earth and an apple is very close, the question that Newton would like to know is to find the gravitational force between a particle and a big ball.[37] It took so long for Newton to solve this problem that he even had to postpone publishing his epic work, *Principia* [38] [Hsiang]. The law which Newton discovered is as follows:

> For a ball of mass M with density function given as a function of the distance to the center of the ball, the gravitational force between the ball and a point P outside the ball is equal to the gravitational force between a particle of mass M at the center of the ball and P.

Let's derive the Newton's law.[39]

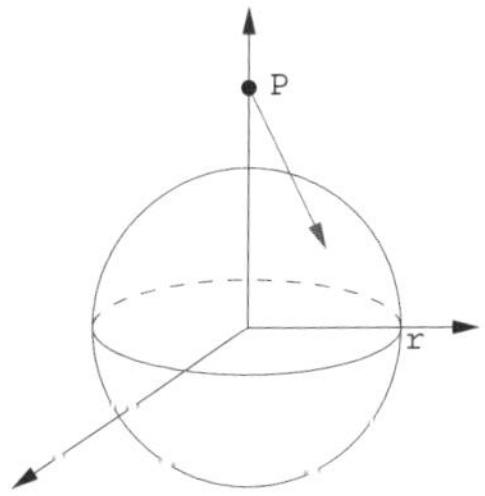

For convenience, set the coordinates so that the center of the given ball B is at the origin and a particle (P, m) outside the ball lies at $(0, 0, R)$. If we let r be the radius of the ball, then $0 < r < R$. Also let μ be the density function of the ball. Then the mass of the ball is given by[40]

---

[34] The average distance between sun and earth is 149,450,000 km and the equatorial radius of sun is 695,553 km.

[35] As the area of sphere is proportional to the square of radius....

[36] I confess that I didn't know that the **proportional constant** is the most interesting thing in a proportional expression until I got old.

[37] A ball in mathematics is always the one packed. The empty one is called a sphere.

[38] Of course Newton had to develop calculus by himself.

[39] There are several different ways to derive the Newton's equation. [Feynman, 1963, Ch. 13-4], [Hsiang]. Once we study the divergence theorem, we can prove it very easily. We may prove it using inversion.

[40] The "area" of the sphere of radius ρ is the same as the side area of the circumscribed cylinder $4\pi\rho^2$. It was one of the discoveries which Archimedes of Syracuse was proud of. The Roman general Marcus Claudius Marcellus who defeated Syracuse let a sphere circumscribed by a cylinder engraved on Archimedes' tomb in accordance with Archimedes's will. Archimedes's tomb had been forgotten until Cicero as the quaestor in western Sicily

$$M := \iiint_B \mu\, dV = \int_0^{2\pi}\int_0^{\pi}\int_0^{r} \mu(\rho)\rho^2 \sin\varphi\, d\rho\, d\varphi\, d\theta = \int_0^r (4\pi\rho^2)\mu(\rho)\, d\rho.$$

Now a particle (P, m) is attracted by a point (x, y, z) of this ball with the force

$$\mathbf{F}(x,y,z) = \frac{Gm\mu(x,y,z)}{|(x,y,z)-(0,0,R)|^2} \frac{(x,y,z)-(0,0,R)}{|(x,y,z)-(0,0,R)|}.$$

We can integrate each component of the force $\mathbf{F}$ to get the force acting on the particle P by the whole ball. Here from the symmetry of the given problem we can immediately know that the x- and y-components are zero. Therefore there remains only the z-component of the gravitational force, and

$$\begin{aligned}
&\iiint_B \mathbf{F}\cdot\mathbf{k}\, dV \\
&= -\iiint_B \frac{Gm\mu(x,y,z)\cdot(R-z)}{|(x,y,z)-(0,0,R)|^3}\, dV \\
&= -\int_0^{2\pi}\int_0^{\pi}\int_0^{r} \frac{Gm\mu(\rho)\cdot(R-\rho\cos\varphi)}{(\rho^2-2R\rho\cos\varphi+R^2)^{3/2}}\rho^2\sin\varphi\, d\rho\, d\varphi\, d\theta \\
&= -Gm\cdot 2\pi\int_0^r \rho^2\mu(\rho)\left(\int_0^{\pi}\frac{(R-\rho\cos\varphi)\sin\varphi\, d\varphi}{(\rho^2+R^2-2R\rho\cos\varphi)^{3/2}}\right) d\rho \\
&\qquad (\text{replace } \sqrt{\rho^2+R^2-2R\rho\cos\varphi} =: u\ ) \\
&= -Gm\cdot 2\pi\int_0^r \rho^2\mu(\rho)\left(\int_{R-\rho}^{R+\rho}\frac{-\rho^2+R^2+u^2}{2Ru^3}\frac{u\, du}{R\rho}\right) d\rho \\
&= -Gm\cdot 2\pi\int_0^r \rho^2\mu(\rho)\left(\frac{1}{2R^2\rho}\int_{R-\rho}^{R+\rho}\frac{u^2+R^2-\rho^2}{u^2}\, du\right) d\rho
\end{aligned}$$

---

discovered the tomb and had it cleaned up in 75 BC but again it has been lost until a tomb on which a sphere circumscribed by a cylinder was engraved was found during foundation work to build a hotel in Syracuse in 1965 [Lee]. Plutarch wrote in "Parallel Lives" as follows [Kim,Kim]:

> Archimedes possessed so high a spirit, so profound a soul, and such treasures of scientific knowledge, that though these inventions had now obtained him the renown of more than human sagacity, he yet would not deign to leave behind him any commentary or writing on such subjects; but, repudiating as sordid and ignoble the whole trade of engineering, and every sort of art that lends itself to mere use and profit, he placed his whole affection and ambition in those purer speculations where there can be no reference to the vulgar needs of life.

$$\begin{aligned}
&= -Gm\cdot 2\pi\int_0^r \rho^2\mu(\rho)\left(\frac{1}{2R^2\rho}\left[u-\frac{R^2-\rho^2}{u}\right]_{R-\rho}^{R+\rho}\right)d\rho\\
&= -Gm\cdot 2\pi\int_0^r \rho^2\mu(\rho)\left(\frac{1}{2R^2}\cdot 4\rho\right)d\rho\\
&= -\frac{Gm}{R^2}\cdot 4\pi\int_0^r \rho^2\mu(\rho)\,d\rho\\
&= -\frac{Gm}{R^2}M.
\end{aligned}$$

□

5.2.1 Basic Exercise

1. Find the gravitational force between a ball and a point inside the ball when the density function of the ball is given by the function of distance from the center.

2. For the **solid angle vector field** $\mathbf{A}_P(X) := \dfrac{X-P}{|X-P|^3}$, show that if we let

$$\mathbf{F}(X) := \iiint_{|P|\le r} \mathbf{A}_P(X)\,dV(P),$$

then

$$\mathbf{F}(X) = \begin{cases} \frac{4\pi r^3}{3|X|^3}X & (|X|\ge r)\\ \frac{4\pi}{3}X & (|X|\le r). \end{cases}$$

5.3 Moments of Inertia

In the plane, the motion of a particle of mass m which moves along the circle of radius r with the center at the origin can be expressed as

$$\mathbf{r}(t) = r\big(\cos\theta(t), \sin\theta(t)\big).$$

Here the velocity vector and speed of the particle are

$$\mathbf{v} = r\dot\theta\big(-\sin\theta, \cos\theta\big), \qquad v = r\left|\dot\theta\right|$$

and so the kinetic energy is

$$\frac{1}{2}mv^2 = \frac{1}{2}mr^2\dot\theta^2.$$

Here

$$I := mr^2$$

is called the **moment of inertia** "about the origin" [41]

Similarly, let's describe the motion of a point $\mathbf{x}_0 = (x_0, y_0, z_0)$ of mass m which rotates about the z-axis in the space.

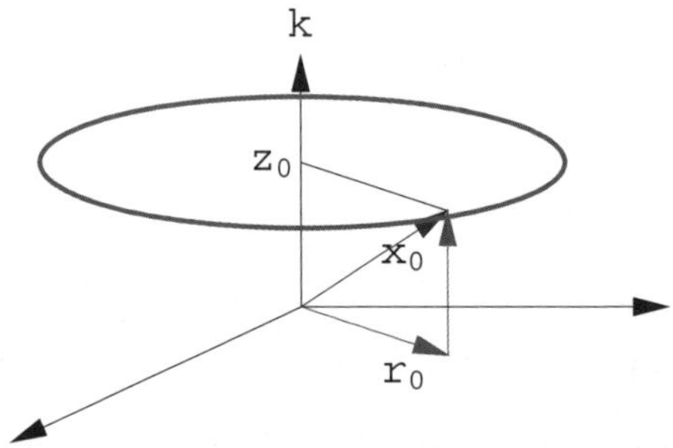

If we project the vector $\mathbf{x}_0$ onto the z-axis, we get $z_0\,\mathbf{k}$. If we let

$$\mathbf{r}_0 := \mathbf{x}_0 - z_0\mathbf{k},$$

the rotation of the point $\mathbf{x}_0$ about the z-axis is given by[42]

$$\mathbf{x}(t) = \mathbf{r}_0 \cos\theta(t) + \mathbf{k} \times \mathbf{r}_0 \sin\theta(t) + z_0\mathbf{k}, \qquad \mathbf{x}(0) = \mathbf{x}_0.$$

Here the velocity vector is

$$\dot{\mathbf{x}} = (-\mathbf{r}_0 \sin\theta + \mathbf{k} \times \mathbf{r}_0 \cos\theta)\,\dot{\theta}$$

and thus the speed is[43]

$$|\dot{\mathbf{x}}| = |\mathbf{r}_0| \left|\dot{\theta}\right| =: r_0 \left|\dot{\theta}\right|.$$

Therefore in this motion, the kinetic energy is

$$\frac{1}{2}m|\dot{\mathbf{x}}|^2 = \frac{1}{2}mr_0^2\dot{\theta}^2 =: \frac{1}{2}I\dot{\theta}^2.$$

From now on, we call

$$I = mr_0^2 = m(x_0^2 + y_0^2)$$

the **moment of inertia** of the particle $(\mathbf{x}_0, m)$ about the z-axis.

In general, if the distance between a point $\mathbf{x}_0$ and a line l is r, then the **moment of inertia** of the particle $(\mathbf{x}_0, m)$ about the line l is given by mr^2.

---

[41] Therefore the kinetic energy is expressed as $\frac{1}{2}I\dot{\theta}^2$.

[42] We let $\theta(t)$ be the angle of rotation $\theta(t)$. If we rotate the vector $\mathbf{r}_0$ about the z-axis by $90°$, we get $\mathbf{k} \times \mathbf{r}_0$.

[43] In other words, in rotation, the speed is given by the product of **angular speed** and the "distance to the axis of rotation".

5.3.1 Moments of Inertia of Solids

When a solid with density function μ rotates about the line l, the kinetic energy is

$$\iiint \frac{1}{2}\mu(\mathbf{x})r(\mathbf{x})^2\dot{\theta}^2\, dV(\mathbf{x}) =: \frac{1}{2}I\dot{\theta}^2.$$

Here r is the distance function between a point of the solid and the line l, and

$$I = \iiint \mu(\mathbf{x})r(\mathbf{x})^2\, dV(\mathbf{x})$$

is the **moment of inertia** of the given solid about the line l.[44]

5.3.2 Example

When the density of a box

$$R := \left[-\frac{a}{2}, \frac{a}{2}\right] \times \left[-\frac{b}{2}, \frac{b}{2}\right] \times \left[-\frac{c}{2}, \frac{c}{2}\right]$$

is given by a constant μ, the mass of the box is

$$M := a \cdot b \cdot c \cdot \mu = abc\mu,$$

and the moment of inertia about the x-axis is

$$\begin{aligned}
I_x &= \int_{-c/2}^{c/2}\int_{-b/2}^{b/2}\int_{-a/2}^{a/2} (y^2+z^2)\mu\, dx\, dy\, dz \\
&= \mu \cdot a \cdot 2 \cdot 2 \int_0^{c/2}\int_0^{b/2} (y^2+z^2)\, dy\, dz \\
&= \mu \cdot 4a \int_0^{c} \left(\frac{1}{3}\left(\frac{b}{2}\right)^3 + \frac{b}{2}z^2\right) dz \\
&= \mu \cdot a \left(\frac{1}{3}\frac{b^3}{2}\frac{c}{2} + \frac{1}{3}\frac{b}{2}\frac{c^3}{2}\right) \\
&= \mu \cdot (abc)(b^2+c^2)/12 \\
&= M(b^2+c^2)/12
\end{aligned}$$

Similarly, the moments of inertia about the x- and z-axis are

$$\begin{aligned}
I_y &= M(a^2+c^2)/12 \\
I_z &= M(a^2+b^2)/12,
\end{aligned}$$

respectively.

---

[44]The moment of inertia is related to the energy required to rotate a body, or make a rotating body stop.

5.3.3 Basic Exercise

When M is the mass of a homogeneous box

$$R = [0, a] \times [0, b] \times [0, c],$$

find its moments of inertia I_x, I_y, I_z.

5.3.4 Moments of Inertia of Plane Regions

For a region R in the plane with density function μ, its **moment of inertia** about the line l is defined by

$$I_l := \iint_R \mu\, r^2\, dV_2,$$

where r is the distance between a point in R and the line l.

5.3.5 Example

Suppose that R is a region in the plane enclosed by

$$y = 0, \qquad x = 1, \qquad y = 2x$$

and the density function of R is

$$\mu(x, y) := 6(x + y + 1).$$

The moment of inertia of R about the x-axis is given by

$$\begin{aligned} I_x &= \int_0^1 \int_0^{2x} y^2 \mu(x, y)\, dy\, dx = \int_0^1 \int_0^{2x} (6xy^2 + 6y^3 + 6y^2)\, dy\, dx \\ &= \int_0^1 \left[2xy^3 + \frac{3}{2} y^4 + 2y^3 \right]_{y=0}^{y=2x} dx = \int_0^1 (40x^4 + 16x^3)\, dx \\ &= \left[8x^5 + 4x^4 \right]_0^1 = 12 \end{aligned}$$

and the moment of inertia of R about the y-axis is

$$I_y = \int_0^1 \int_0^{2x} x^2 \mu(x, y)\, dy\, dx = \frac{39}{5}.$$

5.3.6 Moments of Inertia of Curves

For a curve $X \subset \mathbb{R}^3$ with density function μ, the moment of inertia about the x-axis is given by

$$I_x \;:=\; \int_X (y^2 + z^2)\mu \; ds$$

and the moments of inertia about the y- and z-axis are given by

$$\begin{aligned} I_y &:= \int_X (z^2 + x^2)\mu \ ds, \\ I_z &:= \int_X (x^2 + y^2)\mu \ ds, \end{aligned}$$

respectively.

5.3.7 Example

For the helix

$$X(t) = (\cos 4t)\mathbf{i} + (\sin 4t)\mathbf{j} + t\,\mathbf{k}, \qquad 0 \le t \le 2\pi,$$

we have

$$ds = |X'(t)|\,dt = \sqrt{4^2 + 1}\,dt = \sqrt{17}\,dt.$$

So when the density function of X is 1, the moment of inertia about the z-axis is

$$I_z = \int_X (x^2 + y^2)\,ds = \int_0^{2\pi} (\cos^2 4t + \sin^2 4t)\sqrt{17}\,dt = 2\pi\sqrt{17}.$$

5.3.8 Basic Exercise

Find the moment of inertia of the curve

$$X(t) := (t\cos t)\mathbf{i} + (t\sin t)\mathbf{j} + \frac{2\sqrt{2}}{3}t^{3/2}\,\mathbf{k}, \qquad 0 \le t \le 1$$

about the z-axis when the density function is 1.

5.3.9 Moments of Inertia of Surfaces

Even though we will deal with integrals over surfaces in Chapter 16, we will just state the definition of moments of inertia of surfaces. For the surface $S \subset \mathbb{R}^3$ with density function μ, its moment of inertia about the x-axis is given by

$$I_x := \iint_S (y^2 + z^2)\mu \ dS,$$

and its moments of inertia about the y and z-axis are given by

$$\begin{aligned} I_y &:= \iint_S (z^2 + x^2)\mu \ dS, \\ I_z &:= \iint_S (x^2 + y^2)\mu \ dS, \end{aligned}$$

respectively.

Exercise: Chapter 14 Section 5

1. When a cube bounded by

$$x = \pm 1, \qquad y = \pm 1, \qquad z = 4 \pm 1$$

has the uniform density μ, find the moments of inertia I_x, I_y, I_z.

2. Find the mass M and the center of mass $(\bar{x}, \bar{y}, \bar{z})$ of the region R, and find its moments of inertia about each axis.

$$R\ :\ 0 \le x \le 1,\ 0 \le y \le 1,\ 0 \le z \le 1, \qquad \mu(x, y, z) = x + y + z + 1$$

3. When the curve

$$X(t) := t\,\mathbf{i} + \frac{2\sqrt{2}}{3}t^{3/2}\,\mathbf{j} + \frac{t^2}{2}\,\mathbf{k}, \qquad 90 \le t \le 2$$

has the density function

$$\mu(t) = \frac{1}{t+1},$$

find the moments of inertia about each axis.

4. For the homogeneous surface with mass M

$$x^2 + y^2 = R^2, \qquad -a \le z \le a,$$

show that the moment of inertia is $I_z = MR^2$.

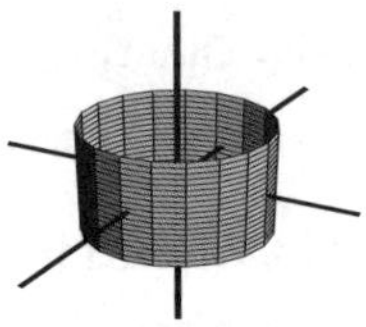

5. Show that the moment of inertia of the interior of the homogeneous cylinder with mass M

$$x^2 + y^2 \le R^2, \qquad -a \le z \le a$$

is $I_z = \dfrac{1}{2}MR^2$.

6. Show that the moment of inertia of the homogeneous solid with mass M given by

$$R_1^2 \le x^2 + y^2 \le R_2^2, \qquad -a \le z \le a$$

is $I_z = \dfrac{1}{2}M(R_2^2 + R_1^2)$.

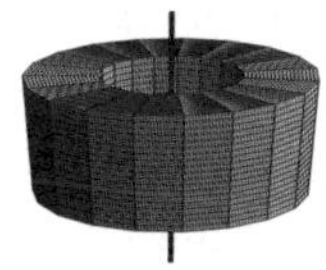

7. Show that for a homogeneous line segment of mass M and of length L, its moment of inertia about the axis through the center of the line is $\frac{1}{12}ML^2$. Also show that if the axis passes through the endpoint of the line segment, then the moment of inertia is $\frac{1}{3}ML^2$.

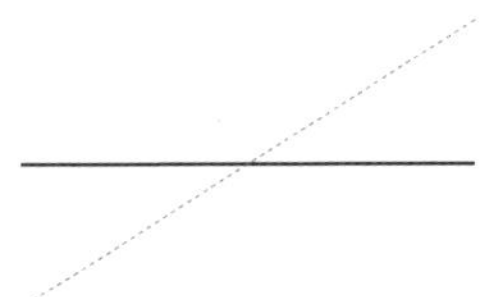

8. Show that the moment of inertia of a homogeneous ball with mass M and radius R about the axis through the center of the ball is $\frac{2}{5}MR^2$.

9. Show that the moment of inertia of a homogeneous sphere with mass M and radius R about the axis through the center of the sphere is $\frac{2}{3}MR^2$.

Chapter 14 Advanced Exercise

1. Make the equation for the annual hours of sunlight at each point of earth and guess the latitude of the place where is most like a desert.

2. Euler's **gamma function** is defined by[45]

$$\Gamma(x) := \int_0^\infty t^{x-1}e^{-t}\,dt \qquad (x>0).$$

   (1) Show that $\Gamma(1)=1$.

   (2) Show that $\Gamma(x+1)=x\Gamma(x)$ for $x>0$,.

   (3) Show that $\Gamma(n+1)=n!$ for $n\in\mathbb{N}$.

   (4) Show that $\Gamma(1/2)=\sqrt{\pi}$.

   (5) Show that $\Gamma'(1)=-\gamma$. (Here, γ is Euler's constant.)

3. For any positive number t, show that[46]

$$\int_{\mathbb{R}^n}\frac{1}{(4\pi t)^{n/2}}e^{-(x_1^2+\cdots+x_n^2)/4t}\,dx_1\cdots dx_n = 1.$$

4. From Fubini's theorem, derive the Leibniz's rule (for differentiation of integrals) introduced in Chapter 11 Section 1.

---

[45] To show the convergence of this improper integral, it suffices to divide the interval into two subintervals and show

$$\int_0^1 t^{x-1}e^{-t}\,dt \le \int_0^1 t^{x-1}\,dt = \frac{1}{x}\left[t^x\right]_{t=0}^{t=1} = \frac{1}{x}$$

and

$$\int_1^\infty t^{x-1}e^{-t}\,dt < \infty.$$

The last inequality holds since there exists c such that

$$t^{x-1}e^{-t} \le \frac{c}{t^2} \qquad (1\le t<\infty).$$

(Recall $\lim_{t\to\infty} t^{x+1}e^{-t}=0$.)

[46] In this problem, the integrand is called the **elementary solution for the heat equation**. A function $u(\mathbf{x},t)$ which satisfies the equation

$$\left(\frac{\partial}{\partial t}-\left(\frac{\partial^2}{\partial x_1^2}+\cdots+\frac{\partial^2}{\partial x_n^2}\right)\right)u(\mathbf{x},t)=0$$

is called a solution for the heat equation.

5. It is important in probability theory or statistics to estimate the value of $\int_T^\infty e^{-x^2/2}\,dx$ for a real number T. If we denote its value by A, we get

$$\begin{aligned} A &= \int_T^\infty e^{-x^2/2}\,dx \\ &= \int_T^\infty \left(-xe^{-x^2/2}\right)\cdot\left(-\frac{1}{x}\right)dx = \left[\frac{e^{-x^2/2}}{-x}\right]_T^\infty - \int_T^\infty \frac{e^{-x^2/2}}{x^2}\,dx \\ &= \frac{e^{-T^2/2}}{T} - \int_T^\infty \frac{e^{-x^2/2}}{x^2}\,dx. \end{aligned}$$

On the other hand, since

$$\int_T^\infty \frac{e^{-x^2/2}}{x^2}\,dx \le \int_T^\infty \frac{e^{-x^2/2}}{T^2}\,dx = \frac{A}{T^2},$$

we get the inequality

$$A \ge \frac{e^{-x^2/2}}{T} - \frac{A}{T^2},$$

and thus

$$\frac{e^{-T^2/2}}{T\left(1+\frac{1}{T^2}\right)} \le \int_T^\infty e^{-x^2/2}\,dx \le \frac{e^{-T^2/2}}{T}.$$

For arbitrary real numbers a and T, show that

$$\frac{e^{-aT^2}}{2aT\left(1+\frac{1}{2aT^2}\right)} \le \int_T^\infty e^{-ax^2}\,dx \le \frac{e^{-aT^2}}{2aT}.$$

6. Show the following integral over a ball $B\ :\ x^2+y^2+z^2 \le 4$

(1) $\displaystyle\iiint_B \frac{dx\,dy\,dz}{\sqrt{(x-1)^2+y^2+z^2}} = \frac{22\pi}{3}$

(2) $\displaystyle\iiint_B \frac{dx\,dy\,dz}{\sqrt{(x-3)^2+y^2+z^2}} = \frac{32\pi}{9}$

(3) $\displaystyle\iiint_B \frac{dx\,dy\,dz}{\sqrt{(x-9)^2+(y-2)^2+z^2}} = \frac{32\sqrt{85}\pi}{255}$

Chapter 15

Divergence and Green's Theorem

If we think of a vector field as a velocity vector of fluid flow, the divergence of the vector field represents the net rate of expansion per unit volume of the fluid. In this chapter, we will study the divergence theorem for vector fields in the plane and use it to derive Green's theorem. Green's theorem states that the integral of "differentiated vector field" $d\mathbf{F}$ on a plane region D is equal to the integral of the vector field $\mathbf{F}$ over the boundary ∂D:

$$\iint_D d\mathbf{F} = \int_{\partial D} \mathbf{F} \cdot d\mathbf{s}.$$

Such a principle holds in general dimension. Green's theorem is extended to the **divergence theorem** in dimension of three or more. In one dimension, Green's theorem is nothing but the **fundamental theorem of calculus** which we already knew very well.

1 Vector Fields and Divergence

1.1 Divergence Function

For a $\mathcal{C}^1$ vector field defined on an open set $U \subset \mathbb{R}^2$

$$\mathbf{F}(x, y) = P(x, y)\,\mathbf{i} + Q(x, y)\,\mathbf{j},$$

the **divergence** of $\mathbf{F}$ at a point X in U is defined by

$$\operatorname{div}\mathbf{F}(X) := \frac{\partial P}{\partial x}(X) + \frac{\partial Q}{\partial y}(X)$$

and the **divergence (function)**[1] of $\mathbf{F}$ is defined by the function

$$\operatorname{div}\mathbf{F} := \frac{\partial P}{\partial x} + \frac{\partial Q}{\partial y}$$

In general, for a vector field in $\mathbb{R}^n$

$$\mathbf{F}(x_1, \ldots, x_n) = f_1(x_1, \ldots, x_n)\,\mathbf{e}_1 + \cdots + f_n(x_1, \ldots, x_n)\,\mathbf{e}_n,$$

the **divergence (function)** of $\mathbf{F}$ is defined by[2]

$$\operatorname{div}\mathbf{F} := D_1 f_1 + \cdots + D_n f_n = \frac{\partial f_1}{\partial x_1} + \cdots + \frac{\partial f_n}{\partial x_n}.$$

1.1.1 Example

The divergence of the position vector field

$$\mathbf{r}(x, y) = x\,\mathbf{i} + y\,\mathbf{j}$$

in the plane $\mathbb{R}^2$ is

$$\operatorname{div}\mathbf{r}(x, y) = \frac{\partial x}{\partial x} + \frac{\partial y}{\partial y} = 1 + 1 = 2.$$

[1]The term "divergence" which we use here is nothing to do with "convergence-divergence" of a sequence.

[2]In some book, they let

$$\nabla := \left(\frac{\partial}{\partial x_1}, \ldots, \frac{\partial}{\partial x_n}\right)$$

and denote the divergence of a vector field $\mathbf{F}$ by $\nabla \cdot \mathbf{F}$ instead of $\operatorname{div}\mathbf{F}$.

The divergence of the position vector field

$$\mathbf{r}(x, y, z) = x\,\mathbf{i} + y\,\mathbf{j} + z\,\mathbf{k}$$

in $\mathbb{R}^3$ is identically[3]

$$\operatorname{div}\mathbf{r} = 3.$$

1.1.2 Example

For the angle element vector field

$$\mathbf{a}(x, y) = -\frac{y}{x^2 + y^2}\,\mathbf{i} + \frac{x}{x^2 + y^2}\,\mathbf{j}$$

the divergence is

$$\operatorname{div}\mathbf{a} = \frac{\partial}{\partial x}\left(-\frac{y}{x^2 + y^2}\right) + \frac{\partial}{\partial y}\left(\frac{x}{x^2 + y^2}\right) = \frac{2xy}{(x^2 + y^2)^2} - \frac{2xy}{(x^2 + y^2)^2} = 0.$$

1.2 Meaning of Divergence

A vector field describes the motion of points in the space[4] and it is the divergence that tells how the volume (or area if the dimension is two) changes as some region moves.

For example, think of the position vector field

$$\mathbf{r}(x, y) = x\,\mathbf{i} + y\,\mathbf{j}$$

as the velocity vector field of a fluid flow.[5] That is, assume that an arbitrary point $X_0 = (x_0, y_0)$ in the plane moves satisfying the equation of motion

$$X'(t) = \mathbf{r}(X(t)), \qquad X(0) = X_0.$$

If we solve this equation of motion,[6] we get

$$X(t) = \left(x_0 e^t, y_0 e^t\right).$$

---

[3] It is equal to the dimension of the domain and also the sum of diagonal entries of the identity map.

[4] Every vector field can be thought of as the velocity vector field of a fluid. The motion described by a vector field is called a **dynamical system**. Vector fields we usually deal with are vector fields independent of time. A vector field $\mathbf{F}_t(X)$ depending on time can be thought of as a vector field independent of time in "the space of one dimension higher".

[5] Here, the flow is generated at the origin.

[6] In coordinates, we can denote it by

$$x'(t) = x(t), \quad x(0) = x_0, \qquad y'(t) = y(t), \quad y(0) = y_0.$$

Therefore if a small disk

$$D_r := \{X \in \mathbb{R}^2 \mid |X - X_0| \le r\}$$

near the point X_0 flows along the vector field $\mathbf{r}$, after time t it becomes a disk[7] of radius re^t

$$D_r(t) = \{Y \in \mathbb{R}^2 \mid |Y - X(t)| \le re^t\}.$$

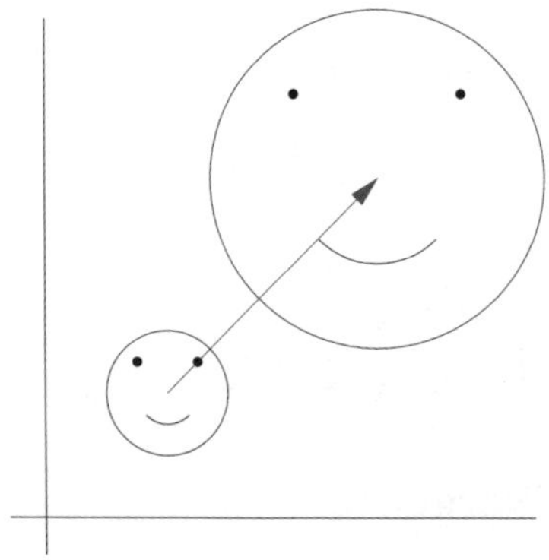

The area of this disk is

$$a(t) := \text{area}\, D_r(t) = \pi(re^t)^2.$$

Therefore the ratio of the area of the disk after time t per the area of the original disk is

$$\frac{a(t)}{a(0)} = \frac{\pi(re^t)^2}{\pi r^2} = e^{2t},$$

and the instantaneous rate of change at $t = 0$ is[8]

$$\left.\frac{d}{dt}\right|_0 \frac{a(t)}{a(0)} = 2 = \operatorname{div} \mathbf{r}(x_0, y_0).$$

We will see in the last section that the principle as in above always holds for general vector fields.

---

[7] If we let Y be the position of the point X after t seconds which satisfies $|X - X_0| = r$, $Y = Xe^t$ and thus $|Y - X_0e^t| = re^t$.

[8] $\dfrac{a'(t_0)}{a(t_0)} = \left.\dfrac{d}{dt}\right|_{t_0} \log a(t)$ will be called the **standard derivative** or **standard net rate of change**. The term "coefficient of expansion" is sometimes used in engineering. If $a(t)$ is the sum of deposit and interest,

$$r = \lim_{\Delta t \to 0} \frac{a(t_0 + \Delta t) - a(t_0)}{a(t_0)} = \frac{a'(t_0)}{a(t_0)}$$

means the **instantaneous interest rate**.

1.2.1 Basic Exercise

1. Show that the divergence of a vector field

$$\mathbf{F}(x,y) := -y\,\mathbf{i} + x\,\mathbf{j}$$

in the plane is identically zero. Also describe the equation of motion given by this vector field, and show that the net rate of change in area near each point is given by the divergence.[9]

2. Explain that the divergence of the vector field given in the following picture is positive at each point.

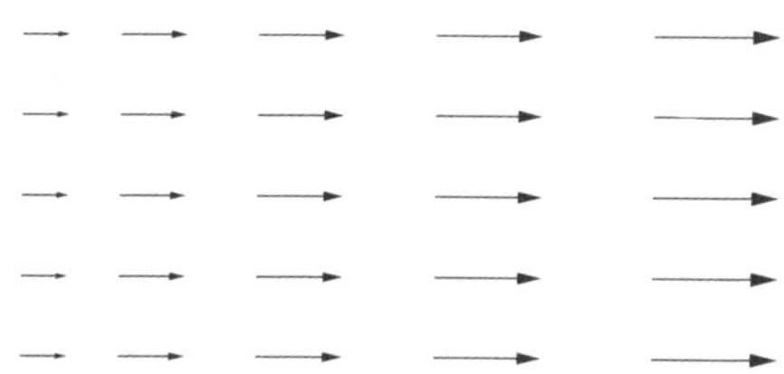

Explain that the divergence of the vector field given in the following picture is negative at each point..

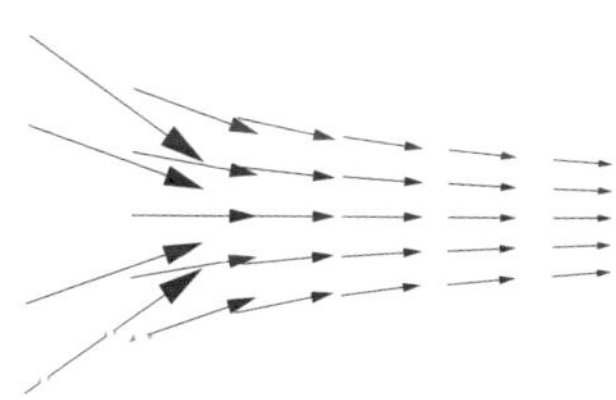

1.3 Incompressible Vector Fields

Like the angle element vector field, a vector field whose divergence is identically zero is called **incompressible**.

1.3.1 Basic Exercise

Suppose that a function $f(x,y)$ in the plane depends only on the distance from the origin to the point (x,y). Show that the vector field $\mathbf{F}(x,y) := f(x,y)(-y\,\mathbf{i}+x\,\mathbf{j})$ is incompressible.

[9]Refer to Chapter 2 Section 4 Exercise 8.

Exercise: Chapter 15 Section 1

1. Find the divergence of the following vector field.

   (1) $\mathbf{F}(x,y) = (x^2 - y^2)\,\mathbf{i} + 2xy\,\mathbf{j}$

   (2) $\mathbf{F}(x,y,z) = yz\,\mathbf{i} + zx\,\mathbf{j} + xy\,\mathbf{k}$

   (3) $\mathbf{F}(x,y,z) = \left(x^2 + y^2 + z^2\right)(3\,\mathbf{i} + 4\,\mathbf{j} + 5\,\mathbf{k})$

2. For vector fields $\mathbf{F}$, $\mathbf{G}$ and a function h defined on an open set in $\mathbb{R}^n$, show that[10]

$$\begin{aligned}\operatorname{div}(\mathbf{F} + \mathbf{G}) &= \operatorname{div}\mathbf{F} + \operatorname{div}\mathbf{G} \\ \operatorname{div}(h\mathbf{F}) &= \operatorname{grad} h \cdot \mathbf{F} + h \operatorname{div}\mathbf{F}.\end{aligned}$$

3. Describe the motion given by the vector field $\mathbf{F}(x) = x^2$ in $\mathbb{R}$, and show that the rate of change of length is the very derivative $\mathbf{F}' = \operatorname{div}\mathbf{F}$ of $\mathbf{F}$.

4. Show that the solid angle vector field

$$\mathbf{A}(X) = \frac{X}{|X|^3}, \qquad X \in \mathbb{R}^3 - \{O\}$$

   is incompressible.

5. Determine the sign of the divergence of the following vector field.

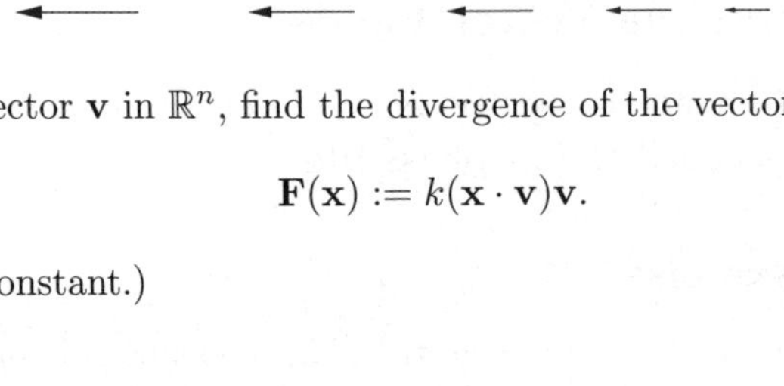

6. For a unit vector $\mathbf{v}$ in $\mathbb{R}^n$, find the divergence of the vector field

$$\mathbf{F}(\mathbf{x}) := k(\mathbf{x} \cdot \mathbf{v})\mathbf{v}.$$

   (Here, k is constant.)

[10]The last formula is sort of **Leibniz's rule** for differentiation.

2 Plane Vector Fields and Divergence Theorem

There are two ways to assign to each point of C a unit vector perpendicular to a given regular plane curve C, that is, a **unit normal vector**.

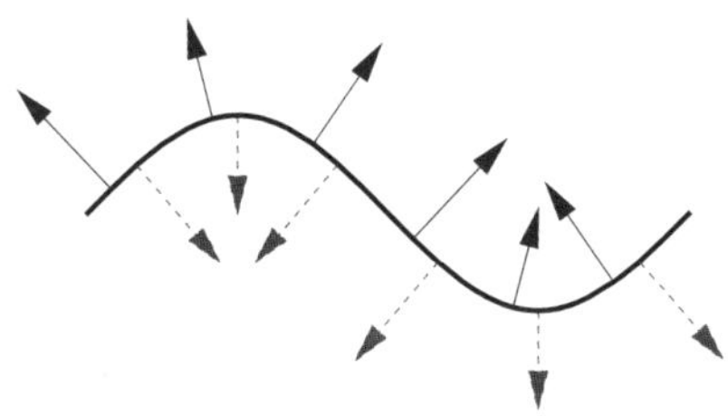

If $\mathbf{n}$ is a unit normal vector field of C, then so is $-\mathbf{n}$.

For a unit normal vector field $\mathbf{n}$ of a curve C and a vector field $\mathbf{F}(x, y)$, the line integral

$$\int_C \mathbf{F} \cdot \mathbf{n}\, ds$$

is called the "**flux** of the vector field $\mathbf{F}$ across the curve C in the direction of $\mathbf{n}$".[11] Since we integrate not the vector field $\mathbf{F}$ but the function $\mathbf{F} \cdot \mathbf{n}$ in this line integral, it doesn't depend on the orientation of the curve C.

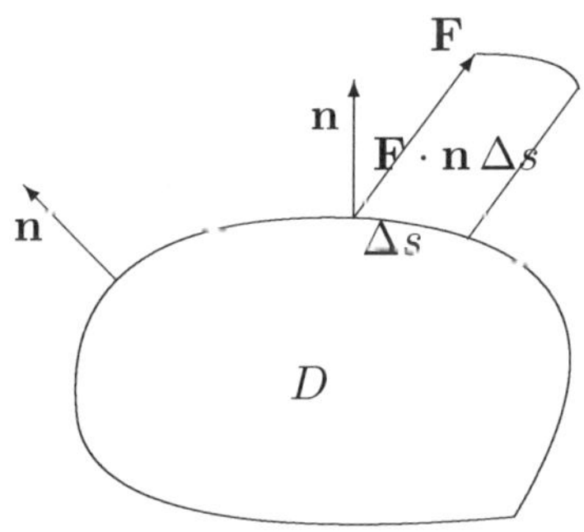

Even though there are two ways to choose a unit normal vector field for a general curve, for the closed curve ∂D which is the boundary of a region D, we usually choose **the outward direction** as the unit normal vector field $\mathbf{n}$.

[11] If $\mathbf{F}$ is the velocity vector field of a fluid, the amount of fluid that flows through a small part of C in unit time is $\mathbf{F} \cdot \mathbf{n}\, ds$, and so the amount of fluid that flows through the whole curve is $\int_C \mathbf{F} \cdot \mathbf{n}\, ds$. Strictly speaking, it would be precise to express it as "the net rate of change in the amount of fluid that flows through" rather than "the amount of fluid that flows through" or "the amount of fluid that flows through in unit time".

The following divergence theorem says that the "net rate of change of the amount that flows out of a region by a vector field" is equal to the "net rate of change of the area of the region."[12] (The amount that flows in the region is thought of as the amount that flows out of the region being negative.)

Theorem 2.0.1 (Divergence Theorem) *For a C^1 vector field $\mathbf{F}(x,y)$ defined on a plane region D,*[13]

$$\int_{\partial D} \mathbf{F}\cdot\mathbf{n}\, ds = \iint_D \operatorname{div}\mathbf{F}\, dV_2$$

Proof. We will sketch the proof of the divergence theorem.

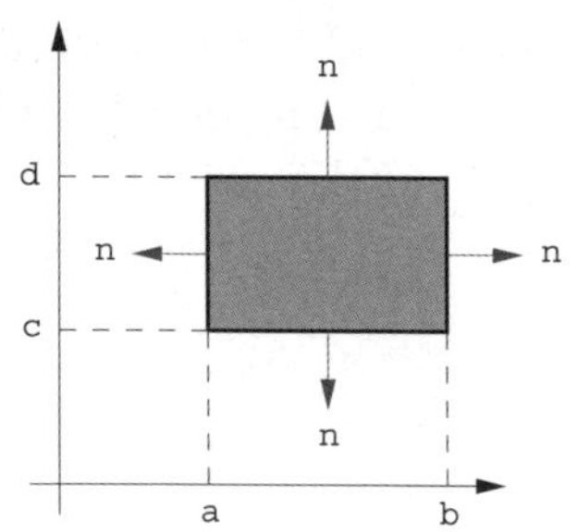

Suppose that the vector field is given by

$$\mathbf{F} = P\mathbf{i} + Q\mathbf{j}$$

and the region is a rectangle

$$D = [a,b]\times[c,d].$$

Then the outward flux of the vector field through the boundary of this region is

$$\int_{\partial D} \mathbf{F}\cdot\mathbf{n}\, ds = -\int_a^b Q(x,c)\,dx + \int_c^d P(b,y)\,dy + \int_a^b Q(x,d)\,dx - \int_c^d P(a,y)\,dy.$$

---

[12] Considering how a vector field moves on a fixed region is equivalent to considering how a region moves while a vector field is fixed. In the last section in this chapter, we will show that

$$\iint_D \operatorname{div}\mathbf{F}\, dV_2 = \left.\frac{d}{dt}\right|_0 \operatorname{area}(D_t).$$

In Chapter 17, we will introduce the divergence theorem in a three-dimensional region.

[13] Here we assume that D is a closed, bounded region and its boundary ∂D is a piecewise regular curve.

On the other hand, since

$$\begin{aligned}\iint_D \operatorname{div}\mathbf{F}\,dV_2 &= \iint_D \left(\frac{\partial P}{\partial x}+\frac{\partial Q}{\partial y}\right)dx\,dy\\ &= \int_c^d\left(\int_a^b \frac{\partial P}{\partial x}\,dx\right)dy+\int_a^b\left(\int_c^d \frac{\partial Q}{\partial y}\,dy\right)dx\\ &= \int_c^d (P(b,y)-P(a,y))\,dy+\int_a^b (Q(x,d)-Q(x,c))\,dx,\end{aligned}$$

we get

$$\int_{\partial D}\mathbf{F}\cdot\mathbf{n}\,ds=\iint_D \operatorname{div}\mathbf{F}\,dV_2.$$

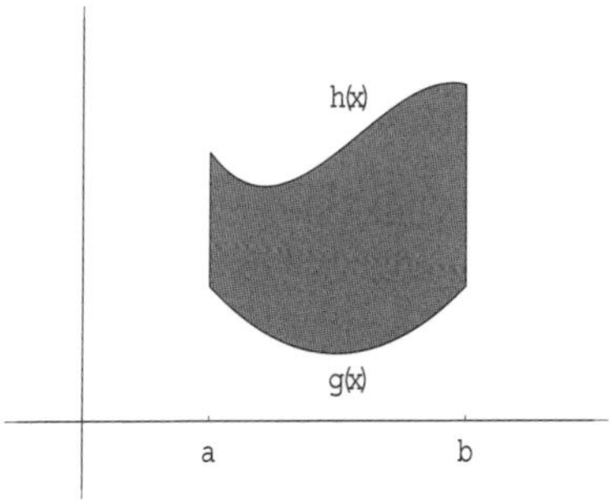

If the region D is of more general shape given by

$$D=\{(x,y)\mid a\le x\le b,\quad g(x)\le y\le h(x)\}$$

for two $\mathcal{C}^1$ functions $g(x)\le h(x)$, we can prove the divergence theorem as we did in the case of rectangle.[14]

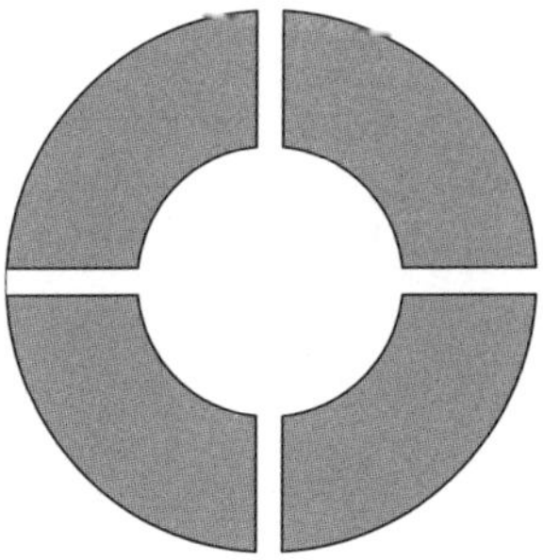

For a general region, we can partition the region D so that D becomes a union of "simple regions" of the type which we just discussed above and then prove the theorem. Here the subtle part is that the new boundaries obtained by dividing the original region become the common boundaries of two adjacent subregions but the directions of their normal vectors $\mathbf{n}$ are opposite to each other and thus the integrals over them cancel each other. □

For example, for the divergences of vector fields $\mathbf{F}(x,y) = x\mathbf{i}$ and $\mathbf{G}(x,y) := y\mathbf{j}$ are both 1, if we apply the divergence theorem to D, we get

[14]Use Leibniz's rule for differentiation under the integral sign introduced in Chapter 11.

$$\text{area}(D) = \int_{\partial D} x\mathbf{i}\cdot\mathbf{n}\, ds = \int_{\partial D} y\mathbf{j}\cdot\mathbf{n}\, ds = \frac{1}{2}\int_{\partial D} \mathbf{r}\cdot\mathbf{n}\, ds$$

In particular, consider the region $D : \dfrac{x^2}{a^2} + \dfrac{y^2}{b^2} \le 1 \ \ (a, b > 0)$ enclosed by an ellipse, we can parameterize its boundary as

$$x = a\cos t, \qquad y = b\sin t \qquad (0 \le t \le 2\pi)$$

and the unit normal vector is

$$\mathbf{n} = (b\cos t, a\sin t)/v(t) \qquad \left(v(t) = \sqrt{b^2\cos^2 t + a^2\sin^2 t}\right).$$

Therefore the area of D is

$$\frac{1}{2}\int_{\partial D} \mathbf{r}\cdot\mathbf{n}\, ds = \frac{1}{2}\int_0^{2\pi} (a\cos t, b\sin t)\cdot(b\cos t, a\sin t)\, dt = \frac{1}{2}\int_0^{2\pi} ab\, dt = ab\pi.$$

2.0.2 Basic Exercise

1. Let D_r be a disk of radius r with the center at P in the plane. For a vector field $\mathbf{F}$ defined near P, show that

$$\operatorname{div}\mathbf{F}(P) = \lim_{r\to 0} \frac{1}{\pi r^2}\int_{\partial D_r} \mathbf{F}\cdot\mathbf{n}\, ds.$$

   From this, explain the meaning of divergence.

2. For the vector field

$$\mathbf{A}(x, y) := \frac{x}{x^2+y^2}\mathbf{i} + \frac{y}{x^2+y^2}\mathbf{j}$$

   defined on the plane except the origin, if we let

$$\mathbf{F}(x, y) = \frac{x}{\sqrt{x^2+y^2}}\mathbf{A}(x, y), \qquad \mathbf{G}(x, y) = \frac{y}{\sqrt{x^2+y^2}}\mathbf{A}(x, y),$$

   show that $\operatorname{div}\mathbf{F} = 0$ and $\operatorname{div}\mathbf{G} = 0$. Also show that

$$\int_S \mathbf{F}\cdot\mathbf{n}\, ds = 0, \qquad \int_S \mathbf{G}\cdot\mathbf{n}\, ds = 0,$$

   for the unit circle

$$S : x^2 + y^2 = 1.$$

2.1 Two-Dimensional Gauss' Law

Consider the vector field

$$\mathbf{A}(x,y) := \frac{x}{x^2+y^2}\,\mathbf{i} + \frac{y}{x^2+y^2}\,\mathbf{j}$$

in the plane. Its norm is the reciprocal of the distance from the origin and its direction is the same as the direction of the position vector.[15] Its divergence is

$$\begin{aligned}
\operatorname{div}\mathbf{A}(x,y) &= \frac{\partial}{\partial x}\left(\frac{x}{x^2+y^2}\right) + \frac{\partial}{\partial y}\left(\frac{y}{x^2+y^2}\right) \\
&= \frac{(x^2+y^2)-x(2x)}{(x^2+y^2)^2} + \frac{(x^2+y^2)-y(2y)}{(x^2+y^2)^2} \\
&= 0.
\end{aligned}$$

Therefore for any **simple closed curve**[16] $C \subset \mathbb{R}^2 - \{O\}$, if the origin is not contained in the interior $\operatorname{int} C$,

$$\int_C \mathbf{A}\cdot\mathbf{n}\,ds = \iint_{\operatorname{int} C} \operatorname{div}\mathbf{A}\,dV_2 = 0.$$

If the origin is contained in the interior of C, then for a very small $\epsilon > 0$, we can apply the divergence theorem to the region

$$D := \big(C \cup \operatorname{int} C\big) - \{(x,y) \mid x^2+y^2 < \epsilon^2\}$$

and get

$$0 = \iint_D \operatorname{div}\mathbf{A}\,dV_2 = \int_C \mathbf{A}\cdot\mathbf{n}\;ds + \int_{x^2+y^2=\epsilon^2} \mathbf{A}\cdot\mathbf{n}\;ds.$$

---

[15]This vector field is the "two-dimensional solid angle vector field". If we rotate this vector field by 90°, it becomes the angle element vector field.

[16]

A simple closed curve is a curve whose two end points are equal and which has no other intersection point. In the plane, the complementary $\mathbb{R}^2 - C$ of a simple closed curve C consists of two **pieces**(**connected components**). One is bounded and the other is unbounded. The former, $\operatorname{int} C$, is called the **interior** of C and the latter, $\operatorname{ext} C$, is called the **exterior** of C: $\mathbb{R}^2 - C = (\operatorname{int} C) \cup (\operatorname{ext} C)$. The curve C is the boundary for both $\operatorname{int} C$ and $\operatorname{ext} C$. Of course, the normal direction $\mathbf{n}$ on C is determined to be the outward direction.

Thus we get

$$\begin{aligned}\int_C \mathbf{A}\cdot\mathbf{n}\,ds &= -\int_{x^2+y^2=\epsilon^2} \mathbf{A}\cdot\mathbf{n}\;ds\\ &= -\int_{x^2+y^2=\epsilon^2} \frac{x\mathbf{i}+y\mathbf{j}}{\epsilon^2}\cdot\left(-\frac{x\mathbf{i}+y\mathbf{j}}{\epsilon}\right)\;ds\\ &= \frac{1}{\epsilon}\int_{x^2+y^2=\epsilon^2} ds\\ &= 2\pi.\end{aligned}$$

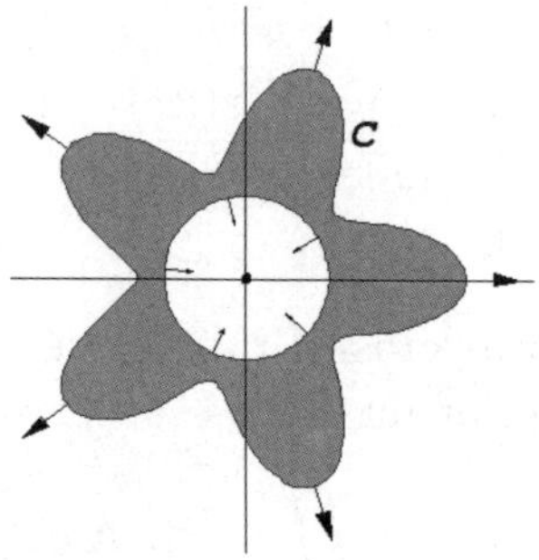

In summary, for any simple closed curve $C \subset \mathbb{R}^2 - \{O\}$,

$$\int_C \mathbf{A}\cdot\mathbf{n}\;ds = \begin{cases} 0 & (O \in \operatorname{ext} C)\\ 2\pi & (O \in \operatorname{int} C).\end{cases}$$

2.1.1 Example

For a point P in the plane, let's call the vector field

$$\mathbf{A}_P(X) := \frac{X-P}{|X-P|^2}$$

"the potential due to a point P at X".[17] Here

$$\operatorname{div}\mathbf{A}_P = 0$$

and hence for a simple closed curve $C \subset \mathbb{R}^2 - \{P\}$,

$$\int_C \mathbf{A}_P(X)\cdot\mathbf{n}\,ds = \begin{cases} 2\pi & (P \in \operatorname{int} C)\\ 0 & (P \in \operatorname{ext} C).\end{cases}$$

[17] Like the magnetic force which is generated by electric current of an electric wire through P perpendicular to the plane

2.1.2 Basic Exercise (Two-dimensional Gauss' Law)

For distinct points $P_1, \ldots, P_k$ in the plane and real numbers $q_1, \ldots, q_k$, let

$$\mathbf{E} := q_1 \mathbf{A}_{P_1} + \cdots + q_k \mathbf{A}_{P_k}.$$

For a simple closed curve $C \subset \mathbb{R}^2 - \{P_1, \ldots, P_k\}$, evaluate[18]

$$\frac{1}{2\pi} \int_C \mathbf{E} \cdot \mathbf{n}\, ds.$$

2.1.3 Example

Let

$$\mathbf{F}(X) := \int_{|P|=r_0} \mathbf{A}_P(X)\, ds(P)$$

be "the potential due to the whole circle $|P| = r_0$ at X".

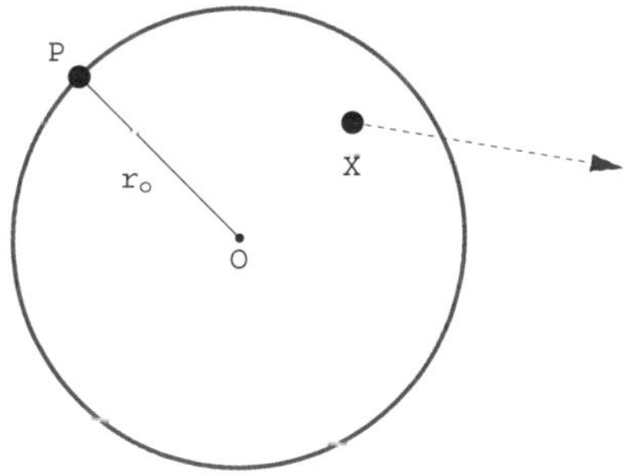

This integral $\mathbf{F}(X)$ is a vector-valued integral whose components are obtained from the integrals of component functions of $\mathbf{A}_P(X)$. In other words,

$$\mathbf{F}(X) \cdot \mathbf{i} = \int_{|P|=r_0} \mathbf{A}_P(X) \cdot \mathbf{i}\, ds(P), \qquad \mathbf{F}(X) \cdot \mathbf{j} = \int_{|P|=r_0} \mathbf{A}_P(X) \cdot \mathbf{j}\, ds(P).$$

The divergence of this vector field $\mathbf{F}(X)$ is zero:[19]

$$\operatorname{div} \mathbf{F}(X) = \int_{|P|=r_0} \operatorname{div} \mathbf{A}_P(X)\, ds(P) = 0.$$

[18] If $q_1 = \cdots = q_k = 1$, it is the number of points among $P_1, \ldots, P_k$ inside the closed curve. In general, it is "the total electric charge" inside the closed curve.

[19] From Leibniz's "rule for differentiation of integrals" and $\operatorname{div} \mathbf{A}_P(X) = 0$.

The vector field $\mathbf{F}(X)$ is defined for points which is not on the circle $|X| = r_0$. Due to the symmetry of the given circumstance, the vector field $\mathbf{F}(X)$ is proportional to the position vector field $\mathbf{r}(X) = X$ and so there exists a function $h(r)$ such that

$$\mathbf{F}(X) = h(r)\mathbf{r} \qquad (r = |X|). \tag{15.1}$$

Now let's show that

$$\mathbf{F}(X) = \begin{cases} 0 & (|X| < r_0) \\ (2\pi r_0)\mathbf{A}_O(X) & (|X| > r_0). \end{cases}$$

In other words, there occurs the same phenomenon as the case where there is no force acting inside of $|X| = r_0$ and the source of force is concentrated at the center outside this circle.

Case I: Consider the case where $|X| = r_1 < r_0$.
Since the divergence of the vector field $\mathbf{F}$ is zero,

$$\int_{|X|=r_1} \mathbf{F}(X) \cdot \mathbf{n}\, ds = \iint_{|X|\le r_1} \operatorname{div} \mathbf{F}(X)\, dV_2. = 0$$

On the other hand, from (15.1) we get

$$\int_{|X|=r_1} \mathbf{F}(X) \cdot \mathbf{n}\, ds = \int_{|X|=r_1} h(r)\mathbf{r} \cdot \mathbf{n}\, ds = h(r_1)r_1 \cdot 2\pi r_1$$

and thus

$$h(r_1)r_1 \cdot 2\pi r_1 = 0.$$

Therefore $h(r_1) = 0$ for $0 < r_1 < r_0$. From this, we get

$$\mathbf{F}(X) = h(r)\mathbf{r} = 0 \qquad (|X| < r_0).$$

Case II: Now consider the case where $|X| = r > r_0$. For convenience, we may assume $X = (r, 0)$.[20] We need to prove

$$\mathbf{F}(r, 0) \cdot \mathbf{i} = 2\pi r_0 \cdot \frac{1}{r}.$$

[20] By taking proper coordinates, we can make X lie on the x-axis.

$$\begin{aligned}
\mathbf{F}(r,0)\cdot\mathbf{i} &= \int_{|P|=r_0} \mathbf{A}_P(r,0)\cdot\mathbf{i}\,ds \\
&= \int_{-\pi}^{\pi} \frac{(r,0)-r_0(\cos\theta,\sin\theta)}{|(r,0)-r_0(\cos\theta,\sin\theta)|^2}\cdot\mathbf{i}\cdot r_0\,d\theta \\
&= r_0\int_{-\pi}^{\pi} \frac{r-r_0\cos\theta}{r^2+r_0^2-2rr_0\cos\theta}\,d\theta \\
&= 2r_0\int_{0}^{\pi} \frac{r-r_0\cos\theta}{r^2+r_0^2-2rr_0\cos\theta}\,d\theta \\
&\qquad \left(t:=\tan\frac{\theta}{2},\quad dt=\frac{1+t^2}{2}\,d\theta,\quad \cos\theta=\frac{1-t^2}{1+t^2}\right) \\
&= 2r_0\int_0^{\infty} \frac{r-r_0\frac{1-t^2}{1+t^2}}{r^2+r_0^2-2rr_0\frac{1-t^2}{1+t^2}}\,\frac{2\,dt}{1+t^2} \\
&= 4r_0\int_0^{\infty} \frac{r(1+t^2)-r_0(1-t^2)}{(r^2+r_0^2)(1+t^2)-2rr_0(1-t^2)}\,\frac{dt}{1+t^2} \\
&= 4r_0\int_0^{\infty} \frac{(r-r_0)+(r+r_0)t^2}{(r-r_0)^2+(r+r_0)^2t^2}\,\frac{dt}{1+t^2} \\
&\qquad \Bigg(\textbf{Integration by parts}^{21}\textbf{:} \\
&\qquad a^2\neq b^2 \;\Rightarrow\; \frac{a+bT}{(a^2+b^2T)(1+T)} = \frac{\frac{ab}{a+b}}{a^2+b^2T}+\frac{\frac{1}{a+b}}{1+T}\Bigg) \\
&= 4r_0\int_0^{\infty}\left(\frac{\frac{(r-r_0)(r+r_0)}{2r}}{(r-r_0)^2+(r+r_0)^2t^2}+\frac{\frac{1}{2r}}{1+t^2}\right)dt \\
&= 4r_0\cdot\frac{1}{2r}\int_0^{\infty}\left(\frac{(r-r_0)(r+r_0)}{(r-r_0)^2+(r+r_0)^2t^2}+\frac{1}{1+t^2}\right)dt \\
&\qquad \left(\Leftarrow \text{ Use} \int\frac{dt}{a^2+b^2t^2}=\frac{1}{ab}\arctan\left(\frac{b}{a}t\right)\right) \\
&= 2r_0\cdot\frac{1}{r}\left(\left[\arctan\frac{r+r_0}{r-r_0}\,t\right]_0^{\infty}+\left[\arctan t\right]_0^{\infty}\right) \\
&= 2r_0\cdot\frac{1}{r}\left(\frac{\pi}{2}+\frac{\pi}{2}\right) \\
&= 2\pi r_0\cdot\frac{1}{r}.
\end{aligned}$$

So we are done.

---

[21] As for relative primes p and q there exist integers a, b such that $\frac{1}{pq}=\frac{a}{p}+\frac{b}{q}$, if polynomials $p(x)$ and $q(x)$ are relatively prime polynomials, then there exist polynomials

When $|X| > r_0$, we can use the **inversion** to give the more elegant proof. First, let

$$X_* := \frac{r_0^2}{|X|^2} X$$

be the "inversion" of a point X with respect to the circle $|P| = r_0$. Then for any point P on the circle, $\triangle OXP$ and $\triangle OPX_*$ are similar with the similar ratio $|X| : r_0$.

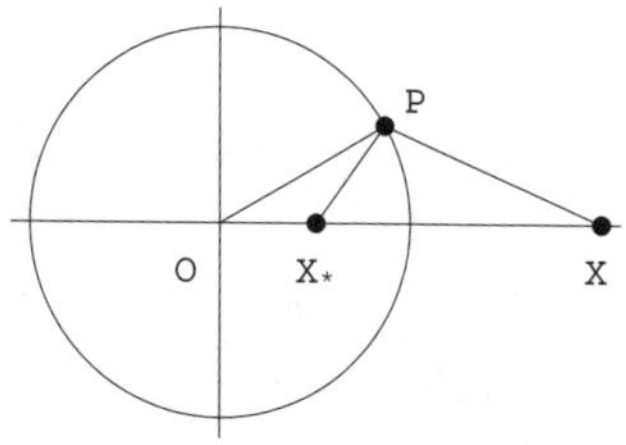

Therefore, if $|P| = r_0$, $\mathbf{u}(X) = X/|X|$ and $\angle OXP = \theta = \angle OPX_*$, then we have

$$\mathbf{A}_P(X) \cdot \mathbf{u}(X) = \frac{\cos\theta}{|X-P|} = \frac{r_0}{|X|}\frac{\cos\theta}{|P-X_*|} = \frac{r_0}{|X|}\mathbf{A}_{X_*}(P) \cdot \mathbf{u}(P).$$

So by the divergence theorem, we get

$$\begin{aligned}
&\mathbf{F}(X) \cdot \mathbf{u}(X) \\
= &\left(\int_{|P|=r_0} \mathbf{A}_P(X)\, ds(P)\right) \cdot \mathbf{u}(X) \\
= &\int_{|P|=r_0} \mathbf{A}_P(X) \cdot \mathbf{u}(X)\, ds(P) \\
= &\int_{|P|=r_0} \frac{r_0}{|X|}\mathbf{A}_{X_*}(P) \cdot \mathbf{u}(P)\, ds(P) \\
= &\frac{r_0}{|X|}\int_{|P-X_*|=\epsilon} \mathbf{A}_{X_*}(P) \cdot \mathbf{n}(P)\, ds(P) \qquad \left(\mathbf{n}(P) := \frac{P-X_*}{|P-X_*|}\right) \\
= &\frac{r_0}{|X|} \cdot 2\pi.
\end{aligned}$$

Therefore

$$\mathbf{F}(X) = (\mathbf{F}(X) \cdot \mathbf{u}(X))\mathbf{u}(X) = \frac{2\pi r_0}{|X|}\mathbf{u}(X) = 2\pi r_0 \mathbf{A}_O(X).$$

$a(x),\ b(x)$ such that $\dfrac{1}{p(x)q(x)} = \dfrac{a(x)}{p(x)} + \dfrac{b(x)}{q(x)}$.

Exercise: Chapter 15 Section 2

1. Suppose that the divergence of a plane vector field $\mathbf{F}(x, y)$ is zero. And suppose that $\mathbf{F}$ is defined on the boundary ∂D of some region D and

$$\int_{\partial D} \mathbf{F} \cdot \mathbf{n}\, ds \neq 0.$$

Then prove that there exists a point in D at which $\mathbf{F}$ is not well defined.

2. If the gradient vector field of a function $h(x, y)$ defined on the plane except at the origin is proportional to the angle element vector field $\mathbf{a}(x, y)$, then show that

$$\mathbf{F}(x, y) := h(x, y) \left(\frac{x}{x^2 + y^2}\, \mathbf{i} + \frac{y}{x^2 + y^2}\, \mathbf{j} \right)$$

is divergence free.[22]

3. For a point P in the plane, let

$$\mathbf{A}_P(X) := \frac{X - P}{|X - P|^2} = \text{grad } \log |X - P|.$$

For a positive number r_0, consider the vector field

$$\mathbf{F}(X) := \int_{|P| = r_0} \mathbf{A}_P(X)\, ds.$$

Show that the function

$$\varphi(X) := \int_{|P| = r_0} \log |X - P|\, ds \qquad (|X| \neq r_0)$$

is a potential function of $\mathbf{F}$ and use this to show that

$$\varphi(X) = \begin{cases} 2\pi r_0 \log r_0 & (|X| < r_0) \\ 2\pi r_0 \log |X| & (|X| > r_0). \end{cases}$$

4. For a vector field $\mathbf{F}$ defined on an open set in the plane, show that if

$$\int_{\partial D} \mathbf{F} \cdot \mathbf{n}\, ds = 0$$

for any closed disk D contained in the domain, then $\operatorname{div} \mathbf{F} = 0$.

---

[22] It means that $\operatorname{div} \mathbf{F} = 0$.

3 Plane Vector Fields and Rotation

For a $\mathcal{C}^1$ vector field

$$\mathbf{F}(x, y) = P(x, y)\,\mathbf{i} + Q(x, y)\,\mathbf{j}$$

defined on an open set in the plane, the **rotation** of $\mathbf{F}$ at a point X in the domain is defined by

$$\operatorname{rot}\mathbf{F}(X) := \frac{\partial Q}{\partial x}(X) - \frac{\partial P}{\partial y}(X)$$

Also $X \mapsto \operatorname{rot}\mathbf{F}(X)$ is called " **rotation (function)** of $\mathbf{F}$".

3.0.1 Example

For the position vector field

$$\mathbf{r}(x, y) = x\,\mathbf{i} + y\,\mathbf{j},$$

we have

$$\operatorname{rot}\mathbf{r} = \frac{\partial y}{\partial x} - \frac{\partial x}{\partial y} = 0$$

and so its rotation is identically zero.

3.0.2 Example

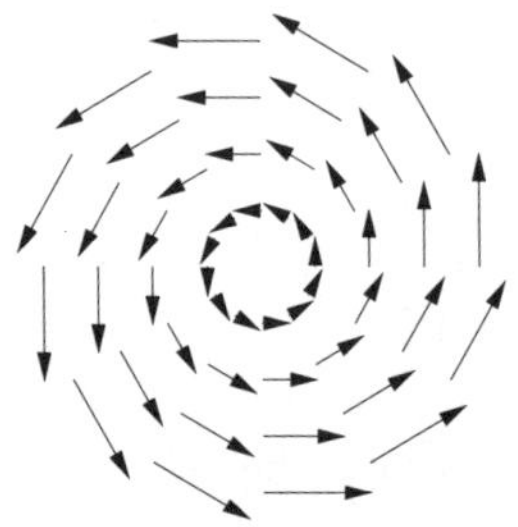

"angular speed"[23] ω

For a real number ω, the rotation of the vector field

$$\mathbf{F}(x, y) := \omega(-y\,\mathbf{i} + x\,\mathbf{j})$$

is

$$\operatorname{rot}\mathbf{F}(x, y) = \omega\left(\frac{\partial x}{\partial x} - \frac{\partial(-y)}{\partial y}\right) = 2\omega.$$

In the plane, if we consider the rotation with

$$X(t) = r(\cos\omega t, \sin\omega t),$$

[23]The **angular velocity** of this rotation is

$$\boldsymbol{\omega} := \omega\,\mathbf{k}.$$

The angular velocity is a vector of which the direction is equal to the direction of the rotational axis and the norm is equal to the **angular speed**.

we get

$$X'(t) = \omega r(-\sin \omega t, \cos \omega t) = \mathbf{F}(X(t))$$

and thus rot $\mathbf{F}$ is twice the angular speed of rotation described by $\mathbf{F}$.[24]

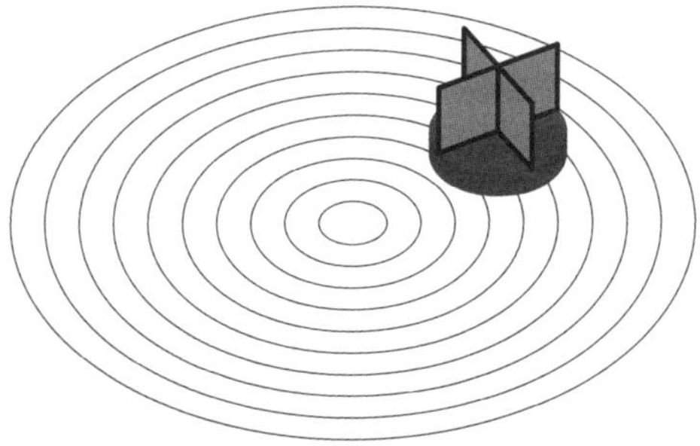

3.0.3 Example

The rotation of the angle element vector field

$$\mathbf{a}(x, y) = -\frac{y}{x^2 + y^2}\,\mathbf{i} + \frac{x}{x^2 + y^2}\,\mathbf{j}$$

is identically zero:

$$\begin{aligned} \text{rot}\,\mathbf{a} &= \frac{\partial}{\partial x}\left(\frac{x}{x^2+y^2}\right) - \frac{\partial}{\partial y}\left(-\frac{y}{x^2+y^2}\right) \\ &= \frac{(x^2+y^2) - x(2x)}{(x^2+y^2)^2} + \frac{(x^2+y^2) - y(2y)}{(x^2+y^2)^2} \\ &= 0. \end{aligned}$$

3.1 Meaning of Rotation

For a plane vector field

$$\mathbf{F}(x, y) = P(x, y)\,\mathbf{i} + Q(x, y)\,\mathbf{j},$$

its rotation

$$\text{rot}\,\mathbf{F} = \frac{\partial Q}{\partial x} - \frac{\partial P}{\partial y}$$

has the following meaning.[25]

[24]The rotation measure not the **revolution** around the axis but the **rotation**.

[25]The **meaning** could be various like in the story "blind men and an elephant." In the above situation, the vector field is thought of as "force."

For convenience, assume that $\mathbf{F}$ is defined near the origin O. Consider a small circle near the origin

$$C_r(t) := r(\cos t, \sin t), \qquad 0 \le t \le 2\pi,$$

and let the unit tangent vector field of this circle be

$$\mathbf{t}(t) := (-\sin t, \cos t).$$

The rotational force of a point on C_r rotating around the origin by the vector field $\mathbf{F}$ is[26]

$$\begin{aligned}
& r\mathbf{F}(C_r(t)) \cdot \mathbf{t}(t) \\
= \;& r\big[\cos t\; Q(r\cos t, r\sin t) - \sin t\; P(r\cos t, r\sin t)\big] \\
= \;& r\big[\cos t\{Q(O) + r\cos t\; Q_x(O) + r\sin t\; Q_y(O) + o(r)\} \\
& -\sin t\{P(O) + r\cos t\; P_x(O) + r\sin t\; P_y(O) + o(r)\}\big] \\
= \;& r\big[\cos t\; Q(O) + r\cos^2 t\; Q_x(O) + r\cos t\sin t\; Q_y(O) \\
& -\sin t\; P(O) - r\sin t\cos t\; P_x(O) - r\sin^2 t\; P_y(O) + o(r)\big].
\end{aligned}$$

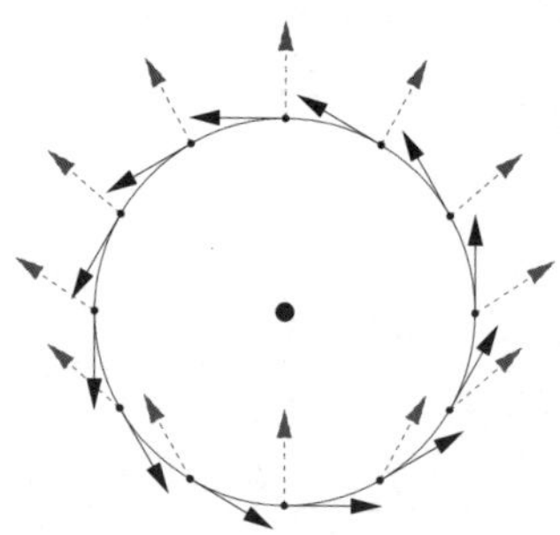

Therefore the rotational force acting on the whole circle C_r is

$$\begin{aligned}
& \int_{C_r} r\mathbf{F}(C_r(t)) \cdot \mathbf{t}(t)\, ds \\
= \;& \int_0^{2\pi} r\big[\cos t\, Q(O) + r\cos^2 t\, Q_x(O) + r\cos t\sin t\, Q_y(O) \\
& -\sin t\, P(O) - r\sin t\cos t\, P_x(O) - r\sin^2 t\, P_y(O) + o(r)\big]\, r\, dt \\
= \;& r^2 \int_0^{2\pi} \big(r\cos^2 t\, Q_x(O) - r\sin^2 t\, P_y(O) + o(r)\big)\; dt \\
= \;& r^3\pi\,(Q_x(O) - P_y(O)) + o(r^3) \\
= \;& r^3\pi\, \mathrm{rot}\, \mathbf{F}(O) + o(r^3).
\end{aligned}$$

[26]We let $Q_x = \dfrac{\partial Q}{\partial x}$, $Q_y = \dfrac{\partial Q}{\partial y}$ and so on. This force is equal to $C_r(t) \times \mathbf{F} \cdot \mathbf{k}$. (We saw it in Chapter 8.)

From this we get[27]

$$\operatorname{rot}\mathbf{F}(O)=\lim_{r\to 0}\frac{1}{\pi r^2}\int_{C_r}\mathbf{F}\cdot\mathbf{t}\,ds.$$

This is the reason why we call rot $\mathbf{F}$ the rotation.

3.2 Irrotational Fields

From now on, If a vector field has the identically zero rotation, it is called an **irrotational vector field**. The following theorem comes immediately from the Commutativity of Partial Derivatives.

Theorem 3.2.1 *If a C^1 vector field $\mathbf{F}(x,y)$ defined on an open set in the plane has a potential function, then $\mathbf{F}(x,y)$ is irrotational.*

Proof. If the vector field

$$\mathbf{F}(x,y)=P(x,y)\mathbf{i}+Q(x,y)\mathbf{j}$$

is the gradient vector field of a function φ, then

$$P=\frac{\partial\varphi}{\partial x},\qquad Q=\frac{\partial\varphi}{\partial y}$$

and thus

$$\operatorname{rot}\mathbf{F}=\frac{\partial Q}{\partial x}-\frac{\partial P}{\partial y}=\frac{\partial}{\partial x}\left(\frac{\partial\varphi}{\partial y}\right)-\frac{\partial}{\partial y}\left(\frac{\partial\varphi}{\partial x}\right)=0.$$

□

The converse of this theorem is true locally, which is called Poincaré's lemma. (Refer to Chapter 13.)

[27]Such a principle holds not only for the origin but also for any point in the domain of $\mathbf{F}$. If we study Green's theorem later, we will get the more general formula for more general curves.

4 Boundary and Orientation

4.1 Plane and Orientation

Consider the rotational motion around a point in the plane. A counter-clockwise rotation is called **positively-oriented** and a clockwise rotation[28] is called **negatively-oriented**.

When two linearly independent vectors $\mathbf{v}, \mathbf{w}$ are given in the plane, we can rotate $\mathbf{v}$ through an angle less than 180° in the positive direction or in the negative direction to make it point the same direction as $\mathbf{w}$. If the direction in which we rotate $\mathbf{v}$ is positive, then the pair $(\mathbf{v}, \mathbf{w})$ is called "**positively-oriented**" and the orientation of $(\mathbf{v}, \mathbf{w})$ is called the "**standard orientation**".

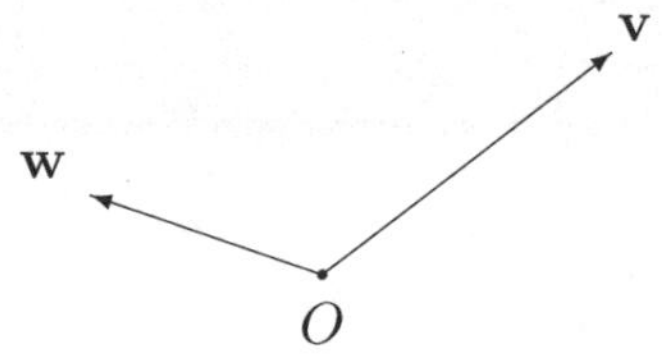

For example, the standard basis $(\mathbf{i}, \mathbf{j})$ is positively-oriented.

Two linearly independent vectors $(\mathbf{v}, \mathbf{w})$ is positively-oriented if and only if[29]

$$\det(\mathbf{v}, \mathbf{w}) > 0.$$

The reason for this property is the following: First, let $\mathbf{v}^*$ be the vector obtained by rotating a plane vector $\mathbf{v}$ by 90° in the positive direction. That is, if $\mathbf{v} = (a, b)$, then $\mathbf{v}^* = (-b, a)$. Then for any plane vector $\mathbf{w}$, we know that[30]

$$\det(\mathbf{v}, \mathbf{w}) = \mathbf{v}^* \cdot \mathbf{w}.$$

Now if a unit vector $\mathbf{w}$ is obtained by rotating $\mathbf{v}$ in the positive direction, there exists $\theta \in (0, \pi)$ such that $\mathbf{w} = \mathbf{v}\cos\theta + \mathbf{v}^*\sin\theta$, and thus

$$\det(\mathbf{v}, \mathbf{w}) = \mathbf{v}^* \cdot (\mathbf{v}\cos\theta + \mathbf{v}^*\sin\theta) = \sin\theta > 0.$$

---

[28] The direction in which a shadow of a stick (by the sun) on the Northern Hemisphere moves

[29] The orientation of a pair of vectors can be "defined" by this "property".

[30] $\det\begin{pmatrix} a & c \\ b & d \end{pmatrix} = \begin{pmatrix} -b \\ a \end{pmatrix} \cdot \begin{pmatrix} c \\ d \end{pmatrix}$.

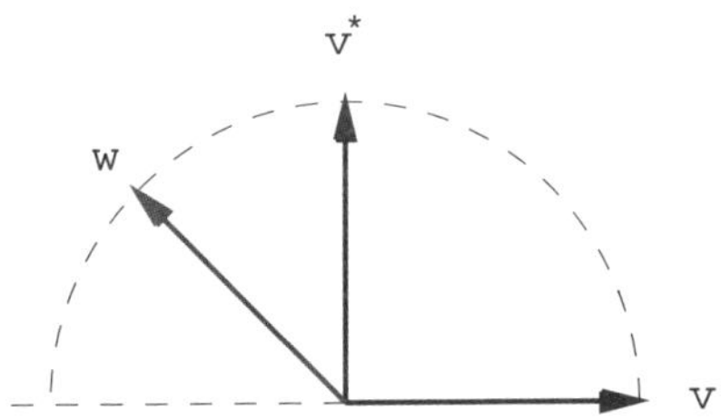

If $\mathbf{v}, \mathbf{w}$ are not unit vectors, we may consider the unit vectors of the same directions as $\mathbf{v}, \mathbf{w}$.

If we obtain $\mathbf{w}$ by rotating $\mathbf{v}$ in the negative direction (through the angle less than 180°), $(\mathbf{v}, \mathbf{w})$ is called **negatively-oriented**. The pair $(\mathbf{v}, \mathbf{w})$ is negatively-oriented if and only if $(\mathbf{w}, \mathbf{v})$ is positively-oriented.

4.1.1 Basic Exercise

Examine whether the following pair of vectors $(\mathbf{v}, \mathbf{w})$ is positively-oriented or negatively-oriented.

(1) $\mathbf{v} = (1, 1), \quad \mathbf{w} = (-1, 1)$ (2) $\mathbf{v} = (1, 2), \quad \mathbf{w} = (-1, -3)$

4.2 Orientation of Boundary

Suppose that the boundary ∂D of D consists of regular curves. Then the direction of boundary ∂D is defined as follows.

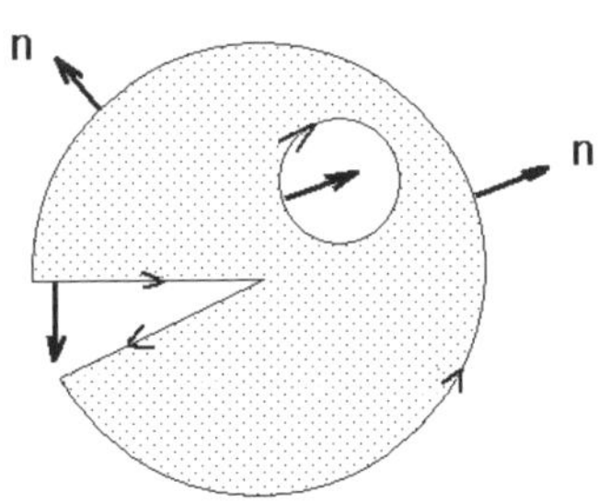

First, at each point on the boundary consider a **unit normal vector**, that is, a unit vector perpendicular to the boundary. There are two such vectors, the inward and outward unit normal vectors. We will denote the outward normal vector by $\mathbf{n}$, and call it the **standard unit normal vector** on the boundary of the region.

Now, if we rotate the standard unit normal vector **n** by 90° at each point on the boundary,[31] we get a vector **t** tangent to the boundary. The direction of this **standard unit tangent vector t** is called the **orientation of the boundary**.

> If we walk along the boundary in the direction of the orientation,[32] the interior is always on the left.

If the boundary of the region is not a regular curve but a piecewise regular curve, there is a **discrete** set of points where the tangent direction is not defined, and we won't think of the normal direction at such **singular points**. Discrete singular points don't have any effect on line integrals.

[31] Unless stated otherwise, we always rotate a vector in the positive direction (counterclockwise).

[32] No hand-standing, please.

5 Green's Theorem

Green's theorem[33] says that the integral of the rotation of a vector field on a region[34] is equal to the circulation of the vector field along the boundary.

Theorem 5.0.1 (Green's Theorem) *Let* $\mathbf{F}(x,y)$ *be a* C^1 *vector field defined on a region* D *in* $\mathbb{R}^2$. *Then*[35]

$$\int_{\partial D} \mathbf{F} \cdot d\mathbf{s} = \iint_D \operatorname{rot} \mathbf{F} \, dV_2$$

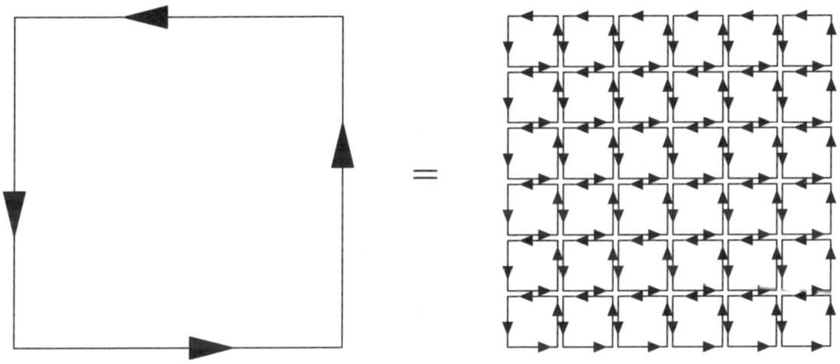

[33]George Green (1793–1841). English mathematician and physicist. He taught himself mathematics, but in 1828 he published "An essay on the application of mathematical analysis to the theories of electricity and magnetism" on which he introduced an important concept, later called **Green function**, in the boundary value problems of partial differential equations and the theory of integral transform. The term "potential" was first used in this paper. (In 1744, Euler called the strain energy of a spring *vis potentialis.*) Green's work was hardly known until W. Thomson (who later became Lord Kelvin) located a copy of Green's essay and realized its significance, and so had it reprinted in 1850 [Birkhoff].

[34]Still we assume that it is a bounded closed region and its boundary is a piecewise regular curve.

[35]When we integrate a vector field along a curve, usually we integrate the function given by the inner product of the vector field and the unit tangent vector of the curve. If a given curve is a **plane curve**, when we integrate a plane vector field along this curve, we can integrate the function given by the inner product of the vector field and the unit normal vector. Integrating the function given by the inner product with the tangent vector is an **intrinsic** concept, independent of how the given curve lies in the space and the coordinate system. However, integrating the function given by the inner product with the normal vector is a **relative** concept, not just depending on the curve but depending on how the given curve lies in the plane. Sometimes an intrinsic concept is important and sometimes a relative one is useful.

If we express a vector field $\mathbf{F}(x, y)$ as a differential form

$$\mathbf{F}(x, y) = P(x, y)\, dx + Q(x, y)\, dy,$$

Green's theorem is expressed as

$$\int_{\partial D} P(x, y)\, dx + Q(x, y)\, dy = \iint_D \left(\frac{\partial Q}{\partial x} - \frac{\partial P}{\partial y} \right) dxdy$$

Proof. Let

$$\mathbf{F} = P\,\mathbf{i} + Q\,\mathbf{j},$$

be the given vector field, and $\mathbf{n}$ and $\mathbf{t}$ be the standard unit normal and tangent vector fields on the boundary ∂D of the region D, respectively. If we rotate the vector field

$$\mathbf{G} := Q\,\mathbf{i} - P\,\mathbf{j}$$

by 90° then we get $\mathbf{F}$, and if we rotate the normal vector field $\mathbf{n}$ by 90°, we get the tangent vector field $\mathbf{t}$. Therefore

$$\mathbf{F} \cdot \mathbf{t} = \mathbf{G} \cdot \mathbf{n}$$

and thus[36]

$$\begin{aligned} \int_{\partial D} \mathbf{F} \cdot ds &= \int_{\partial D} \mathbf{F} \cdot \mathbf{t}\, ds = \int_{\partial D} \mathbf{G} \cdot \mathbf{n}\, ds \\ &= \iint_D \operatorname{div} \mathbf{G}\, dV_2 = \iint_D \left(\frac{\partial Q}{\partial x} - \frac{\partial P}{\partial y} \right) dV_2 \\ &= \iint_D \operatorname{rot} \mathbf{F}\, dV_2. \end{aligned}$$

□

5.0.2 Example

Along the counterclockwise curve $C : \; x^2 + y^2 = 4$, evaluate the integral

$$\int_C \left(2y + \sqrt{9 + x^3} \right) dx + \left(5x + e^{\arctan y} \right) dy$$

[36] In the plane, the divergence theorem is equivalent to Green's theorem. There is a natural reason why we prove Green's theorem first and then derive the divergence theorem.

Solution. By Green's theorem, its value is equal to

$$\iint_{x^2+y^2\le 4} (5-2)\,dxdy = 3\cdot 4\pi = 12\pi.$$

Corollary 5.0.3 *The area of the region D in the plane is*

$$\frac{1}{2}\int_{\partial D} -ydx + xdy = -\int_{\partial D} ydx = \int_{\partial D} xdy.$$

Proof. By Green's theorem,

$$-\int_{\partial D} ydx = \iint_D dV_2 = \int_{\partial D} xdy$$

and we are done. □

5.0.4 Example

Use the above Corollary to find the area enclosed by the ellipse

$$D: \quad \frac{x^2}{a^2} + \frac{y^2}{b^2} \le 1. \qquad \text{(Here, } a, b > 0.)$$

Solution. If we parameterize the ellipse by

$$\partial D: \quad x = a\cos t, \quad y = b\sin t \qquad (0 \le t \le 2\pi)$$

we get[37]

$$\begin{aligned} \text{area}(D) &= \int_{\partial D} xdy = \int_0^{2\pi} (a\cos t)(b\cos t)\,dt \\ &= ab\int_0^{2\pi} \cos^2 t\,dt = \pi ab. \end{aligned}$$

[37] Since $\cos^2 t + \sin^2 t = 1$ and $\int_0^{2\pi} \cos^2 t\,dt = \int_0^{2\pi} \sin^2 t\,dt$, we get $\int_0^{2\pi} \cos^2 t\,dt = \pi$.

5.0.5 Basic Exercise

For a point P in the domain of a plane vector field $\mathbf{F}(x, y)$, show that

$$\operatorname{rot} \mathbf{F}(P) = \lim_{D \to P} \frac{1}{\text{area} D} \int_{\partial D} \mathbf{F} \cdot d\mathbf{s}.$$

(Here D is a small connected region containing P, and $D \to P$ means that

$$\sup\{|X - P| \mid X \in D\}$$

converges to zero.)

Exercise: Chapter 15 Section 5

1. If C is the unit circle $x^2+y^2=1$ oriented counterclockwise, evaluate the line integral

$$\int_C \left(e^{\arctan x}+(xy+1)e^{xy}\right)dx+\left(x^3+3xy^2+x^2e^{xy}\right)dy.$$

2. Use Green's theorem to evaluate the following integral.

(1) $\displaystyle\int_C (x+y^2)dx+(y+x^2)dy,$

C is the square with four vertices at $(\pm1,\pm1)$

(2) $\displaystyle\int_C (x^2+y^2)dx-2xy\,dy,$

C is the triangle enclosed by $x=0,\ y=0,\ x+y=1$

(3) $\displaystyle\int_C (y+e^x)dx+(2x^2+\cos y)dy,$

C is the triangle with vertices $(0,0)$, $(1,1)$, $(2,0)$

(4) $\displaystyle\int_C (x^2-y^2)dx+xy\,dy,$

C is the boundary of the region enclosed by $y=x$ and $y=x^2$

(5) $\displaystyle\int_C \left(-y^2+e^{e^x}\right)dx+(\arctan y)dy,$

C is the boundary of the region enclosed by $y=x^2$ and $x=y^2$

(6) $\displaystyle\int_C x^2\,dy-y^2\,dx,$

C is the curve given by $r=1+\cos\theta$ in polar coordinates

3. Use Green's theorem to find the area of the region enclosed by the x-axis and the cycloid

$$x=a(t-\sin t),\quad y=a(1-\cos t)\qquad (0\le t\le 2\pi).$$

4. Find the area of the region enclosed by curves

$$x(t)=\cos^3 t,\qquad y(t)=\sin^3 t\qquad (0\le t\le 2\pi).$$

5. Find the area of the region enclosed by the Folium of Descartes $x^3+y^3=3xy$ in the first quadrant.

6. Show that the center $(\bar{x},\bar{y})$ of the region D in the plane is given by

$$\bar{x}=\frac{1}{2A}\int_{\partial D} x^2\,dy,\qquad \bar{y}=-\frac{1}{2A}\int_{\partial D} y^2\,dx.$$

Here A denotes the area of D, and we assume that the boundary of D is piecewise C^1.

7. For a function $Q(x,y) := e^{-y^2} \arctan x$ defined on the region
$$D = \{(x,y) \in \mathbb{R}^2 \mid y \geq 0\},$$
show that
$$\int_{\partial D} Q(x,y)\, dy \neq \iint_D \frac{\partial Q}{\partial x}\, dx\, dy.$$
From this, we can see that Green's theorem is not true for an unbounded region.

8. For the angle element vector field $\mathbf{a}$, show that if the interior D of a simple closed plane curve C contains the origin, then[38]
$$\int_C \mathbf{a} \cdot d\mathbf{s} = 2\pi, \qquad \iint_D \operatorname{rot} \mathbf{a}\, dV_2 = 0.$$
And explain why this doesn't contradict to Green's theorem.

9. Find the rotation of the vector field consisting of real and imaginary parts of the following complex function.

   (1) $f(z) = z = x + yi$
   (2) $f(z) = z^2 = (x^2 - y^2) + 2xyi$
   (3) $f(z) = z^3 = (x^3 - 3xy^2) + (3x^2y - y^3)i$
   (4) $f(z) = e^z = e^x(\cos y + i \sin y)$

   Also show that the complex conjugate function[39] $\bar{f}(z)$ of the above function gives a vector field that is irrotational and incompressible.

[38]Here the orientation of $C = \partial D$ is the standard orientation.

[39]The **complex conjugate** of a complex number $z = x + yi$ is $\bar{z} = x - yi$.

6 Divergence and Change of Volume

We will explain that the divergence of a vector field indicates the expansion of volume of the region. For convenience, we will consider the case where the dimension is $n = 3$. When $n = 2$ the proof could be a little bit easier but the case of general dimension is essentially the same.

To understand the meaning of divergence, let's consider the following lemma.

Lemma 6.0.1 *If a square matrix*

$$A(t) = \begin{pmatrix} a_{11}(t) & a_{12}(t) & a_{13}(t) \\ a_{21}(t) & a_{22}(t) & a_{23}(t) \\ a_{31}(t) & a_{32}(t) & a_{33}(t) \end{pmatrix}$$

as a function of time t is the identity matrix at $t = 0$, then[40]

$$\left.\frac{d}{dt}\right|_0 \det A(t) = \operatorname{tr} \left.\frac{d}{dt}\right|_0 A(t)$$

Proof. Since $a_{ij}(0) = \delta_{ij}$, for a 3-permutation σ, we get

$$\begin{aligned}
&\left.\frac{d}{dt}\right|_0 \det A(t) \\
=\ &\left.\frac{d}{dt}\right|_0 \sum_\sigma (\operatorname{sgn}\sigma) a_{1\sigma(1)}(t)\, a_{2\sigma(2)}(t)\, a_{3\sigma(3)}(t) \\
=\ &\sum_\sigma (\operatorname{sgn}\sigma) \Big(a'_{1\sigma(1)}(0)\, a_{2\sigma(2)}(0)\, a_{3\sigma(3)}(0) + a_{1\sigma(1)}(0)\, a'_{2\sigma(2)}(0)\, a_{3\sigma(3)}(0) \\
&+ a_{1\sigma(1)}(0)\, a_{2\sigma(2)}(0)\, a'_{3\sigma(3)}(0) \Big) \\
=\ &a'_{11}(0) + a'_{22}(0) + a'_{33}(0) \\
=\ &\operatorname{tr} \left.\frac{d}{dt}\right|_0 A(t).
\end{aligned}$$

□

Now fix a point P in the domain of the vector field

$$\mathbf{F}(x, y, z) = f_1\,\mathbf{i} + f_2\,\mathbf{j} + f_3\,\mathbf{k},$$

[40] Here we assume that $a_{ij}(t)$ is differentiable. This lemma is true not only for 3×3 square matrix but for any $n \times n$ square matrix. If you feel uneasy with $n = 3$, it is easier to consider it when $n = 2$.

and let D be a small region containing P and D_t be the region onto which D flows in the direction of the vector field $\mathbf{F}$ in t second.

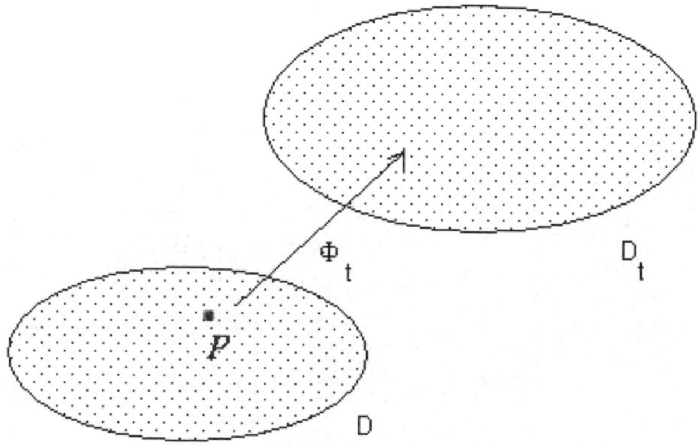

In other words, a point in D_t is the image of the map

$$\Phi_t : D \to D_t,$$

which satisfies the following condition:

$$\frac{d}{dt}\Phi_t(X) = \mathbf{F}\left(\Phi_t(X)\right), \qquad \Phi_0(X) = X.$$

Since Φ_0 is the identity map and its derivative[41] $\Phi_0'(P)$ is also the identity map,

$$\det \Phi_0'(P) = 1 > 0.$$

Therefore, by continuity, if t has very small absolute value, then

$$\det \Phi_t'(P) > 0.$$

By the inverse function theorem, Φ_t is a C^1 locally invertible map near P. Therefore by change of variables, we get

$$\text{vol}\,(D_t) = \iiint_{D_t} dx\,dy\,dz = \iiint_D \det \Phi_t'\,dx\,dy\,dz.$$

[41]Here the derivative means the derivative with respect to the space variable (x, y, z), that is, the Jacobian matrix of Φ_0, not the derivative with respect to the time variable t.

From the above lemma, the net rate of change of volume of D is

$$\begin{aligned}
\left.\frac{d}{dt}\right|_0 \mathrm{vol}\,(D_t) &= \iiint_D \left.\frac{d}{dt}\right|_0 \det \Phi_t'\, dx\, dy\, dz \\
&= \iiint_D \mathrm{tr}\left(\left.\frac{d}{dt}\right|_0 \Phi_t'\right) dx\, dy\, dz \\
&= \iiint_D \mathrm{tr}\left(\left(\left.\frac{d}{dt}\right|_0 \Phi_t\right)'\right) dx\, dy\, dz \\
&= \iiint_D \mathrm{tr}\,\mathbf{F}'\, dx\, dy\, dz
\end{aligned}$$

and so[42]

$$\left.\frac{d}{dt}\right|_0 \mathrm{vol}\,(D_t) = \iiint_D \mathrm{div}\,\mathbf{F}\, dx\, dy\, dz$$

If we apply the mean value theorem in this equation, there exists a point P^* in D such that[43]

$$\left.\frac{d}{dt}\right|_0 \mathrm{vol}\,(D_t) = \mathrm{vol}(D)\mathrm{div}\,\mathbf{F}(P^*).$$

Therefore we have

$$\mathrm{div}\,\mathbf{F}(P^*) = \left.\frac{d}{dt}\right|_0 \frac{\mathrm{vol}\,(D_t)}{\mathrm{vol}(D)}.$$

If D converges to P, we get

$$\mathrm{div}\,\mathbf{F}(P) = \lim_{D\to P} \left.\frac{d}{dt}\right|_0 \frac{\mathrm{vol}\,(D_t)}{\mathrm{vol}(D)}$$

Therefore, the divergence $\mathrm{div}\,\mathbf{F}(P)$ of a vector field $\mathbf{F}$ at a point P indicates the **net rate of change of volume** near P.

---

[42]The divergence of a vector field $\mathbf{F}$ is the trace of the derivative of the vector-valued function $\mathbf{F}$ (that is, the **Jacobian matrix** of $\mathbf{F}$):

$$\mathbf{F}' = \begin{pmatrix} \dfrac{\partial f_1}{\partial x_1} & \cdots & \dfrac{\partial f_1}{\partial x_n} \\ \vdots & \ddots & \vdots \\ \dfrac{\partial f_n}{\partial x_1} & \cdots & \dfrac{\partial f_n}{\partial x_n} \end{pmatrix}, \qquad \mathrm{div}\,\mathbf{F} = \mathrm{tr}\,\mathbf{F}'.$$

[43]For convenience, we assume that the region D is connected.

Corollary 6.0.2 *If we let D be a small region containing P and D_t be the region onto which D flows in the direction of the vector field* $\mathbf{F}$ *in t second,*

$$\frac{\mathrm{vol}\,(D_t)}{\mathrm{vol}\,(D)} \approx 1 + t\,\mathrm{div}\,\mathbf{F}(P) \qquad (|t| \ll 1).$$

6.0.3 Basic Exercise

Show that if the divergence $\mathrm{div}\,\mathbf{F}$ of a vector field $\mathbf{F}$ is constant, then

$$\mathrm{div}\,\mathbf{F} = \left.\frac{d}{dt}\right|_0 \frac{\mathrm{vol}(D_t)}{\mathrm{vol}(D)}$$

for any region D.

Chapter 15 Advanced Exercise

1. For a function $h(x,y)$ defined on $\mathbb{R}^2 - \{O\}$, let

$$\mathbf{F}(x,y) = h(x,y)\mathbf{A}(x,y), \qquad \mathbf{A}(x,y) := \frac{(x,y)}{x^2+y^2}.$$

Show that if $\operatorname{div}\mathbf{F} = 0$ then $h(x,y)$ is constant.

2. For two functions $P(x,y)$, $Q(x,y)$ defined on an open set in $\mathbb{R}^2$ and two vectors $\mathbf{v}$, $\mathbf{w} \in \mathbb{R}^2$, let the vector field $\mathbf{F}$ be

$$\mathbf{F}(x,y) = P(x,y)\,\mathbf{v} + Q(x,y)\,\mathbf{w}.$$

Show that

$$\operatorname{div}\mathbf{F} = D_{\mathbf{v}}P + D_{\mathbf{w}}Q.$$

3. For a vector field

$$\mathbf{F}(x,y) = P(x,y)\,\mathbf{i} + Q(x,y)\,\mathbf{j}$$

defined on an open set U in $\mathbb{R}^2$, show the following.

(i) If $\displaystyle\int_C \mathbf{F}\cdot d\mathbf{s} = 0$ for any simple closed curve $C \subset U$, then $\dfrac{\partial P}{\partial y} = \dfrac{\partial Q}{\partial x}$.

(ii) If $\displaystyle\int_C \mathbf{F}\cdot\mathbf{n}\;ds = 0$ for any simple closed curve $C \subset U$, then $\dfrac{\partial P}{\partial x} = -\dfrac{\partial Q}{\partial y}$.

(iii) If $\displaystyle\int_C \mathbf{F}\cdot d\mathbf{s} = 0 = \int_C \mathbf{F}\cdot\mathbf{n}\,ds$ for any simple closed curve $C \subset U$, then

$$\left(\frac{\partial^2}{\partial x^2} + \frac{\partial^2}{\partial y^2}\right)P = 0, \qquad \left(\frac{\partial^2}{\partial x^2} + \frac{\partial^2}{\partial y^2}\right)Q = 0.$$

4. Find an example of a plane vector field $\mathbf{F}(x,y)$ such that

$$\operatorname{div}\mathbf{F} = 0, \qquad \operatorname{rot}\mathbf{F} = 0.$$

5. For a point (x,y) in $\mathbb{R}^2$, let $(x,y)^*$ be the point obtained by rotating (x,y) by $90°$ around the origin:

$$(x,y)^* := (-y,x).$$

Similarly, for a plane vector field $\mathbf{F}(x,y) = f_1(x,y)\,\mathbf{i} + f_2(x,y)\,\mathbf{j}$, let $\mathbf{F}^*$ be the vector field obtained by rotating $\mathbf{F}$ by $90°$:

$$\mathbf{F}^* = -f_2\,\mathbf{i} + f_1\,\mathbf{j}.$$

For a function $f(x,y)$ and a vector field $\mathbf{F}(x,y)$, show that

$$\operatorname{rot}(f\mathbf{F}) = -\operatorname{grad} f\cdot\mathbf{F}^* + f\cdot\operatorname{rot}\mathbf{F}.$$

6. For the angle element vector field **a** and a $\mathcal{C}^1$ function $f(r)$, show that if the vector field $\mathbf{F}(x,y) = f(\sqrt{x^2+y^2})\,\mathbf{a}(x,y)$ is irrotational, then f is constant.

7. For a vector $(t_1,\ldots,t_n)$ and an $n\times n$ orthogonal matrix[44] (a_{ij}), assume that in $\mathbb{R}^n$ is given a new coordinate system

$$y_i = \sum_{j=1}^{n} a_{ij}x_j + t_i \qquad (i = 1,\ldots,n).$$

Show that the divergence of a vector field **F** is independent of coordinate system.

[44] A square matrix A is called an **orthogonal matrix** if $AA^{\mathrm{t}} = A^{\mathrm{t}}A = I$. If we think of a matrix as a linear transformation on $\mathbb{R}^n$, then it is a linear transformation which preserves the norm of vectors.

Part VI

Surface Integrals, Divergence Theorem and Stokes' Theorem

Chapter 16

Surfaces and Surface Integrals

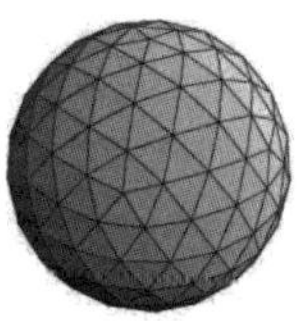

When I am working on a problem, I never think about beauty but when I have finished, if the solution is not beautiful, I know that it is wrong.

— Buckminster Fuller (1895–1983), American architect

In this chapter, we introduce the concept of surface, and study the way to find the surface area. Also we study how to integrate functions and vector fields over a surface. we integrate a vector field sometimes along a curve, but sometimes over a surface. For integral, we need both a "function to integrate" and a "region over which we integrate". We integrate a "one-dimensional function" over a one-dimensional region like an interval or a curve, a "two-dimensional function" over a two-dimensional region like a surface, and in general an "n-dimensional function" over an n-dimensional region. Here a "one-dimensional function" means a vector field(or 1-form).

As there is one-to-one correspondence between lines through the origin and planes perpendicular to that line in the three-dimensional space, we may think of "vector fields" as "two-dimensional functions", and thus we can integrate a vector field over a surface and the value of this integral is called the **flux** of the vector field. When we integrate a vector field, we should always consider the orientation.

Over a 0-dimensional region like a point, we integrate a "0-dimensional function", which is a real-valued function in the usual sense. If we integrate a function $f(X)$ over a point P, we get $f(P)$.

1 Surfaces

1.1 Parametric Surfaces

We can define a surface in various ways. Usually we deal with parametric surfaces. A (parametric) **surface** in the space $\mathbb{R}^3$ is a continuous map defined on a region $D \subset \mathbb{R}^2$

$$X : D \to \mathbb{R}^3, \quad (u, v) \mapsto X(u, v) \in \mathbb{R}^3.$$

1.1.1 Example

The following maps

$$\begin{aligned} X(u,v) &= (\cos u, \sin u, v) \\ Y(r,\theta) &= (r\cos\theta, r\sin\theta, r) \\ S(\varphi,\theta) &= (\cos\theta\sin\varphi, \sin\theta\sin\varphi, \cos\varphi) \end{aligned}$$

are parametrizations of the **cylinder**, **circular cone** and **sphere**, respectively.

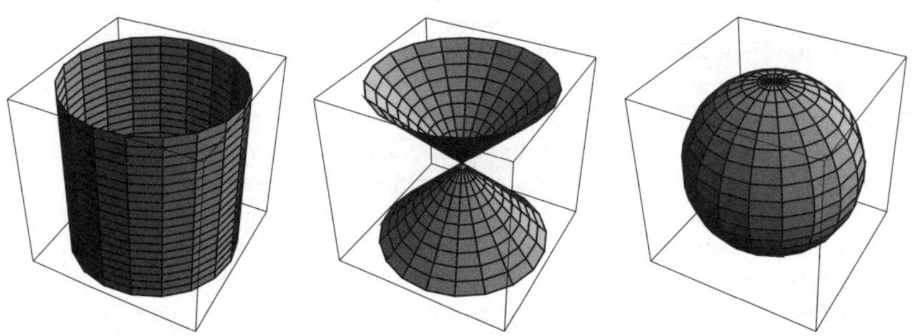

1.2 Graphs and Surfaces

We can easily get a surface from the graph of a function

$$z = f(x, y)$$

of two variables:

$$G_f(x, y) := (x, y, f(x, y)).$$

If $f(x, y) = e^{-(x^2+y^2)/3} \cos(x^2 + y^2)$, then the graph is as follows.

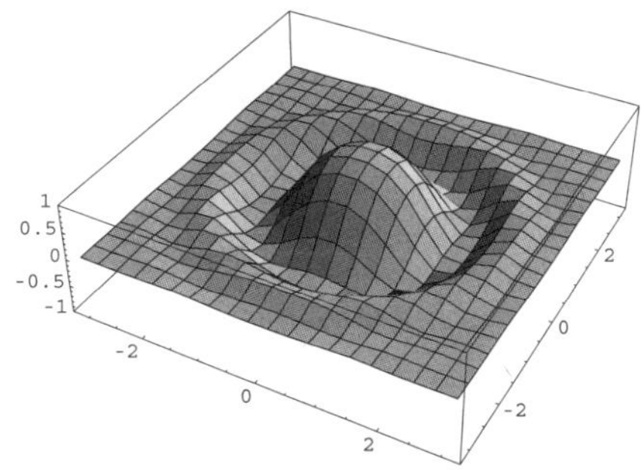

1.3 Level Surfaces

Often a surface is given by an equation.

Suppose that there is $\mathcal{C}^1$ function $f(x, y, z)$ so that a point $P = (x_0, y_0, z_0)$ satisfies an equation

$$f(x, y, z) = 0. \tag{16.1}$$

If

$$\frac{\partial f}{\partial z}(P) \neq 0,$$

then by the implicit function theorem, points which satisfy the equation (16.1) can be expressed locally of the form

$$z = g(x, y).$$

Therefore near P the level surface

$$Z(f) := \{(x, y, z) \mid f(x, y, z) = 0\}$$

is a surface given as the graph of some function $z = g(x, y)$.

Similarly, by the implicit function theorem, if

$$\frac{\partial f}{\partial x}(P) \neq 0,$$

then near P the level surface $Z(f)$ can be expressed as a graph of a function

$$x = h(y, z).$$

And if $\dfrac{\partial f}{\partial y}(P) \neq 0$, then near P the level surface $Z(f)$ can be expressed as a graph of some function

$$y = k(x, z).$$

Therefore if

$$\operatorname{grad} f(P) \neq \mathbf{0}$$

then one of $\dfrac{\partial f}{\partial x}(P)$, $\dfrac{\partial f}{\partial y}(P)$, $\dfrac{\partial f}{\partial z}(P)$ is not zero and thus we get the following conclusion:

Theorem 1.3.1 *Suppose that $f : U \to \mathbb{R}$ is a C^1 function defined on a region U in the space. If* $\operatorname{grad} f(P) \neq 0$ *at a point P on the level surface*[1]

$$Z(f) = \{P \in U \mid f(P) = 0\},$$

then the level surface can be locally parameterized near P.

1.3.2 Basic Exercise

Parameterize the **hyperboloid of one sheet**[2]

$$x^2 + y^2 - z^2 = 1$$

and the **hyperboloid of two sheets**

$$x^2 - y^2 - z^2 = 1.$$

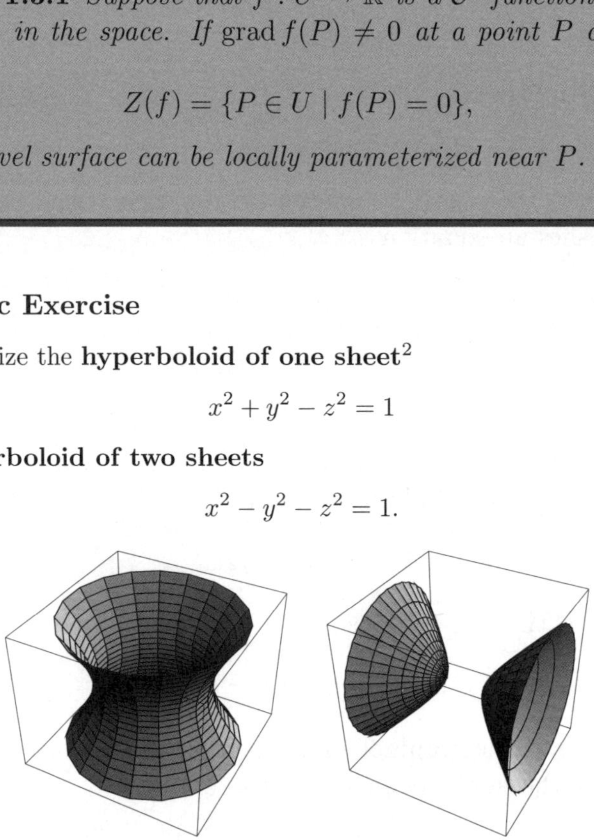

Picture: Hyperboloids

[1]If $\operatorname{grad} f(P) \neq 0$ at every point P on the level surface, it is called a **regular level surface**.

[2]There are several ways to parameterize a surface. In this problem, it is enough to find just one parametrization. For example, the hyperboloid of one sheet can be parameterized by

$$x = \cosh u \cos v, \qquad y = \cosh u \sin v, \qquad z = \sinh u.$$

1.4 Regular Surfaces

Unless stated otherwise, we will assume that a (parametric) surface $X : D \to \mathbb{R}^3$ is $\mathcal{C}^1$. Then two vectors

$$X_u := \frac{\partial X}{\partial u}, \qquad X_v := \frac{\partial X}{\partial v}$$

are tangent to the "surface".

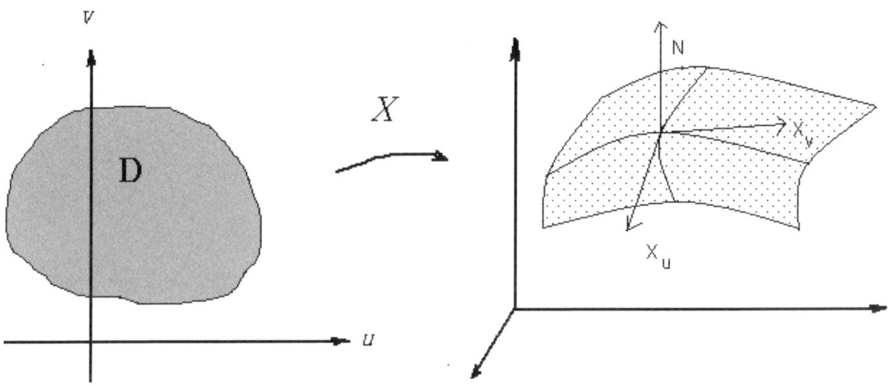

If these two vectors are linearly independent at each point of the surface, that is

$$\mathbf{N} := X_u \times X_v \neq \mathbf{0},$$

then we call X a **regular surface**. On a regular surface, every vector on the tangent plane is expressed as a linear combination of X_u and X_v. And $\mathbf{N}$ is a vector field perpendicular to the tangent plane, i.e. a **normal vector field**.

We will assume that a surface is regular unless stated otherwise. We can think of the **tangent plane** on a regular surface, which is a plane perpendicular to a normal vector.

1.4.1 Basic Exercise

Explain that a **surface of revolution** obtained by rotating a curve $(0, r(t), z(t))$ in the yz-plane around the z-axis can be parameterized by

$$X(\theta, t) = (r(t)\cos\theta, r(t)\sin\theta, z(t)).$$

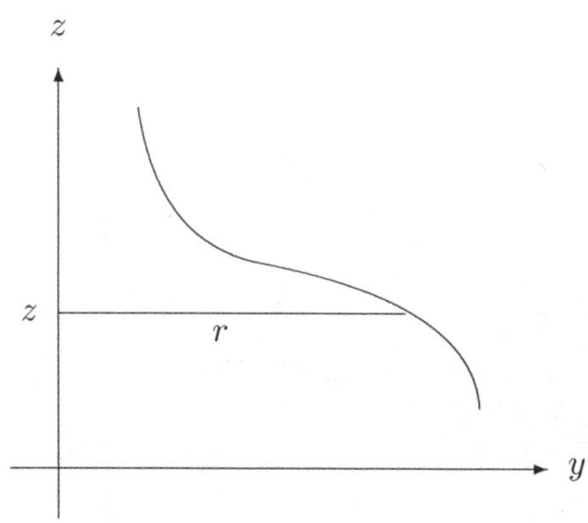

Show that the tangent vector X_t of the

$$\textbf{meridian} \quad t \mapsto X(\theta_0, t)$$

and the tangent vector X_θ of the

$$\textbf{parallel} \quad \theta \mapsto X(\theta, t_0)$$

are mutually perpendicular. In other words, show that

$$X_t(\theta_0, t_0) \cdot X_\theta(\theta_0, t_0) = 0.$$

Exercise: Chapter 16 Section 1

1. Parameterize the ellipsoid $z = x^2 + y^2$.
2. Parameterize the following surface.

   (1) $\dfrac{x^2}{a^2} + \dfrac{y^2}{b^2} + \dfrac{z^2}{c^2} = 1$ (2) $z = x^2 + 2y^2$

   (3) $x^2 + 2y^2 - 3z^2 = 0$ (4) $x^2 + 2y^2 - 3z^2 = 1$

   (5) $x^2 - 2y^2 - 3z^2 = 1$
3. Parameterize the **torus** obtained by rotating a circle of radius r with the center at $(0, R, 0)$ in the yz-plane around the z-axis. (Here, $0 < r < R$.)

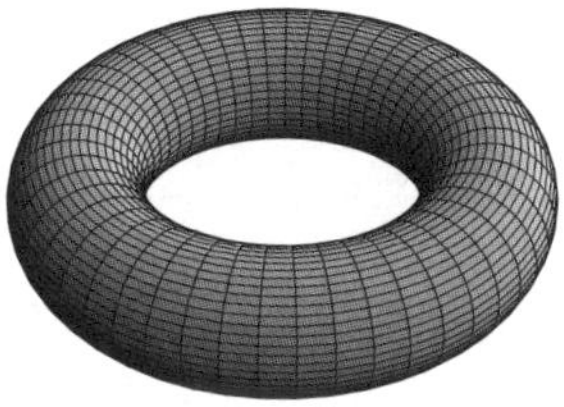

4. (**Möbius' strip**[3]) Sketch the surface given by

$$X(u, v) := (\cos u, \sin u, 0) + v\left((0, 0, 1)\sin\frac{u}{2} + (\cos u, \sin u, 0)\cos\frac{u}{2}\right)$$

   for $0 \le u \le 2\pi$, $-\dfrac{1}{2} \le v \le \dfrac{1}{2}$.

[3]August Ferdinand Möbius, (1790–1868) German astronomer and mathematician.

2 Surface Area

Finding the area of a surface is a little more difficult than finding the length of a curve. To find the length of a curve, we can pick some points on the curve and add up the lengths of **polygonal paths** (straight line segments) joining successive points, and then if we take the limit as the norm of the partition of the curve tends to zero we can get the length of the curve.

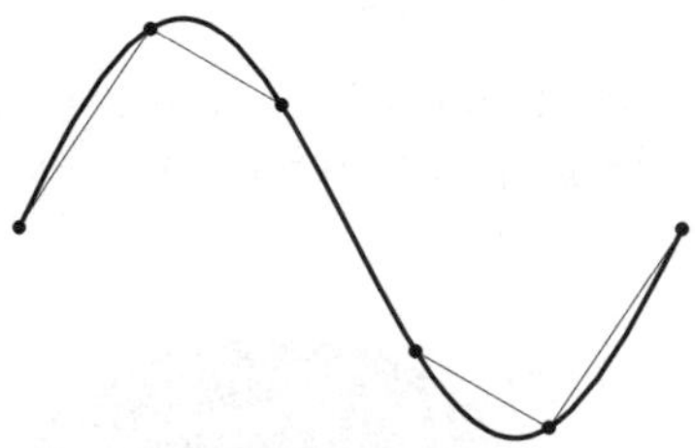

However to find the area of a surface, if we partition a surface,[4] find the area of the polyhedron with edges joining points from the partition and take the limit as the partition gets finer and finer, we can not succeed.[5]

Now let's see how to get the surface area. For a regular surface

$$X(u,v) = \big(x(u,v), y(u,v), z(u,v)\big), \qquad (u,v) \in D$$

defined on a region D in the uv-plane, since X_u and X_v are linearly independent, the vector

$$\mathbf{N} := X_u \times X_v \neq \mathbf{0}$$

is a nonzero vector perpendicular to the surface.

[4] To be precise, we partition the region on which the parameters of the surface are defined.

[5]

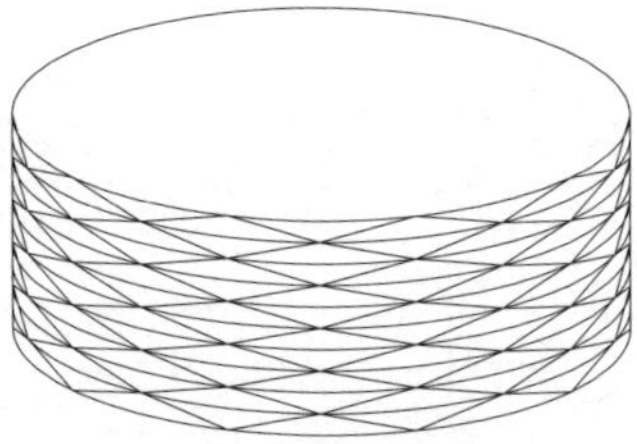

For example, if we partition a cylinder as in the accompanying figure, the area of the polyhedron we have tends to the infinity as the partition gets finer. In general, we cannot say that the polyhedron obtained by partitioning a surface is close to the tangent plane of the surface. As the tangent lines of two curves should be close in order that their lengths are close, the tangent planes of two surfaces should be close in order that their surface areas are close .

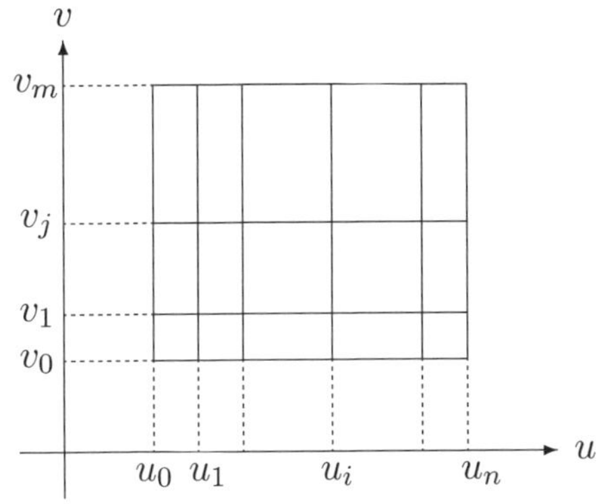

In order to get the surface area, we divide the region D into small rectangles and then a small piece on the surface

$$\begin{aligned} X_{ij} &:= \{X(u_i + s\Delta u, v_j + t\Delta v) \mid 0 \le s, t \le 1\} \\ &\approx \{X(u_i, v_j) + sX_u(u_i, v_j)\Delta u + tX_v(u_i, v_j)\Delta v \mid 0 \le s, t \le 1\} \end{aligned}$$

is approximately a parallelogram. So the area of X_{ij} is approximated by

$$\text{area}(X_{ij}) \approx |X_u(u_i, v_j) \times X_v(u_i, v_j)|\ \Delta u\, \Delta v = |\mathbf{N}(u_i, v_j)| \Delta u\, \Delta v.$$

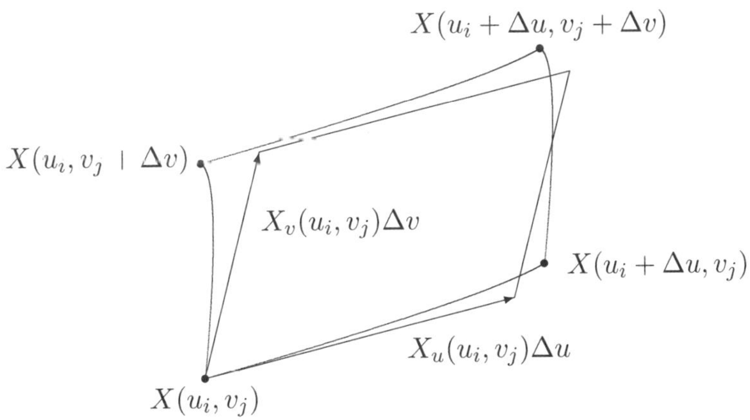

Therefore the area of X is

$$\text{area}(X) = \sum_{i,j} \text{area}\,(X_{ij}) \approx \sum_{i,j} |\mathbf{N}(u_i, v_j)|\ \Delta u\, \Delta v.$$

If we take the limit as we refine the partition of D, we get

$$\boxed{\text{area}(X) = \iint_D |\mathbf{N}(u,v)|\, dudv}$$

Of course, we have

$$|\mathbf{N}| = \sqrt{|X_u|^2|X_v|^2 - (X_u \cdot X_v)^2}$$

and

$$\boxed{d\mathrm{S} := |\mathbf{N}(u,v)|\, du\, dv}$$

is called the **area element** of the surface.[6]

2.0.1 Example

For a surface X given by the graph of a function $f(x,y)$,

$$X(x,y) := (x, y, f(x,y)),$$

we have

$$X_x = \left(1, 0, \frac{\partial f}{\partial x}\right), \qquad X_y = \left(0, 1, \frac{\partial f}{\partial y}\right)$$

and the normal vector field is[7]

$$\mathbf{N} = X_x \times X_y = \left(-\frac{\partial f}{\partial x}, -\frac{\partial f}{\partial y}, 1\right) = \text{grad}(z - f(x,y)).$$

Therefore the area element is[8]

$$d\mathrm{S} = \sqrt{\left(\frac{\partial f}{\partial x}\right)^2 + \left(\frac{\partial f}{\partial y}\right)^2 + 1}\, dx\, dy =: \sqrt{{f_x}^2 + {f_y}^2 + 1}\ dx\, dy.$$

---

[6] It is hard to distinguish ds from $d\mathrm{S}$ on the board,. However since ds appears with $\int$ and $d\mathrm{S}$ with $\iint$, it won't cause much confusion. Often the area element is denoted by $d\sigma$ or dA.

[7] Because the gradient vector is perpendicular to the level surface $z - f(x,y) = 0$.

[8] If we project a parallelogram which makes an angle θ with the xy-plane to the xy-plane, the area of the resulting projection is $\cos\theta$ times the original area. On the other hand, since the cosine of the angle between the parallelogram perpendicular to $\mathbf{N} = (-f_x, -f_y, 1)$ and the xy-plane is

$$\frac{\mathbf{N}}{|\mathbf{N}|} \cdot \mathbf{k} = \frac{1}{\sqrt{{f_x}^2 + {f_y}^2 + 1}},$$

we get $dx\, dy = \dfrac{1}{\sqrt{{f_x}^2 + {f_y}^2 + 1}}\, d\mathrm{S}$, that is,

$$d\mathrm{S} = \sqrt{{f_x}^2 + {f_y}^2 + 1}\ dx\, dy.$$

2.0.2 Example

If we let A be the area of the graph of the function

$$f(x,y) = \cosh\sqrt{x^2+y^2} = \cosh r \qquad (D\ :\ x^2+y^2 \le 1),$$

then

$$\begin{aligned} A &= \iint_D \sqrt{f_x{}^2 + f_y{}^2 + 1}\ dx\,dy \\ &= \iint_D \sqrt{\left(\frac{x}{r}\sinh r\right)^2 + \left(\frac{y}{r}\sinh r\right)^2 + 1}\ dx\,dy \\ &= \iint_D \cosh r\ dx\,dy \\ &= \int_0^{2\pi}\int_0^1 \cosh r \cdot r\,dr\,d\theta \\ &= 2\pi\left(\big[r\sinh r\big]_0^1 - \int_0^1 \sinh r\,dr\right) \\ &= 2\pi\big(\sinh 1 - (\cosh 1 - 1)\big) \\ &= 2\pi\left(-\frac{1}{e} + 1\right). \end{aligned}$$

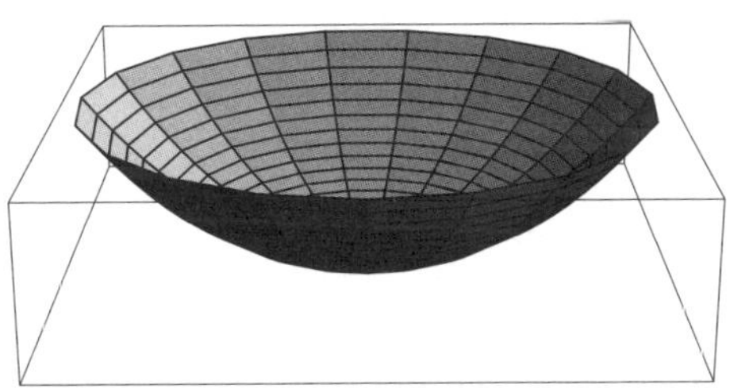

2.0.3 Basic Exercise

Find the area of the catenoid

$$\cosh z = r = \sqrt{x^2+y^2}, \quad z_1 \le z \le z_2.$$

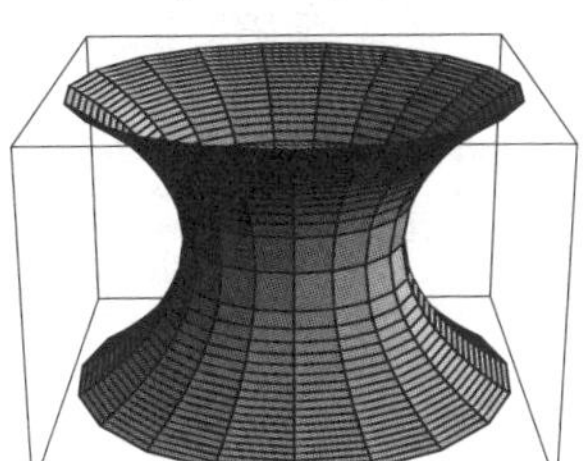

2.1 Area of Sphere

Archimedes showed that the area of the sphere is equal to the area of the cylinder circumscribed to the sphere and moreover he found out that the area of the portion of the sphere between two parallel planes perpendicular to the central axis of this cylinder is equal to the area of the portion of the cylinder between those two planes.[9] [Hsiang].

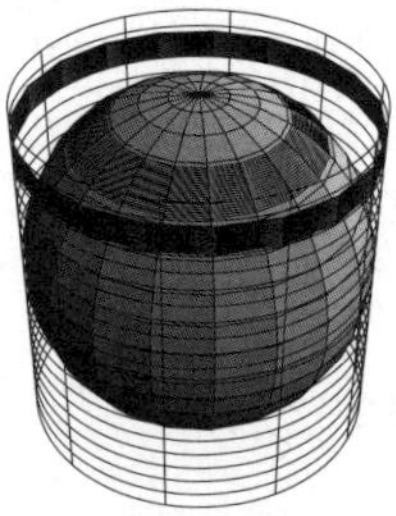

[9]The Archimedes Palimpsest is a medieval parchment manuscript of 174 pages. It contains seven treatises by Archimedes, "Equilibrium of Planes", "Spiral Lines", "The Measurement of the Circle", "Sphere and Cylinder", "On Floating Bodies", "The Method of Mechanical Theorems", and "The Stomachion". It was purchased at a Christie's sale on Oct. 28, 1998, by an anonymous collector for $2,002,500. [New York Times, Washington Post]. Archimedes' work was mentioned by many people in many books, but until 1899 there were only translations of his work and no "original" nor "copy" of his work had been found. Since the materials to write on was rare in the ancient times, the medieval manuscripts were made of parchment, which you can unbind, and scrape off the text, and overwrite with a new text. About one hundred years ago from now, a liturgical book written on parchment was found. Using digital processing of untraviolet and X-ray the text written on it previously and then erased became readable. The text recovered turned out a copy of Archimedes' work by some scribe in the 10th century. It had belonged to a French family from the 1920s, and it was auctioned at "Christie's" in New York on Oct. 28, 1998. The Greek orthodox contended that it had belonged to them and had been stolen.

The following is the last words by Archimedes:

Nolo tubare circulos meos! — Don't touch my circles!

To investigate this principle, let's parameterize the sphere and cylinder by[10]

$$\begin{aligned} X(\varphi,\theta) &= (\cos\theta\sin\varphi, \sin\theta\sin\varphi, \cos\varphi) \\ Y(\varphi,\theta) &= (\cos\theta, \sin\theta, \cos\varphi). \end{aligned}$$

On the sphere,

$$\begin{aligned} X_\theta &= (-\sin\theta\sin\varphi, \cos\theta\sin\varphi, 0) \\ X_\varphi &= (\cos\theta\cos\varphi, \sin\theta\cos\varphi, -\sin\varphi). \end{aligned}$$

Thus

$$|X_\theta| = \sin\varphi, \qquad |X_\varphi| = 1, \qquad X_\theta \cdot X_\varphi = 0$$

and

$$|\mathbf{N}| = \sin\varphi.$$

Therefore the area element of the sphere is

$$\boxed{dS = \sin\varphi\, d\varphi\, d\theta}$$

Similarly, on the cylinder,

$$Y_\varphi = (0, 0, -\sin\varphi), \qquad Y_\theta = (-\sin\theta, \cos\theta, 0)$$

and thus

$$|\mathbf{N}| = \sqrt{|Y_\varphi|^2 |Y_\theta|^2 - (Y_\varphi \cdot Y_\theta)^2} = \sin\varphi.$$

Therefore the area element of the cylinder is

$$dS = \sin\varphi\, d\varphi\, d\theta,$$

which is equal to the area element of the sphere.

From this we get Archimedes' theorem.

In particular, the area of the unit sphere (or cylinder) is

$$\int_0^{2\pi} \int_0^{\pi} \sin\varphi\, d\varphi\, d\theta = 4\pi.$$

In general, the area of the sphere of radius r is[11]

$$\boxed{A = 4\pi r^2}$$

[10] For convenience, think of the sphere of radius 1. Here we let $0 \le \varphi \le \pi$, $0 \le \theta \le 2\pi$.

[11] Would it be possible that the brightness of a point with distance r from the source of light is inversely proportional to the area of the sphere of radius r? In the isotropic three-dimensional space ...

2.1.1 Basic Exercise

Show that the area element of the sphere of radius ρ is expressed as

$$dS = \rho^2 \sin\varphi \, d\varphi \, d\theta$$

in "spherical coordinates" (φ, θ).

2.1.2 Example

Find the area of the surface

$$X(\theta, \varphi) := (\cos\theta \sin\varphi, \sin\theta \sin\varphi, \cos\varphi)$$

for $0 \le \theta \le 2\pi$, $\frac{\pi}{4} \le \varphi \le \frac{\pi}{2}$.

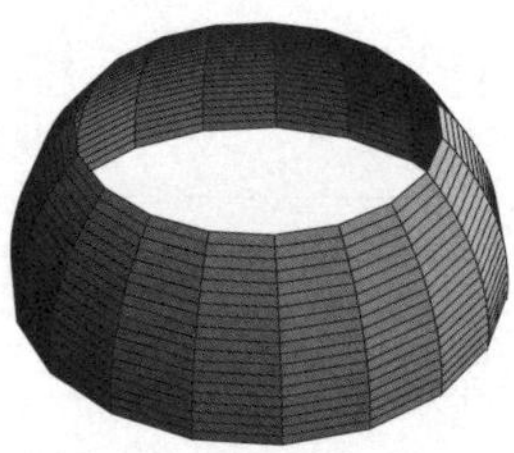

Solution. Since

$$dS = \sin\varphi \, d\varphi \, d\theta,$$

the area we want is[12]

$$\int_{\pi/4}^{\pi/2} \int_0^{2\pi} \sin\varphi \, d\theta \, d\varphi = 2\pi \cdot (-\cos(\pi/2) + \cos(\pi/4)) = \sqrt{2}\pi.$$

[12]Since we want to find the area of a portion of the unit sphere, we can check the answer by comparing it with the surface area of the cylinder with height $\sqrt{2}$ by Archimedes' theorem.

2.2 Solid Angle

The **solid angle** subtended by some region at a point P is the area of the projection of the region onto the unit sphere centered at P.

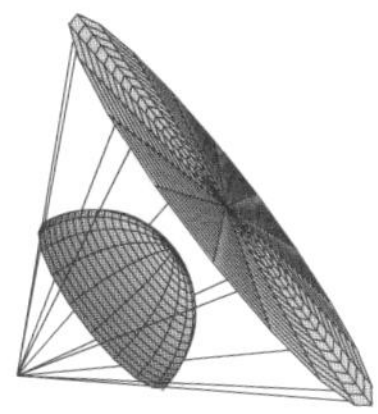

For example, the solid angle of the **first octant**

$$x > 0, \quad y > 0, \quad z > 0$$

at the origin is $4\pi/8 = \pi/2$.

Similarly, the solid angle of a rectangular box at one vertex is $\pi/2$. Also the solid angle of the box at one point on a edge is π and the solid angle at the center of a face is 2π.

2.2.1 Example

At the origin $O := (0, 0, 0)$, the solid angle of the **circular cone** given by

$$C \ : \ z \geq \sqrt{x^2 + y^2}$$

is

$$\int_0^{2\pi} \int_0^{\pi/4} \sin\varphi \, d\varphi \, d\theta = 2\pi \big[-\cos\varphi \big]_0^{\pi/4} = 2\pi \left(1 - \frac{1}{\sqrt{2}} \right).$$

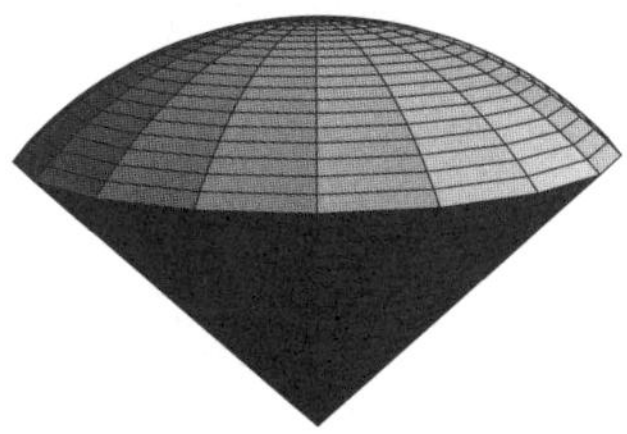

Exercise: Chapter 16 Section 2

1. For four points in $\mathbb{R}^2$,

$$A = (a, b+k), \qquad B = (a+h, b+k), \qquad C = (a, b), \qquad D = (a+h, b)$$

let S_1, S_2, S_3 be the areas of surfaces of revolution obtained by rotating line segments AB, BC, CD around the x-axis respectively.

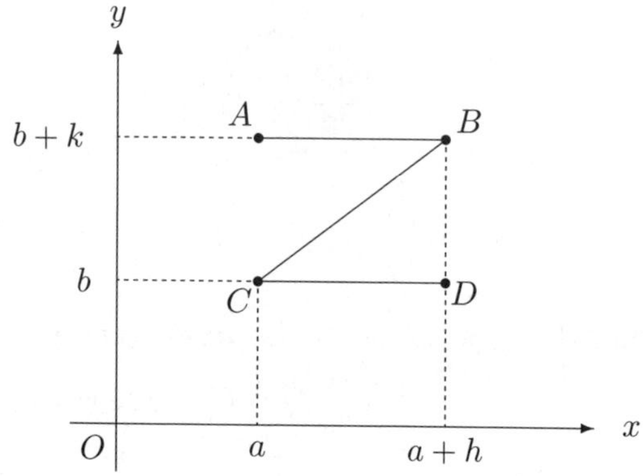

Verify when there holds

$$S_1 \geq S_2 \geq S_3.$$

2. Find the area of the surface

$$x^2 + y^2 + z^2 = 1, \qquad \left(x - \frac{1}{2}\right)^2 + y^2 \leq \frac{1}{4}.$$

3. Show that the area of the surface obtained by rotating the graph of a positive function

$$y = f(x) > 0 \qquad (x_0 \leq x \leq x_1)$$

around the x-axis is

$$\int_{x_0}^{x_1} (2\pi y)\sqrt{1 + \left(\frac{dy}{dx}\right)^2}\, dx.$$

4. Show that the area of the surface obtained by rotating a plane curve

$$(x(t), y(t)) \qquad (t_0 \leq t \leq t_1)$$

around the x-axis is

$$\int_{t_0}^{t_1} (2\pi y)\sqrt{\left(\frac{dx}{dt}\right)^2 + \left(\frac{dy}{dt}\right)^2}\, dt.$$

(Here we assume $y(t) \geq 0$.)

5. Find the area of the surface obtained by rotating the cycloid

$$x(t) = t - \sin t, \qquad y(t) = 1 - \cos t \qquad (0 \le t \le 2\pi)$$

around the x-axis.

6. (**Pappus' Theorem II**) Let d be the distance between the center of the curve given by the equation $y = f(x) > 0$ and the x-axis and let l be the length of the curve.

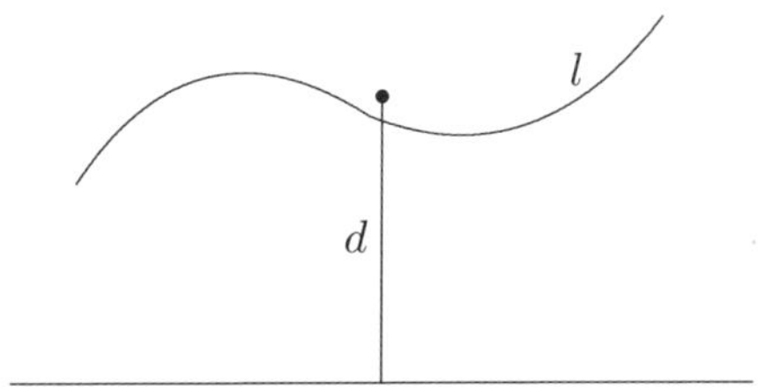

Show that the area of the surface obtained by rotating this curve around the x-axis is $2\pi d \cdot l$. Also use this to show that the surface area of the torus with "big radius" R and "small radius" r is $4\pi^2 rR$.[13]

7. Find the area of the portion of the sphere $x^2 + y^2 + z^2 = 1$ inside the elliptic cylinder $x^2 + \dfrac{y^2}{a^2} \le 1$. (Here $0 < a \le 1$.)

8. For a real number a, show that the solid angle of the circular cone given by

$$C \,:\, z \ge a\sqrt{x^2 + y^2}$$

at the apex O is[14]

$$2\pi\left(1 - \frac{a}{\sqrt{1 + a^2}}\right).$$

9. A triangle on the sphere—that is, a **spherical triangle** is formed by three **great circular arcs**. For a **spherical triangle** ABC with the center at O, let $\theta_A, \theta_B, \theta_C$ be the angles at the vertices and and let $\mathbf{a} := \overrightarrow{OA}$, $\mathbf{b} := \overrightarrow{OB}$, $\mathbf{c} := \overrightarrow{OC}$. Then show that

$$\cos\theta_A = \frac{\mathbf{b}\cdot\mathbf{c}|\mathbf{a}|^2 - (\mathbf{a}\cdot\mathbf{b})(\mathbf{a}\cdot\mathbf{c})}{|\mathbf{a}\times\mathbf{b}|\cdot|\mathbf{a}\times\mathbf{c}|} = \frac{(\mathbf{a}\times\mathbf{b})\cdot(\mathbf{a}\times\mathbf{c})}{|\mathbf{a}\times\mathbf{b}|\cdot|\mathbf{a}\times\mathbf{c}|}.$$

Also show that the solid angle of the spherical triangle ABC at the origin of the sphere is

$$\theta_A + \theta_B + \theta_C - \pi.$$

---

[13]It is a surface of revolution obtained by revolving the circle$x^2 + (y - R)^2 = r^2$ around the x-axis

[14]If this cone has an aperture of α, then $\cos\dfrac{\alpha}{2} = \dfrac{a}{\sqrt{1 + a^2}}$.

10. Suppose that a vertex V of a polyhedron is an **n-hedral angle**,[15] at which n faces $F_1, \ldots, F_n$ meet. And call the sum of the angles between two adjacent faces F_i and F_{i+1} the **sum of dihedral angles** at V:

$$\text{Sum of dihedral angles } := \sum_{i=1}^{n} \angle\{F_i, F_{i+1}\} \qquad (F_{n+1} := F_1).$$

Show that the solid angle of this polyhedron at the vertex V is[16]

$$\text{Sum of dihedral angles } - (n-2)\pi.$$

11. Find the solid angle of a regular tetrahedron at one vertex. Also find the solid angle of a regular octahedron at one vertex.

12. Show that the area of the portion of a sphere S of radius bigger than 1 inside a unit sphere with center at one point of S is always π.

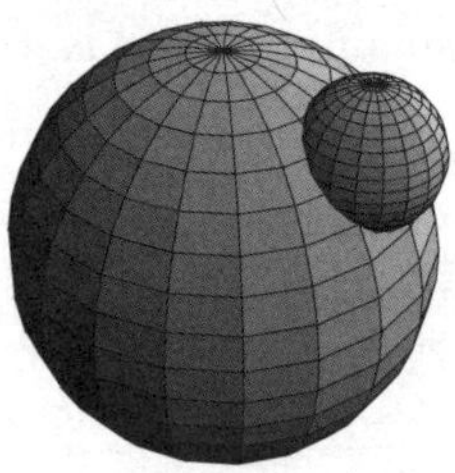

[15]Of course, $n \geq 3$.

[16]First, prove it when $n = 3$. For three points A, B, C on the sphere with center at the origin O, the angle between the faces OAB and OBC is equal to $\angle B$ on the **spherical triangle** ABC. From this, at the trihedral angle O, the sum of dihedral angles is the "sum of interior angles of the spherical triangle"

$$\angle A + \angle B + \angle C = \pi + \text{solid angle}.$$

When O is a **multihedral angle**, if we think of it as a sum of trihedral angles, the sum of dihedral angles is equal to the "sum of interior angles of the spherical polygon". A **convex multihedral angle** has dihedral angles always less than 180°, but **convex multihedral angle** can have a dihedral angle more than π. The sum of exterior angles of a polygon is always 2π on the plane, and $2\pi +$ solid angle on the sphere.

3 Surface Integrals

Think of a surface $X : D \to \mathbb{R}^3$ as a thin metallic surface. If the density at each point is given by the function $f(x, y, z)$, then the mass of the surface X is

$$\iint_D f(X(u,v))\,|X_u \times X_v|\,dudv$$

as in the way to find the surface area.[17]

In general, the **surface integral** of a function $f(x, y, z)$ defined on a surface $X : D \to \mathbb{R}^3$ is defined by

$$\iint_X f\,dS := \iint_D f(X(u,v))\,|X_u \times X_v|\,du\,dv.$$

When an integrable function is called a "density function", its integral is called the **mass**.

3.0.1 Example

The area element of the cone

$$X \;:\; z = \sqrt{x^2+y^2} \qquad (x^2+y^2 \le 1)$$

is

$$\begin{aligned} dS &= \sqrt{1+\left(\frac{\partial z}{\partial x}\right)^2+\left(\frac{\partial z}{\partial y}\right)^2}\,dx\,dy \\ &= \sqrt{1+\left(\frac{x}{z}\right)^2+\left(\frac{y}{z}\right)^2}\,dx\,dy \\ &= \sqrt{2}\,dxdy. \end{aligned}$$

If the density function is given by $f(x, y, z) = z$ at each point on the cone, then the mass of the cone is

$$\begin{aligned} \iint_X f\,dS &= \iint_{x^2+y^2\le 1} z\sqrt{2}\,dx\,dy \\ &= \sqrt{2}\int_0^{2\pi}\int_0^1 r\cdot r\,dr\,d\theta = \sqrt{2}(2\pi)\cdot\frac{1}{3} \\ &= \frac{2\sqrt{2}\pi}{3}. \end{aligned}$$

[17]The mass of a litle piece of the surface is $f\ dS$.

3.0.2 Exercise (Fubini's theorem)

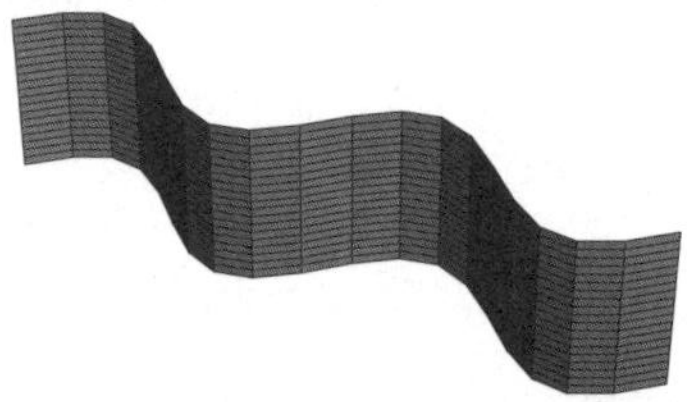

For a plane curve

$$C(t) = (x(t), y(t)) \qquad (t_0 \le t \le t_1)$$

consider the **cylindrical surface**

$$X(t, u) := (x(t), y(t), u) \qquad (u_0 \le u \le u_1).$$

Show that for a function $f(x, y, z)$,

$$\iint_X f \, dS = \int_C \left(\int_{u_0}^{u_1} f(x, y, u) \, du \right) ds.$$

3.1 Area Element and Reparametrization

For a $\mathcal{C}^1$ invertible map $G : \tilde{D} \to D$ between two regions $D, \tilde{D}$ in $\mathbb{R}^2$ and a surface $X : D \to \mathbb{R}^3$ defined on D, the map

$$\tilde{X} := X \circ G : \tilde{D} \to \mathbb{R}^3$$

is called a **reparametrization** of the surface X. If the Jacobian determinant of G, $\det G'$, is always positive (or always negative), then it is called an **orientation-preserving** (or **orientation-reversing**) reparametrization.

Theorem 3.1.1 *The surface integral of a function f defined on a parametric surface $X : D \to \mathbb{R}^3$ doesn't depend on the parametrization, that is, the surface integral doesn't change when the surface is reparameterized.*

Proof. For a map

$$G : \tilde{D} \to D, \qquad G(\tilde{u}, \tilde{v}) = (u, v),$$

let $\tilde{X} := X \circ G : \tilde{D} \to \mathbb{R}^3$ be the reparametrization of X. Then by the change of variables formula, we get

$$\begin{aligned} \iint_X f \, dS &= \iint_D f\big(X(u, v)\big) \, |X_u \times X_v| \, du \, dv \\ &= \iint_{\tilde{D}} f\big(X \circ G(\tilde{u}, \tilde{v})\big) \, |X_u \times X_v| \, \big|\det G'(\tilde{u}, \tilde{v})\big| \, d\tilde{u} \, d\tilde{v}. \end{aligned}$$

It suffices to show that

$$|X_u \times X_v| \left|\det G'(\tilde{u}, \tilde{v})\right| = |(X \circ G)_{\tilde{u}} \times (X \circ G)_{\tilde{v}}| .$$

Since

$$(X \circ G)_{\tilde{u}} = X_u \frac{\partial u}{\partial \tilde{u}} + X_v \frac{\partial v}{\partial \tilde{u}}, \qquad (X \circ G)_{\tilde{v}} = X_u \frac{\partial u}{\partial \tilde{v}} + X_v \frac{\partial v}{\partial \tilde{v}},$$

we have

$$(X \circ G)_{\tilde{u}} \times (X \circ G)_{\tilde{v}} = (X_u \times X_v) \det \frac{\partial(u, v)}{\partial(\tilde{u}, \tilde{v})} = (X_u \times X_v) \det G'$$

and so we are done. □

3.2 Average of Function

The average of a function f defined on a surface X is defined by

$$\bar{f} := \frac{1}{\operatorname{area} X} \iint_X f \, dS.$$

Also if we let $\bar{x}$ $\bar{y}$ $\bar{z}$ be the averages of the x-, y-, z-component functions of X, respectively, then the point $\bar{X} := (\bar{x}, \bar{y}, \bar{z})$ is called the (geometric) **center** of X .

3.2.1 Basic Exercise

Show that the center $\bar{X} := (\bar{x}, \bar{y}, \bar{z})$ of a surface X satisfies the equations[18]

$$\iint_X (x - \bar{x}) \, dS = 0, \qquad \iint_X (y - \bar{y}) \, dS = 0, \qquad \iint_X (z - \bar{z}) \, dS = 0.$$

[18]Using the integral of a vector-valued function, these equations are expressed simply as

$$\iint_X (X - \bar{X}) \, dS = \mathbf{0}.$$

3.2.2 Example

Find the center $(\bar{x}, \bar{y}, \bar{z})$ of the upper hemisphere

$$z = \sqrt{a^2 - x^2 - y^2}, \qquad x^2 + y^2 \leq a^2.$$

Solution. From the **symmetry** of the surface, we get

$$\bar{x} = 0, \qquad \bar{y} = 0.$$

On the other hand, since

$$\frac{\partial z}{\partial x} = -\frac{x}{z}, \qquad \frac{\partial z}{\partial y} = -\frac{y}{z},$$

we have

$$dS = \sqrt{1 + \left(\frac{x}{z}\right)^2 + \left(\frac{y}{z}\right)^2}\, dx\, dy = \frac{1}{z}\sqrt{x^2 + y^2 + z^2}\, dx\, dy = \frac{a}{z}\, dx\, dy.$$

and thus

$$\bar{z} = \frac{1}{\text{area}} \iint_{x^2+y^2 \leq a^2} z \cdot \frac{a}{z}\, dx\, dy = \frac{1}{2\pi a^2} \cdot a \cdot \pi a^2 = \frac{a}{2}.$$

Therefore the center of the hemisphere is

$$(\bar{x}, \bar{y}, \bar{z}) = \left(0, 0, \frac{a}{2}\right).$$

3.2.3 Center of Mass

If the density function of the surface X is f, the **mass** of X is defined by

$$M := \iint_X f\, dS,$$

and the **center of mass** $(\bar{x}, \bar{y}, \bar{z})$ of X is defined by

$$\bar{x} = \frac{1}{M} \iint_X xf\, dS, \qquad \bar{y} = \frac{1}{M} \iint_X yf\, dS, \qquad \bar{z} = \frac{1}{M} \iint_X zf\, dS.$$

If the density function is constant, then the center of mass is equal to the geometric center.

3.2.4 Basic Exercise

For a surface with the density function f, show that its center of mass $(\bar{x}, \bar{y}, \bar{z})$ satisfies

$$\iint_X (x-\bar{x})f \, dS = 0, \qquad \iint_X (y-\bar{y})f \, dS = 0, \qquad \iint_X (z-\bar{z})f \, dS = 0.$$

3.2.5 Example

For the surface

$$X \ : \ z = \sqrt{a^2 - x^2 - y^2} \qquad (x^2 + y^2 \le a^2)$$

with the density function $\mu(x, y, z) = z$, find the center of mass $(\bar{x}, \bar{y}, \bar{z})$ and the average density $\bar{\mu}$.

From Example (3.2.2),

$$dS = \frac{a}{z} \, dx \, dy, \qquad \mu \, dS = a \, dx \, dy.$$

Therefore the mass is

$$M := \iint_X \mu \, dS = \iint_{x^2+y^2 \le a^2} a \, dx \, dy = \pi a^3.$$

On the other hand, since the area of X is $A := 2\pi a^2$, the average density is

$$\bar{\mu} = \frac{M}{A} = \frac{\pi a^3}{2\pi a^2} = \frac{a}{2}.$$

Also form the symmetry, $\bar{x} = \bar{y} = 0$ and

$$\begin{aligned}
\bar{z} &= \frac{1}{M} \iint_X z \cdot \mu \, dS \\
&= \frac{1}{\pi a^3} \iint_{x^2+y^2 \le a^2} \sqrt{a^2 - x^2 - y^2} \, a \, dx \, dy \\
&= \frac{1}{\pi a^2} \int_0^{2\pi} \int_0^a \sqrt{a^2 - r^2} \, r \, dr \, d\theta \\
&= \frac{1}{\pi a^2} \left(\int_0^{2\pi} d\theta \right) \left(\int_0^a \sqrt{a^2 - r^2} \, r \, dr \right) \\
&= \frac{2}{a^2} \int_0^a u^2 \, du \qquad (\Leftarrow \sqrt{a^2 - r^2} =: u) \\
&= \frac{2}{3} a.
\end{aligned}$$

Therefore the center of mass of X is

$$(\bar{x}, \bar{y}, \bar{z}) = \left(0, 0, \frac{2}{3}a\right).$$

3.2.6 Example

For a sphere of radius ρ, $S : x^2 + y^2 + z^2 = \rho^2$, find the integral[19]

$$V(t) := \iint_S \frac{dS}{\sqrt{x^2 + y^2 + (z-t)^2}} \qquad (-\rho < t < \rho).$$

If we parameterize the sphere as

$$x = \rho\cos\theta\sin\varphi, \qquad y = \rho\sin\theta\sin\varphi, \qquad z = \rho\cos\varphi,$$

on the region $0 \le \theta \le 2\pi,\ 0 \le \varphi \le \pi$, then $dS = \rho^2 \sin\varphi\, d\varphi\, d\theta$ and

$$\begin{aligned} V(t) &= \int_0^{2\pi}\int_0^{\pi} \frac{\rho^2\sin\varphi\, d\varphi\, d\theta}{\sqrt{\rho^2\sin^2\varphi + (\rho\cos\varphi - t)^2}} \\ &= \rho^2 \cdot 2\pi \cdot \int_0^{\pi} \frac{\sin\varphi\, d\varphi}{\sqrt{\rho^2 + t^2 - 2\rho t\cos\varphi}} \\ &= \rho^2 \cdot 2\pi \cdot \int_{\rho-t}^{\rho+t} \frac{u\, du/\rho t}{u} \qquad (\Leftarrow\ u := \sqrt{\rho^2 + t^2 - 2\rho t\cos\varphi}\ ) \\ &= \rho^2 \cdot 2\pi \cdot \frac{2}{\rho} \\ &= 4\pi\rho. \end{aligned}$$

3.2.7 Basic Exercise

For the sphere $S(4) : x^2 + y^2 + z^2 = 4^2$, show that

$$\iint_{S(4)} \frac{dS}{\sqrt{(x-1)^2 + (y-2)^2 + (z-3)^2}} = 4\pi \cdot 4.$$

---

[19]This function is one of the most important function in potential theory [Kellog], [Davis, Snider]. Once we study the divergence theorem later, we can easily compute the integral. This example tells us that the potential energy of the gravitational field is constant inside the sphere. Therefore the gravitational field is zero inside the sphere.

Exercise: Chapter 16 Section 3

1. Over the unit sphere $S : x^2+y^2+z^2=1$, integrate the following function:

$$x, \quad y, \quad z, \quad x^2, \quad y^2, \quad z^2, \quad xy, \quad yz, \quad zx, \quad x^2y, \quad xyz$$

2. Find the area and center of the following surface.
(Here $r=\sqrt{x^2+y^2}$, $\rho=\sqrt{x^2+y^2+z^2}$.)

   (1) Ellipsoid $z=r^2$, $r \le a$

   (2) Cone $z=r$, $r \le a$

   (3) The portion of the sphere $\rho=a$ inside the cone $z=r$

   (4) The portion of $x^2+y^2+z^2=4$ inside the cylinder $x^2+y^2=2x$, satisfying $z \ge 0$

3. When the density function of the surface S is given by f, find the mass, average density and center of mass.

   (1) $f(x,y,z)=xyz$, S is the portion of $x+y+z=1$ in the first octant

   (2) $f(x,y,z)=x+y$, S is the portion of $z=2x+3y$ inside the cylinder $x^2+y^2=9$, satisfying $x \ge 0$, $y \ge 0$

   (3) $f(x,y,z)=z$, S is given by $z=r^2$ inside the cylinder $r=2$ in cylindrical coordinates

   (4) $f(x,y,z)=(x^2+y^2)z$, S is the portion of $\rho=1$ satisfying $z \ge 0$ in spherical coordinates

   (5) $f(x,y,z)=z^2$, S is given by $x=\cos\theta$, $y=\sin\theta$, $z=z$, $0 \le \theta \le 2\pi$, $0 \le z \le 2\pi$

4. Suppose that a continuous function f defined on a surface X attains the maximum and minimum values, and let $\bar{f}$ be the average of f. Show that if X is connected then there exists a point X_0 on the surface such that

$$\bar{f}=f(X_0).$$

5. For a sphere

$$S(\rho) \;:\; x^2+y^2+z^2=\rho^2$$

with the constant density μ, let

$$M := \iint_{S(\rho)} \mu \, dS = 4\pi\mu\rho^2$$

be the mass. For a real number $t>\rho$, show that

$$\iint_{S(\rho)} \frac{\mu \, dS}{\sqrt{x^2+y^2+(z-t)^2}} = \frac{M}{t}.$$

6. Evaluate the following integral.

(1) $$\iint_{S(4)} \frac{dS}{\sqrt{(x-1)^2+y^2+z^2}}$$

(2) $$\iint_{S(4)} \frac{dS}{\sqrt{x^2+(y-3)^2+z^2}}$$

(3) $$\iint_{S(5)} \frac{dS}{\sqrt{(x-2)^2+(y-3)^2+z^2}}$$

7. (**Pappus' theorem III**) When the density function of a plane curve $C \subset \mathbb{R}^2$ is given by μ,

$$m := \int_C \mu \, ds$$

is the mass of the curve C and

$$M := \iint_S \mu \, dS$$

is the mass of a surface of revolution S obtained by rotating the curve C around the straight line $l \subset \mathbb{R}^2$.[20] If we let d be the distance from the center of mass of C to the line l, show that

$$M = 2\pi dm.$$

[20] Assume that l and C don't intersect.

4 Vector Fields and Surface Integrals

4.1 Surface and Orientation

As the orientation of a curve is important when we integrate a vector field along a curve, the orientation of a surface is also important when we integrate a vector field over a surface.

An **orientation** of a surface in the three-dimensional space is a way to determine the normal direction of the surface at each point of the surface that varies continuously.

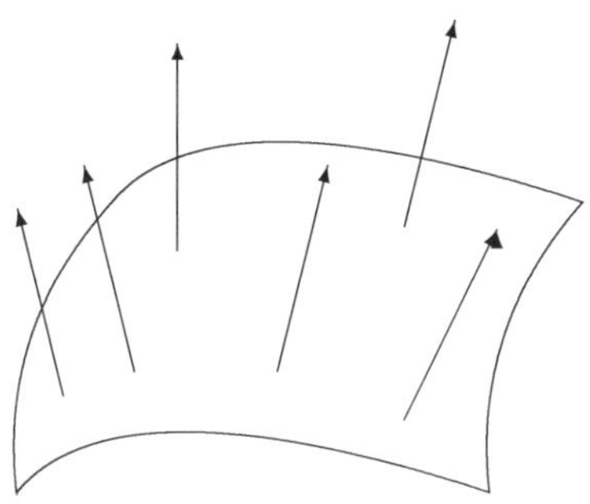

When a parametrization $X(u,v)$ for a surface S is given, a **unit normal vector field**

$$\mathbf{n} := \frac{X_u \times X_v}{|X_u \times X_v|}$$

determines an orientation of the surface. However if we reparameterize S as

$$\tilde{X}(\tilde{u},\tilde{v}) := X(\tilde{v},\tilde{u}),$$

then

$$\tilde{X}_{\tilde{u}} \times \tilde{X}_{\tilde{v}} = -X_u \times X_v$$

and thus the orientation is reversed.

On a parametric surface, the orientation doesn't change by an orientation-preserving reparametrization, but becomes opposite by an orientation-reversing reparametrization.[21]

Since every surface can be parameterized locally, there always exists an orientation for any surface locally.

[21]Every surface is not orientable. For example, Möbius strip is not orientable.

4.2 Closed Surfaces

Like a sphere or a torus, a bounded, connected and closed surface without boundary is called a **closed surface.**[22]

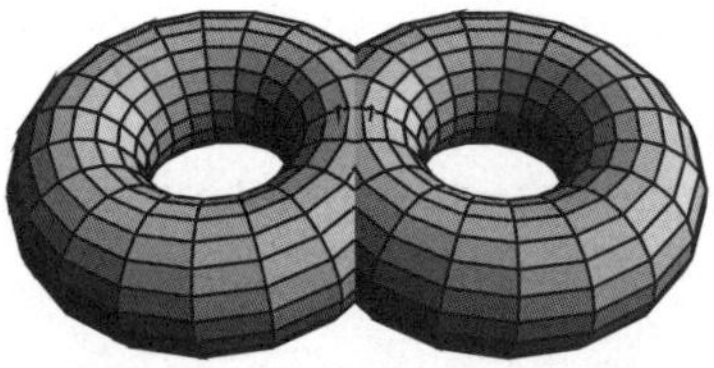 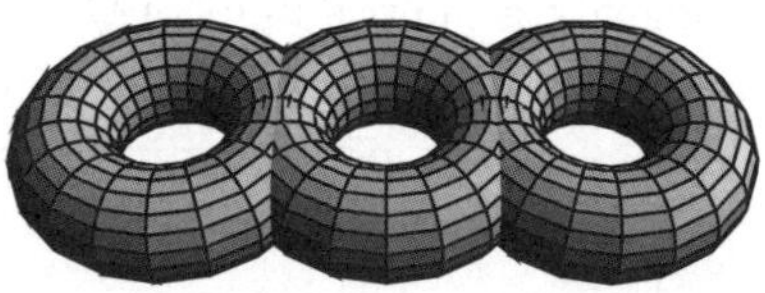

The complementary of a closed surface S in the three-dimensional space consists of two connected components. One is bounded and the other is unbounded. The former $\operatorname{int} S$ is called the **interior** of S, and the latter $\operatorname{ext} S$ is called the **exterior** of S:[23]

$$\mathbb{R}^3 - S = (\operatorname{int} S) \cup (\operatorname{ext} S).$$

On a closed surface, unless stated otherwise, we assume that the **outward unit normal vector** gives the orientation of the surface.

4.3 Vector Fields and Surface Integrals

For a vector field $\mathbf{F}(x, y, z) \in \mathbb{R}^3$ defined at every point of an **oriented surface** — that is, a surface S with an orientation, the surface integral

$$\iint_S \mathbf{F} \cdot d\mathbf{S} := \iint_S \mathbf{F} \cdot \mathbf{n} \, dS$$

is called the "integral of the vector field $\mathbf{F}$ over the surface S". The surface integral of a vector field changes its sign if the orientation of the surface is reversed.

If a vector field $\mathbf{F}$ represents the velocity of a fluid flow, the above integral is the net quantity of fluid to flow across the surface S per unit time, and this integral is called the " **flux** of a vector field $\mathbf{F}$ across the oriented surface S".

[22] compact surface without boundary

[23] Here, S is the boundary of both $\operatorname{int} S$ and $\operatorname{ext} S$.

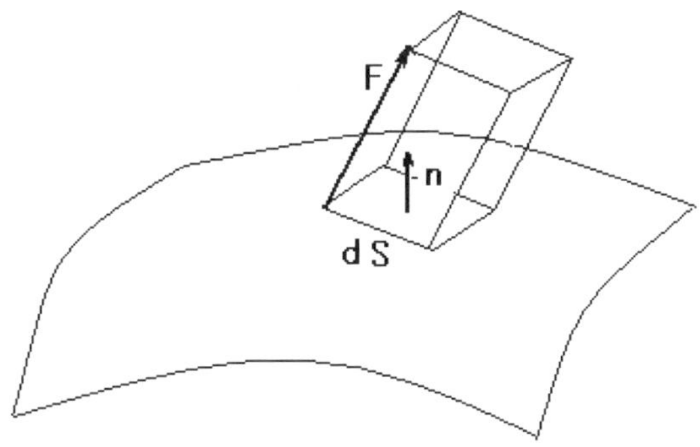

If S is parameterized by $X : D \to S \subset \mathbb{R}^3$ on a plane region D,

$$\mathbf{N} := X_u \times X_v, \qquad \mathbf{n} = \frac{\mathbf{N}}{|\mathbf{N}|}, \qquad d\mathrm{S} = |\mathbf{N}|\, du\, dv, \qquad \mathbf{n}\, d\mathrm{S} = \mathbf{N}\, du\, dv$$

and so we know that

$$\iint_S \mathbf{F} \cdot d\mathbf{S} = \iint_S \mathbf{F} \cdot \mathbf{n}\, d\mathrm{S} = \iint_D \mathbf{F} \cdot \mathbf{N}\, du\, dv.$$

4.3.1 Example

Find the flux of the vector field

$$\mathbf{F} = x\,\mathbf{i} + y\,\mathbf{j} + z\,\mathbf{k}$$

out of the region enclosed by the ellipsoid $z = x^2 + y^2$ and the plane $z = 4$.

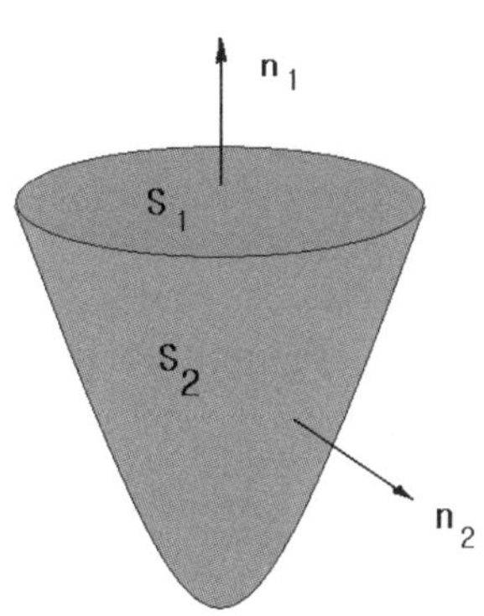

Solution. Let S_1 be the upper disk, and

$$\mathbf{n}_1 := \mathbf{k}$$

be the outward unit normal vector. Also let S_2 be the portion of the ellipsoid, and $\mathbf{n}_2$ be the outward unit normal vector, where $\mathbf{n}_2$ is obtained by "normalizing"

$$\mathbf{N} := -X_x \times X_y = -\left(1, 0, \frac{\partial z}{\partial x}\right) \times \left(0, 1, \frac{\partial z}{\partial y}\right) = \left(\frac{\partial z}{\partial x}, \frac{\partial z}{\partial y}, -1\right) = (2x, 2y, -1)$$

Since S_1 is of radius 2,

$$\iint_{S_1} \mathbf{F} \cdot \mathbf{n}_1 \, dS = \iint_{S_1} z \, dS = \iint_{S_1} 4 \, dS = 4 \cdot \left(\pi \cdot 2^2\right) = 16\pi.$$

On the other hand, we get

$$\begin{aligned}
\iint_{S_2} \mathbf{F} \cdot \mathbf{n}_2 \, dS &= \iint_{x^2+y^2 \le 4} \mathbf{F} \cdot \mathbf{N} \, dx \, dy \\
&= \iint_{x^2+y^2 \le 4} (2x^2 + 2y^2 - (x^2 + y^2)) \, dx \, dy \\
&= \int_0^{2\pi} \int_0^2 r^2 \cdot r \, dr \, d\theta = 2\pi \cdot \left(\frac{1}{4} \cdot 2^4\right) \\
&= 8\pi.
\end{aligned}$$

Therefore, the flux is

$$16\pi + 8\pi = 24\pi.$$

4.3.2 Basic Exercise

For the solid angle vector field $\mathbf{A} = \frac{\mathbf{r}}{r^3}$, find the integral out of the surface[24]

S_1 : the portion of the surface $z = 1 - x^2 - y^2$ satisfying $z \ge 0$.

Here the orientation is given by the unit normal vector $\mathbf{n}$ satisfying $\mathbf{n} \cdot \mathbf{k} \ge 0$.

---

[24] Answer: 2π

Exercise: Chapter 16 Section 4

1. Find the absolute value of the integral $\iint_S \mathbf{F} \cdot d\mathbf{S}$ over the following surface S.

   (1) $\mathbf{F} = x\mathbf{i} + y\mathbf{j} + z\mathbf{k}$, S is the portion of the plane $2x + 2y + z = 3$ in the first octant

   (2) $\mathbf{F} = 2y\mathbf{i} + 3z\mathbf{k}$, S is the portion of $z = 3x + 2$ inside the cylinder $x^2 + y^2 = 4$

   (3) $\mathbf{F} = z\mathbf{k}$, S is the upper hemisphere $x^2 + y^2 + z^2 = 2^2$, $z \geq 0$

   (4) $\mathbf{F} = y\mathbf{i} - x\mathbf{j}$, S is the portion of $z = r$ inside the cylinder $r = 3$ in cylindrical coordinates

   (5) $\mathbf{F} = 2x\mathbf{i} + 2y\mathbf{j} + 3\mathbf{k}$, S is the portion of the ellipsoid $z = 4 - x^2 - y^2$ above the xy-plane

   (6) $\mathbf{F} = 3xy^2\,\mathbf{i} + 3x^2y\,\mathbf{j} + z^3\,\mathbf{k}$, S is the sphere $x^2 + y^2 + z^2 = 1$

2. Let S be the boundary of the region enclosed by the coordinate planes and the plane $2x + 3y + z = 6$. Find the flux of the vector field $\mathbf{F} = x\mathbf{i} - y\mathbf{j}$ out of the surface S.

3. Let S be the boundary of the region enclosed by an ellipsoid $z = 4 - x^2 - y^2$ and the xy-plane. Find the flux of the vector field $\mathbf{F} = 2x\mathbf{i} + 2\mathbf{j} + 3\mathbf{k}$ out of the surface S.

4. If we let S be the boundary of the region enclosed by two ellipsoids $z = x^2 + y^2$ and $z = 18 - x^2 - y^2$, find the flux of the vector field $\mathbf{F} = z^2\mathbf{k}$ out of the surface S.

5. Let $\mathbf{N}(u, v)$ be a normal vector field of a regular surface $X(u, v)$ and $\tilde{\mathbf{N}}(\tilde{u}, \tilde{v})$ be the normal vector field of an orientation-preserving reparametrization $\tilde{X}(\tilde{u}, \tilde{v})$ of X. Show that

$$\frac{\mathbf{N}}{|\mathbf{N}|} = \frac{\tilde{\mathbf{N}}}{|\tilde{\mathbf{N}}|}.$$

6. Let $\mathbf{F}$ be a vector field defined on the space, S be a sphere with center at the origin, and $\mathbf{r}$ be the position vector field. Show that

$$\iint_S \mathbf{r} \times \mathbf{F} \cdot d\mathbf{S} = 0.$$

Chapter 17

Divergence Theorem

For a vector field defined on a bounded region in the three-dimensional space, the divergence theorem says that the flux of the vector field through the boundary of the region is equal to the integral of the divergence of the vector field over the interior of the region. We already considered such a principle for plane vector fields in Chapter 15 and it is similar to the fundamental theorem of calculus over a real interval. We study Gauss' Law as an application of the divergence theorem. For further reading, refer to [Schey], [Davis, Snider], [Danielson].

1 Divergence Theorem

The **divergence theorem**[1] states:

> The net outward flux of a vector field through a closed surface in space is equal to the triple integral of the divergence of the vector field over the region enclosed by the surface (that is, the net rate of change of volume of the region).

Theorem 1.0.1 (Divergence Theorem) *For a C^1 vector field* $\mathbf{F}$ *defined on a region*[2] *R in $\mathbb{R}^3$,*

$$\iint_{\partial R} \mathbf{F} \cdot d\mathbf{S} = \iiint_R \operatorname{div} \mathbf{F} \, dV.$$

The proof of the divergence theorem is in the last section.[3]

1.0.2 Example

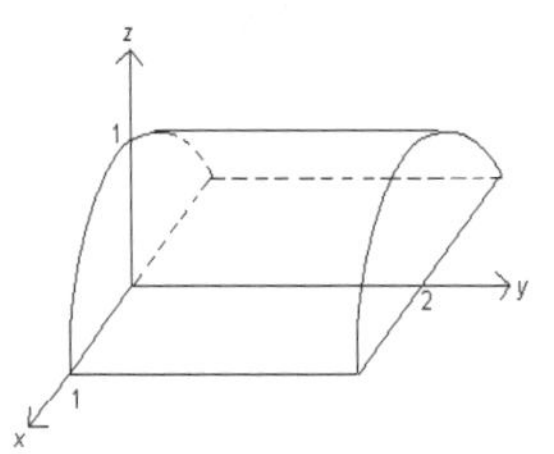

Let S be the boundary of the region R enclosed by three planes $z = 0$, $y = 0$, $y = 2$ and the parabolic cylinder $z = 1 - x^2$. For a vector field

$$\mathbf{F} = (x + \cos y)\,\mathbf{i} + (y + \sin z)\,\mathbf{j} + (z + e^x)\,\mathbf{k},$$

find $\displaystyle\iint_S \mathbf{F} \cdot d\mathbf{S}$.

---

[1] Refer to the last section in Chapter 15. This theorem was discovered by Gauss in Germany and Michel Ostrogradski (1810–1861) in Russia in the 1830s.

[2] Here we assume that a region is a closed, bounded region whose boundary is a piecewise regular surface. A surface made of a union of regular surfaces $S_1, \ldots, S_n$ whose boundaries are piecewise regular curves is called a **piecewise regular surface** if

$$S_i \cap S_j \subset \partial S_i \cap \partial S_j.$$

And the orientation of the surface ∂R is determined by the outward unit normal vector $\mathbf{n}$ over the boundary ∂R of the region R.

[3] Before we consider the general case, we may consider and prove an easier case where the region is a rectangular box. This case can be proved in a similar way to the two-dimensional case. It would be useful to consider some easy case first.

Solution. Since $\operatorname{div}\mathbf{F} = 3$,

$$\begin{aligned}\iint_S \mathbf{F}\cdot d\mathbf{S} &= \iiint_R \operatorname{div}\mathbf{F}\,dV = \int_{-1}^{1}\int_0^2\int_0^{1-x^2} 3\,dz\,dy\,dx \\ &= 3\cdot 2\cdot 2\int_0^1 (1-x^2)\,dx = 8.\end{aligned}$$

1.0.3 Example

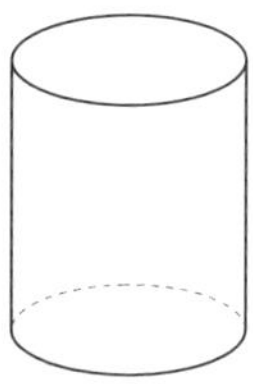

Let R be the region enclosed by two planes $z = 0$, $z = 3$ and the cylinder $x^2 + y^2 = 4$ and let S be the boundary of R. Find the flux of the vector field

$$\mathbf{F} = (x^2 + y^2 + z^2)(x\,\mathbf{i} + y\,\mathbf{j} + z\,\mathbf{k})$$

over the surface S.

Solution. Since

$$\operatorname{div}\mathbf{F} = 5(x^2 + y^2 + z^2),$$

we get

$$\begin{aligned}\iint_S \mathbf{F}\cdot d\mathbf{S} &= \iiint_R 5(x^2+y^2+z^2)\,dx\,dy\,dz \\ &= \int_0^{2\pi}\int_0^2\int_0^3 5(r^2+z^2)\,r dz\,dr\,d\theta \qquad \text{(cylindrical coordinates)} \\ &= 5\left(\int_0^{2\pi} d\theta\right)\int_0^2 (3r^3 + 9r)\,dr = 10\pi\cdot\left(\frac{3}{4}\cdot 2^4 + \frac{9}{2}\cdot 2\right) \\ &= 300\pi.\end{aligned}$$

1.0.4 Example

For the boundary S of the region R, show that the volume of R is

$$\operatorname{vol}(R) = \frac{1}{3}\iint_S (x\mathbf{i} + y\mathbf{j} + z\mathbf{k})\cdot d\mathbf{S}.$$

Solution. It immediately comes from the divergence theorem and the fact that $\operatorname{div}(x\mathbf{i} + y\mathbf{j} + z\mathbf{k}) = 3$.

1.0.5 Example

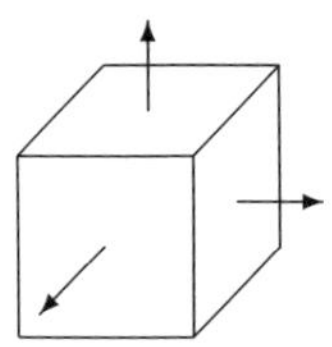

Find the flux of the solid angle vector field

$$\mathbf{A} := \frac{\mathbf{r}}{r^3}$$

out of the cube of side length 1 with the center at the origin.[4]

Solution. Let R be the given cube. Then the flux is

$$\iint_{\partial R} \mathbf{A} \cdot d\mathbf{S}.$$

If we let $\mathbf{B}$ be a very small ball of radius r with the center at the origin, then the vector field $\mathbf{A}$ is well-defined on the region

$$R - \mathbf{B}$$

and $\operatorname{div} \mathbf{A} = 0$. Therefore from the divergence theorem, we get

$$0 = \iiint_{R-\mathbf{B}} \operatorname{div} \mathbf{A}\, dV = \iint_{\partial R} \mathbf{A} \cdot d\mathbf{S} - \iint_{\partial \mathbf{B}} \mathbf{A} \cdot d\mathbf{S}.$$

Therefore, we have

$$\begin{aligned} \iint_{\partial R} \mathbf{A} \cdot d\mathbf{S} &= \iint_{\partial \mathbf{B}} \mathbf{A} \cdot d\mathbf{S} \\ &= \iint_{\partial \mathbf{B}} \frac{\mathbf{r}}{r^3} \cdot \frac{\mathbf{r}}{r}\, dS = \iint_{\partial \mathbf{B}} \frac{1}{r^2}\, dS \\ &= \frac{1}{r^2} \cdot \left(4\pi r^2\right) = 4\pi. \end{aligned}$$

1.0.6 Basic Exercise

1. Find the flux of the solid angle vector field $\mathbf{A}$ out of the sphere of radius 1 with the center at $(0, 3, 2)$.[5]
2. Find the integral of the solid angle vector field $\mathbf{A}$ out of the following surface.[6] Here the orientation is given by the unit normal vector $\mathbf{n}$ satisfying $\mathbf{n} \cdot \mathbf{k} \geq 0$.

   (1) $S_2 := \{x^2 + y^2 + z^2 = 4, \quad z \geq 0\}$
   (2) $S_3 := \{x^2 + y^2 + z^2 = 9, \quad z \leq 0\}$

---

[4] It can be any closed surface which contains the origin in its interior.

[5] Answer: 0

[6] Answer: 2π, -2π

Exercise: Chapter 17 Section 1

1. Show that the divergence theorem holds for the following case.

   (1) $\mathbf{F} = x\mathbf{i} + y\mathbf{j} + z\mathbf{k}, \quad S : x^2 + y^2 + z^2 = 1$

   (2) $\mathbf{F} = |\mathbf{r}|\mathbf{r}, \quad \mathbf{r} = x\mathbf{i} + y\mathbf{j} + z\mathbf{k}, \quad S : x^2 + y^2 + z^2 = 9$

   (3) $\mathbf{F} = x\mathbf{i} + y\mathbf{j} + z\mathbf{k}$, S is the boundary of the region enclosed by three coordinate planes and planes $x = 2$, $y = 2$, $z = 2$

   (4) $\mathbf{F} = xy\mathbf{i} + yz\mathbf{j} + zx\mathbf{k}$, S is the surface in the previous problem (3)

   (5) $\mathbf{F} = (x+y)\mathbf{i} + (y+z)\mathbf{j} + (z+x)\mathbf{k}$, S is the boundary of the tetrahedron enclosed by three coordinate planes and the plane $x + y + z = 1$

2. Find the flux of the vector field

$$\mathbf{F} = 3xy^2\,\mathbf{i} + 3x^2y\,\mathbf{j} + z^3\,\mathbf{k}$$

   out of the standard unit sphere

$$S \; : \; x^2 + y^2 + z^2 = 1.$$

3. For the outward unit normal vector field

$$\mathbf{n} = n_1\,\mathbf{i} + n_2\,\mathbf{j} + n_3\,\mathbf{k}$$

   of a closed surface S, show that the volume V of the interior of S is

$$V = \iint_S n_1 x \; dS = \iint_S n_2 y \; dS = \iint_S n_3 z \; dS.$$

4. For a $\mathcal{C}^1$ vector field $\mathbf{F}$ defined on an open region U in the space, show that if $\iint_{\partial B} \mathbf{F} \cdot d\mathbf{S} = 0$ for any ball B contained in U, then $\mathbf{F}$ is incompressible.

5. For an incompressible $\mathbf{F}$ defined on an open set U in the space, show that if

$$\iint_S \mathbf{F} \cdot d\mathbf{S} \neq 0$$

   for some closed surface S contained in U, then there exists a point inside S at which $\mathbf{F}$ is not defined.

2 Gauss' Law

Let $\mathbf{E}_1, \ldots, \mathbf{E}_n$ be the electric fields generated by point charges at $P_1, \ldots, P_n$ with charges $q_1, \ldots, q_n$, respectively.[7] Then from Coulomb's law,

$$\mathbf{E}_i(X) = \frac{q_i}{4\pi}\mathbf{A}_{P_i}(X) \qquad (i = 1, \ldots, n),$$

where

$$\mathbf{A}_P(X) := \frac{X - P}{|X - P|^3}.$$

Therefore each electric field $\mathbf{E}_i$ is an incompressible field defined on $\mathbb{R}^3 - \{P_i\}$ and thus the total electric field

$$\mathbf{E} := \mathbf{E}_1 + \cdots + \mathbf{E}_n$$

is also an incompressible field defined on $\mathbb{R}^3 - \{P_1, \ldots, P_n\}$:

$$\operatorname{div} \mathbf{E} = 0.$$

Theorem 2.0.1 (Gauss' Law) *For a closed surface*

$$S \subset \mathbb{R}^3 - \{P_1, \ldots, P_n\},$$

the total electric charge Q enclosed by the surface S is given by[8]

$$Q = \iint_S \mathbf{E} \cdot d\mathbf{S}.$$

Proof. For convenience, suppose that $P_1, \ldots, P_m$ are inside S and $P_{m+1}, \ldots, P_n$ are outside S. For each $i \in \{1, \ldots, n\}$, let B_i be a ball of radius r with the center at P_i. Here we will pick $r > 0$ small enough so that the whole ball B_i is contained either in the interior of S or in the exterior of S and also B_i contains only P_i among $P_1, \ldots, P_n$.

---

[7] We assume that $P_1, \ldots, P_n$ are all distinct points. The electric field $\mathbf{E}_i$ represents a **repulsion** if $q_i > 0$, and an **attraction** if $q_i < 0$.

[8] If $P_1, \ldots, P_m$ are inside S and $P_{m+1}, \ldots, P_n$ are outside S, then $Q = q_1 + \cdots + q_m$. Gauss' Law can be thought of as a property of electric fields, or a property of the solid angle vector field.

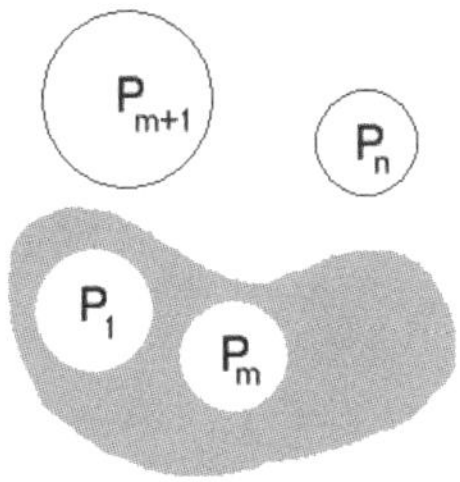

Then the vector field **E** is well-defined on

$$(S \cup \operatorname{int} S) - (B_1 \cup \cdots \cup B_m)$$

and its divergence is zero. So if we let S_i be the boundary of B_i, we get

$$\begin{aligned} 0 &= \iiint_{(S\cup \operatorname{int} S)-(B_1\cup\cdots\cup B_m)} \operatorname{div}\mathbf{E}\, dV \\ &= \left(\iint_S - \iint_{S_1} - \cdots - \iint_{S_m}\right) \mathbf{E}\cdot d\mathbf{S}, \end{aligned}$$

and thus

$$\iint_S \mathbf{E}\cdot d\mathbf{S} = \iint_{S_1} \mathbf{E}\cdot d\mathbf{S} + \cdots + \iint_{S_m} \mathbf{E}\cdot d\mathbf{S}.$$

On the other hand,

$$\iint_{S_i} \mathbf{E}_j \cdot d\mathbf{S} = \begin{cases} \iiint_{B_i} \operatorname{div}\mathbf{E}_j\, dV = 0, & i \neq j \\ q_i, & i = j \end{cases}$$

and so we have

$$\iint_{S_1} \mathbf{E}\cdot d\mathbf{S} + \cdots + \iint_{S_m} \mathbf{E}\cdot d\mathbf{S} = q_1 + \cdots + q_m = Q$$

and we are done. □

Exercise: Chapter 17 Section 2

1. Show that the magnitude of the gravitational force acting on a point $P = (r, 0, 0)$ by the whole z-axis

$$\mathbf{G}(P) = \int_{-\infty}^{\infty} \mathbf{A}_P(0,0,z)\,dz$$

is inversely proportional to r.

2. Show that the gravitational force acting on one point inside a sphere by the whole sphere is zero.

3. Let S be a sphere of radius r and P be a point with the distance R from the center of S. Find the gravitational force acting on P by S

$$\mathbf{G}(P) := \iint_S \mathbf{A}_P\, d\mathrm{S}.$$

4. Gauss' law holds not only when the charge distribution is discrete but also when the charge distribution is continuous. For example, for a continuous function ρ defined on a bounded region[9] R, consider the vector field[10]

$$\mathbf{E}(X) := \frac{1}{4\pi}\iiint_R \rho(P)\mathbf{A}_P(X)\,dV(P) = -\frac{1}{4\pi}\iiint_R \rho(P)\mathbf{A}_X(P)\,dV(P).$$

If $X \notin R$, this integral is well defined as an integral of a continuous vector field. On the other hand, if $X \in R$, it is an improper integral but its value still converges[11] and so $\mathbf{E}$ is well-defined in the whole space.

Gauss' law states that

[9]We assume that the boundary is a piecewise regular surface.

[10]This vector field can be thought of as an electric field in the space with electric charge density function ρ.

[11]For instance, when X is the origin and an interior point of R, if the radius r of the ball with the center at the origin is small, this ball is contained in R and so

$$\begin{aligned}\left|\iiint_{|P|\le r} \rho(P)\mathbf{A}(P)\,dV(P)\right| &\le \left(\max_{|P|\le r}|\rho(P)|\right)\cdot\iiint_{|P|\le r}\frac{1}{|P|^2}\,dV(P)\\ &= \left(\max_{|P|\le r}|\rho(P)|\right)\cdot\left(\int_0^{2\pi}d\theta\right)\cdot\left(\int_0^{\pi}\sin\varphi\,d\varphi\right)\cdot r\\ &< \infty.\end{aligned}$$

(Refer to Exercise in Chapter 14 Section 4.)

Theorem 2.0.2 (Gauss' Law) *The "net electric charge" inside a region R is equal to the electric flux though its boundary ∂R:*

$$\iiint_R \rho\, dV = \iint_{\partial R} \mathbf{E} \cdot d\mathbf{S}.$$

Use Gauss' law and the divergence theorem to derive[12]

$$\operatorname{div} \mathbf{E} = \rho.$$

[12]The classical **electromagnetism** is described by four differential equations, called **Maxwell's equations**. These differential equations describe the interrelationship between electric fields, magnetic fields, electric charges and electric currents. The above equation is one of Maxwell's equations.

3 Proof of Divergence Theorem

Theorem 3.0.1 (Divergence Theorem) *For a $\mathcal{C}^1$ vector field* $\mathbf{F}$ *defined on a region R in $\mathbb{R}^3$,*

$$\iint_{\partial R} \mathbf{F} \cdot d\mathbf{S} = \iiint_R \operatorname{div} \mathbf{F} \, dV.$$

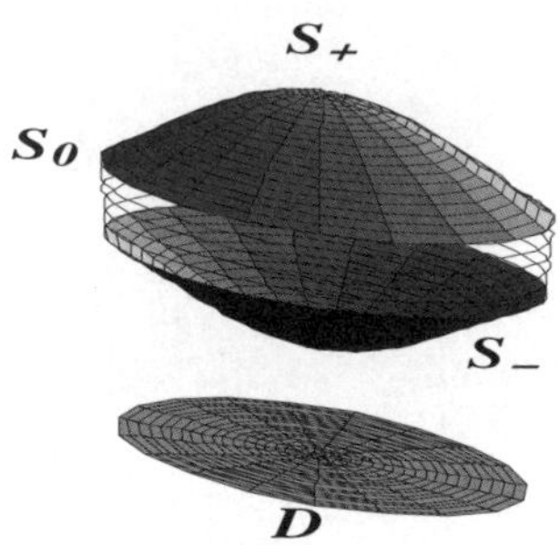

Proof. We will prove the divergence theorem when the region R is given as follows.[13]

$$R: \qquad g(x,y) \le z \le h(x,y), \qquad (x,y) \in D.$$

Here D is a closed region in $\mathbb{R}^2$ whose boundary is a piecewise regular curve.

Let's denote the surfaces given by the graphs of $h(x,y)$ and $g(x,y)$ by

$$S_+: \quad z = h(x,y), \qquad S_-: \quad z = g(x,y),$$

respectively, and the cylindrical surface between S_+ and S_- by S_0. For a given $\mathcal{C}^1$ vector field

$$\mathbf{F} = f_1\mathbf{i} + f_2\mathbf{j} + f_3\mathbf{k},$$

we have

$$\iiint_R \operatorname{div} \mathbf{F} \, dV = \iiint_R \left(\frac{\partial f_1}{\partial x} + \frac{\partial f_2}{\partial y} + \frac{\partial f_3}{\partial z} \right) dx\, dy\, dz \qquad (17.1)$$

$$= \iint_D \left(\int_{g(x,y)}^{h(x,y)} \left(\frac{\partial f_1}{\partial x} + \frac{\partial f_2}{\partial y} + \frac{\partial f_3}{\partial z} \right) dz \right) dx\, dy. \qquad (17.2)$$

---

[13] The general case can be treated by partitioning the region into simple pieces, considering each pieces and then summing them up.

On the other hand, in the above equation, we have[14]

$$\begin{aligned}
&\int_{g(x,y)}^{h(x,y)} \left(\frac{\partial f_1}{\partial x} + \frac{\partial f_2}{\partial y} + \frac{\partial f_3}{\partial z}\right) dz \\
=\ &\left(\frac{\partial}{\partial x}\int_{g(x,y)}^{h(x,y)} f_1\, dz - f_1(x,y,h)\frac{\partial h}{\partial x} + f_1(x,y,g)\frac{\partial g}{\partial x}\right) \\
&+\left(\frac{\partial}{\partial y}\int_{g(x,y)}^{h(x,y)} f_2\, dz - f_2(x,y,h)\frac{\partial h}{\partial y} + f_2(x,y,g)\frac{\partial g}{\partial y}\right) \\
&+\left(f_3(x,y,h) - f_3(x,y,g)\right).
\end{aligned}$$

Therefore from (17.2) we have

$$\iiint_R \operatorname{div} \mathbf{F}\, dV \tag{17.3}$$

$$= \iint_D \left(-f_1(x,y,h)\frac{\partial h}{\partial x} - f_2(x,y,h)\frac{\partial h}{\partial y} + f_3(x,y,h)\right) dx\, dy \tag{17.4}$$

$$+ \iint_D \left(f_1(x,y,g)\frac{\partial g}{\partial x} + f_2(x,y,g)\frac{\partial g}{\partial y} - f_3(x,y,g)\right) dx\, dy \tag{17.5}$$

$$+ \iint_D \left(\frac{\partial}{\partial x}\int_{g(x,y)}^{h(x,y)} f_1\, dz + \frac{\partial}{\partial y}\int_{g(x,y)}^{h(x,y)} f_2\, dz\right) dx\, dy \tag{17.6}$$

and it suffices to show that the equation above is equal to

$$\iint_{S_+} \mathbf{F}\cdot d\mathbf{S} + \iint_{S_-} \mathbf{F}\cdot d\mathbf{S} + \iint_{S_0} \mathbf{F}\cdot d\mathbf{S}.$$

Since the upper boundary S_+ is parameterized by

$$X_+(x,y) := (x, y, h(x,y)) \qquad (x,y) \in D,$$

we may take

$$\mathbf{N} := \frac{\partial X_+}{\partial x} \times \frac{\partial X_+}{\partial y} = -\frac{\partial h}{\partial x}\mathbf{i} - \frac{\partial h}{\partial y}\mathbf{j} + \mathbf{k}$$

as a normal vector field to S_+. In order to have a proper orientation, the third component, that is, the **k**-component of the unit normal vector field in S_+ must be positive, and thus we have

$$\mathbf{n} = \mathbf{N}/|\mathbf{N}|$$

and

$$\mathbf{n}\, dS = \mathbf{N}\, dx\, dy.$$

[14]Refer to Exercise 2 in Chapter 11 Section 1.

Therefore

$$\begin{aligned}
\iint_{S_+} \mathbf{F}\cdot d\mathbf{S} &= \iint_{S_+} \mathbf{F}\cdot\mathbf{n}\, dS \\
&= \iint_D \mathbf{F}(x,y,h)\cdot\mathbf{N}\, dx\, dy \\
&= \iint_D \left(-f_1(x,y,h)\frac{\partial h}{\partial x} - f_2(x,y,h)\frac{\partial h}{\partial y} + f_3(x,y,h)\right) dx\, dy
\end{aligned}$$

and this coincides with (17.4).

Similarly, the lower boundary S_- is parameterized by

$$X_-(x,y) := (x,y,g(x,y)) \qquad (x,y)\in D,$$

and then a normal vector field to S_- is

$$\mathbf{N} := \frac{\partial X_-}{\partial x}\times\frac{\partial X_-}{\partial y} = -\frac{\partial g}{\partial x}\mathbf{i} - \frac{\partial g}{\partial y}\mathbf{j} + \mathbf{k}.$$

Here the unit normal vector field which determines the orientation of S_- is

$$\mathbf{n} = -\mathbf{N}/|\mathbf{N}|$$

and thus

$$\mathbf{n}\, d\mathrm{S} = (-\mathbf{N})\, dx\, dy.$$

Therefore we have

$$\begin{aligned}
\iint_{S_-} \mathbf{F}\cdot d\mathbf{S} &= \iint_{S_-} \mathbf{F}\cdot\mathbf{n}\, d\mathrm{S} \\
&= \iint_D \mathbf{F}(x,y,g)\cdot(-\mathbf{N})\, dx\, dy \\
&= \iint_D \left(f_1(x,y,g)\frac{\partial g}{\partial x} + f_2(x,y,g)\frac{\partial g}{\partial y} - f_3(x,y,g)\right) dx\, dy,
\end{aligned}$$

which coincides with (17.5).

If we denote the unit normal vector field out of the region D in the plane by

$$\mathbf{n}(x,y) = n_1(x,y)\,\mathbf{i} + n_2(x,y)\,\mathbf{j},$$

then

$$\begin{aligned}
\iint_{S_0} \mathbf{F} \cdot d\mathbf{S} &= \int_{\partial D} \left(\int_{g(x,y)}^{h(x,y)} \mathbf{F} \cdot \mathbf{n}\, dz \right) ds \\
&= \int_{\partial D} \left(\int_{g(x,y)}^{h(x,y)} (f_1 n_1 + f_2 n_2)\, dz \right) ds \\
&= \int_{\partial D} \left(n_1 \int_{g(x,y)}^{h(x,y)} f_1\, dz + n_2 \int_{g(x,y)}^{h(x,y)} f_2\, dz \right) ds \\
&= \iint_{D} \left(\frac{\partial}{\partial x} \int_{g(x,y)}^{h(x,y)} f_1\, dz + \frac{\partial}{\partial y} \int_{g(x,y)}^{h(x,y)} f_2\, dz \right) dx\, dy,
\end{aligned}$$

where the divergence theorem in the plane is used in the last step. It is equal to (17.6). So we complete the proof. □

Chapter 17 Advanced Exercise

1. The **Laplacian** of a function $f(x, y, z)$ is defined by[15]

$$\nabla^2 f := \mathrm{div}(\mathrm{grad} f) = \frac{\partial^2 f}{\partial x^2} + \frac{\partial^2 f}{\partial y^2} + \frac{\partial^2 f}{\partial z^2}.$$

For the distance function r from the origin, show that

$$\nabla^2 \left(\frac{1}{r}\right) = 0.$$

Also when we let $r_P(X) := |X - P|$ for a point P in the space, show that

$$\nabla^2 \left(\frac{1}{r_P}\right) = 0.$$

2. For a continuous function μ defined on a region $R \subset \mathbb{R}^3$ and a point (x, y, z) outside R, when we define

$$f(x, y, z) := \iiint_R \frac{\mu(x', y', z')}{\sqrt{(x - x')^2 + (y - y')^2 + (z - z')^2}}\, dx'\, dy'\, dz',$$

show that

$$\nabla^2 f = 0.$$

3. For the outward unit normal vector $\mathbf{n}$ for the boundary S of the region R, if we denote the **n-directional derivative** by

$$\frac{\partial f}{\partial \mathbf{n}} := \mathrm{grad} f \cdot \mathbf{n},$$

show that

$$\iint_S \frac{\partial f}{\partial \mathbf{n}}\, d\mathrm{S} = \iiint_R \nabla^2 f\, dV.$$

Also show that if f is harmonic, that is, f satisfies $\nabla^2 f = 0$, then

$$\iint_S f \frac{\partial f}{\partial \mathbf{n}}\, d\mathrm{S} = \iiint_R |\mathrm{grad} f|^2\, dV.$$

---

[15] $\nabla^2 f$ is also denoted by Δf. Pierre Simon Laplace (1749–1827) showed that in the electric field (or gravitational field) the potential energy function f satisfies **Laplace's equation**

$$\nabla^2 f = 0.$$

Siméon-Denis Poisson (1781–1840) showed that Laplace's equation is satisfied only on a region with no charge and **Poisson's equation**

$$\nabla^2 f = \mu$$

is satisfied on a region where the charge density is μ.

4. Show that if a function f defined on a bounded region R in the space is harmonic in R and identically zero on the boundary of R, then f is identically zero throughout R.

5. (**Green's Identity**) When S be the boundary of a region R in the space, show that following.

$$\text{(1)} \quad \iint_S f\frac{\partial g}{\partial \mathbf{n}}\,d\mathrm{S} = \iiint_R (f\nabla^2 g + \mathrm{grad} f \cdot \mathrm{grad}\, g)\,dV$$

$$\text{(2)} \quad \iint_S \left(f\frac{\partial g}{\partial \mathbf{n}} - g\frac{\partial f}{\partial \mathbf{n}}\right)\,d\mathrm{S} = \iiint_R (f\nabla^2 g - g\nabla^2 f)\,dV$$

6. For a point P in $\mathbb{R}^n$, the n-dimensional **solid angle vector field** is defined as follows:

$$\mathbf{A}_n(X) := \frac{X-P}{|X-P|^n}, \qquad X \in \mathbb{R}^n - \{P\}.$$

Show that

$$\mathrm{div}\,\mathbf{A}_n = 0.$$

Also for $n \neq 2$, show that if we let $r_P(X) := |X-P|$ then

$$\mathbf{A}_n = \mathrm{grad}\left(\frac{1}{2-n}\frac{1}{r_P^{n-2}}\right).$$

7. For a differentiable function $f : (0,\infty) \to \mathbb{R}$, show that the divergence of a vector field

$$\mathbf{F} := f(r)\frac{\mathbf{r}}{r^n}$$

in $\mathbb{R}^n$ is zero if and only if f is constant.

8. (**Continuity Equation**) Let $\rho(x,y,z,t)$ be the density of a fluid at a point (x,y,z) at time t, and $\mathbf{v}(x,y,z,t)$ be the velocity vector. Then the net rate of change of mass in the region R is

$$\frac{d}{dt}\iiint_R \rho(x,y,z,t)\,dV = \iint_R \frac{\partial \rho}{\partial t}\,dV,$$

and it is equal to the flux of the fluid flowing in R through the boundary ∂R:

$$\iiint_R \frac{\partial \rho}{\partial t}\,dV = -\iint_{\partial R} \rho\mathbf{v}\cdot d\mathbf{s}.$$

Use this equation to derive[16]

$$\frac{\partial \rho}{\partial t} = -\mathrm{div}\,(\rho\mathbf{v}).$$

[16]Usually $\mathbf{J} := \rho\mathbf{v}$ is called the **current density**.

9. Let B_r be the ball of radius r with the center at P. Then for a vector field $\mathbf{F}$ defined near P, there exists a point $P^* \in B_r$ such that

$$\iint_{\partial B_r} \mathbf{F} \cdot d\mathbf{S} = \iiint_{B_r} \operatorname{div} \mathbf{F} \, dV = \operatorname{vol}(B_r) \cdot \operatorname{div} \mathbf{F}(P^*).$$

Therefore we get

$$\operatorname{div} \mathbf{F}(P) = \lim_{r \to 0} \frac{1}{\operatorname{vol}(B_r)} \iint_{\partial B_r} \mathbf{F} \cdot d\mathbf{S}.$$

From this equation, explain the reason why $\operatorname{div} \mathbf{F}$ is called the "divergence".

Chapter 18

Curl and Stokes' Theorem

The curl gives us one way to differentiate a vector field As you can guess from the name, the curl measures the circulation of the given vector field. The concept of the curl is first introduced by James Clerk Maxwell (1831–1879) in his book *Treatise on Electricity and Magnetism* (1873). This concept was later generalized by Élie Cartan (1869–1951) and called the **exterior derivative**. Stokes[1] assigned the proof of Stokes' theorem as a part of the examination for the Smith's Prize at the University of Cambridge in 1854. This theorem was first found in the letter to Stokes from Thomson[2] in 1850.

[1]The Navier-Stokes equations, which are the most important equations in fluid mechanics, are also named after Claude-Louis Navier and George Gabriel Stokes (1819–1903) .

[2]William Thomson (1824–1907) became Lord Kelvin in 1892 by Queen Victoria. According to the Guinness Book of Records, a mathematician and physicist Kelvin is the youngest and the fastest graduate of university. Kelvin entered Glasgow University, where his father was a professor of engineering and physics, at the age of 10 on October, 1834 and and graduated at November 14, in the same year. He devised the concept of absolute temperature. Kelvin first introduced the **inversion** with respect to a sphere in the theory of gravity. His name, Kelvin, also used as the scale of absolute temperature (° K).

1 Curl

For a vector field in the three-dimensional space

$$\begin{aligned}\mathbf{F}(x,y,z) &= (f_1(x,y,z),\ f_2(x,y,z),\ f_3(x,y,z)) \\ &= f_1(x,y,z)\,\mathbf{i} + f_2(x,y,z)\,\mathbf{j} + f_3(x,y,z)\,\mathbf{k} \\ &= \sum_{j=1}^{3} f_j(x,y,z)\,\mathbf{e}_j,\end{aligned}$$

the **curl** of **F** is defined by[3]

$$\begin{aligned}\operatorname{curl}\mathbf{F} &:= (D_2f_3 - D_3f_2)\,\mathbf{i} + (D_3f_1 - D_1f_3)\,\mathbf{j} + (D_1f_2 - D_2f_1)\,\mathbf{k} \\ &= \left(\frac{\partial f_3}{\partial y} - \frac{\partial f_2}{\partial z}\right)\mathbf{i} + \left(\frac{\partial f_1}{\partial z} - \frac{\partial f_3}{\partial x}\right)\mathbf{j} + \left(\frac{\partial f_2}{\partial x} - \frac{\partial f_1}{\partial y}\right)\mathbf{k}.\end{aligned}$$

1.0.1 Basic Exercise

Show that the following.[4]

$$\begin{aligned}\operatorname{curl}\mathbf{F} &= \sum_{j=1}^{3}(\operatorname{grad} f_j)\times\mathbf{e}_j = \sum_{i,j=1}^{3}(D_if_j\,\mathbf{e}_i)\times\mathbf{e}_j \\ &= \sum_{i,j,k=1}^{3}\epsilon_{ijk}D_if_j\,\mathbf{e}_k = \sum_{i,j,k=1}^{3}\frac{1}{2}\epsilon_{ijk}\,(D_if_j - D_jf_i)\,\mathbf{e}_k\end{aligned}$$

1.0.2 Example

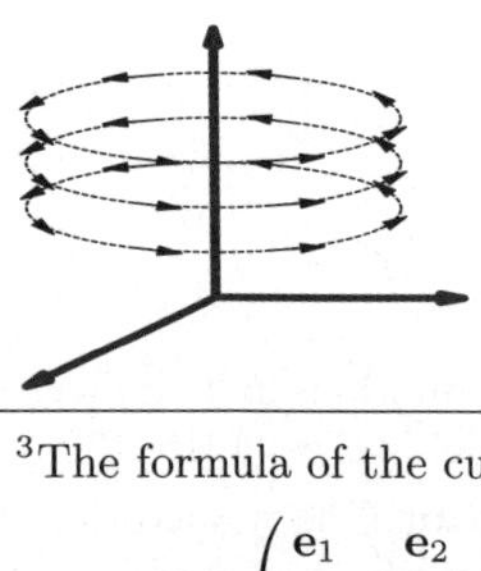

For the vector field

$$\mathbf{F}(x,y,z) = \omega(-y,x,0),$$

its curl is

$$\operatorname{curl}\mathbf{F}(x,y,z) = \omega(0,0,2) = 2\omega\,\mathbf{k}.$$

---

[3]The formula of the curl is like

$$\det\begin{pmatrix}\mathbf{e}_1 & \mathbf{e}_2 & \mathbf{e}_3 \\ D_1 & D_2 & D_3 \\ f_1 & f_2 & f_3\end{pmatrix} = (D_1,D_2,D_3)\times(f_1,f_2,f_3) = \nabla\times(f_1,f_2,f_3)$$

Thus $\operatorname{curl}\mathbf{F}$ is often denoted by $\nabla\times\mathbf{F}$. Instead of $\operatorname{curl}\mathbf{F}$, some use the notation $\operatorname{rot}\mathbf{F}$ as the abbreviation of "rotation" but we use this notation only for the two-dimensional case.

[4]Levi-Civita symbols ϵ_{ijk} appear in Chater 8.

1.0.3 Basic Exercise

Show that if the vector field $\mathbf{F}(x, y, z)$ is obtained from the plane vector field in the xy-plane, that is,

$$\mathbf{F}(x, y, z) = (f_1(x, y),\ f_2(x, y),\ 0)$$

then

$$\operatorname{curl}\mathbf{F} = (0,\ 0,\ D_1 f_2 - D_2 f_1) = \operatorname{rot}(f_1\,\mathbf{i} + f_2\,\mathbf{j})\,\mathbf{k}.$$

The curl of a vector field is independent of the choice of coordinate system: That is, in the space instead of the standard basis $(\mathbf{e}_1, \mathbf{e}_2, \mathbf{e}_3)$ if we use a new orthonormal basis $(\widetilde{\mathbf{e}}_1, \widetilde{\mathbf{e}}_2, \widetilde{\mathbf{e}}_3)$, we can let

$$\widetilde{\mathbf{e}}_j = \sum_{i=1}^{3} a_{ij}\mathbf{e}_i$$

and we know that

$$\sum_{i=1}^{3} a_{ij}a_{ik} = \delta_{jk}.$$

Also for each point, there are the following relations between the original rectangular coordinates (x_1, x_2, x_3) and the new coordinates $(\widetilde{x}_1, \widetilde{x}_2, \widetilde{x}_3)$:

$$x_i = \sum_j a_{ij}\widetilde{x}_j,$$

$$\frac{\partial x_i}{\partial \widetilde{x}_j} = a_{ij}.$$

Now if we let $\widetilde{D}_j = \dfrac{\partial}{\partial \widetilde{x}_j}$, by the chain rule, we have

$$\widetilde{D}_j = \sum_i a_{ij}D_i.$$

Similarly, the vector field $\mathbf{F} = \sum_i f_i\mathbf{e}_i$ can be expressed as $\mathbf{F} = \sum_i \widetilde{f}_i\widetilde{\mathbf{e}}_i$ with

$$\widetilde{f}_j = \sum_i a_{ij}f_i.$$

It is not hard to derive

$$\sum_{i,j} \widetilde{D}_i\widetilde{f}_j(\widetilde{\mathbf{e}}_i \times \widetilde{\mathbf{e}}_j) = \sum_{i,j} D_i f_j(\mathbf{e}_i \times \mathbf{e}_j) = \operatorname{curl}\mathbf{F}.$$

For a connected region D on the plane $z = z_0$ through a point $P = (x_0, y_0, z_0)$, the curl of the vector field $\mathbf{F}(x, y, z) = f_1\,\mathbf{e}_1 + f_2\,\mathbf{e}_2 + f_3\,\mathbf{e}_3$ over the boundary ∂D is

$$\begin{aligned}\int_{\partial D} \mathbf{F} \cdot d\mathbf{s} &= \int_{\partial D} (f_1\,\mathbf{e}_1 + f_2\,\mathbf{e}_2) \cdot d\mathbf{s} \qquad (\Leftarrow\ \mathbf{e}_3 \cdot d\mathbf{s} = 0)\\ &= \iint_D \mathrm{rot}(f_1\,\mathbf{e}_1 + f_2\,\mathbf{e}_2) dV_2 = \iint_D \left(\frac{\partial f_2}{\partial x} - \frac{\partial f_1}{\partial y}\right) dV_2.\end{aligned}$$

Therefore we know that

$$\lim_{D\to P} \frac{1}{\mathrm{area}\, D} \int_{\partial D} \mathbf{F} \cdot d\mathbf{s} = \left(\frac{\partial f_2}{\partial x} - \frac{\partial f_1}{\partial y}\right)(P)$$

is the "z-component" of $\mathrm{curl}\,\mathbf{F}(P)$.

In general, suppose that in the space there is a plane region D whose orientation is given by the unit normal vector $\mathbf{n}$. Then we can introduce a new rectangular coordinate system $(\widetilde{x}, \widetilde{y}, \widetilde{z})$ so that the $\widetilde{z}$-axis points the direction of $\mathbf{n}$. That is, if we let $\widetilde{\mathbf{e}}_1, \widetilde{\mathbf{e}}_2, \widetilde{\mathbf{e}}_3$ be the unit vectors in the directions of axes, then $\widetilde{\mathbf{e}}_3 = \mathbf{n}$. Now

$$\mathbf{F} = \widetilde{f_1}\widetilde{\mathbf{e}}_1 + \widetilde{f_2}\widetilde{\mathbf{e}}_2 + \widetilde{f_3}\widetilde{\mathbf{e}}_3$$

and we can apply the above discussion to the new coordinates to get

$$\int_{\partial D} \mathbf{F}\cdot d\mathbf{s} = \int_{\partial D} \left(\widetilde{f_1}\widetilde{\mathbf{e}}_1 + \widetilde{f_2}\widetilde{\mathbf{e}}_2\right)\cdot d\mathbf{s} = \iint_D \left(\frac{\partial \widetilde{f_2}}{\partial \widetilde{x}} - \frac{\partial \widetilde{f_1}}{\partial \widetilde{y}}\right) dV_2 = \iint_D \mathrm{curl}\mathbf{F}\cdot\mathbf{n} dV_2.$$

1.0.4 Basic Exercise

Let $\mathbf{F}$ be a vector field defined near a point P and D be a region on the plane through P with a unit normal vector $\mathbf{n}$. Show that[5]

$$\lim_{D\to P} \frac{1}{\mathrm{area} D} \int_{\partial D} \mathbf{F} \cdot d\mathbf{s} = \mathrm{curl}\,\mathbf{F}(P) \cdot \mathbf{n}.$$

[5]If we let $\mathbf{F}_D$ be the component of $\mathbf{F}$ with respect to the plane D, the right hand side of the equation is $\mathrm{rot}\mathbf{F}_D$.

1.0.5 Basic Exercise

For vector fields $\mathbf{F}, \mathbf{G}$ and a function h, show that[6]

$$\begin{aligned} \operatorname{curl}(\mathbf{F}+\mathbf{G}) &= \operatorname{curl}\mathbf{F} + \operatorname{curl}\mathbf{G} \\ \operatorname{curl}(h\,\mathbf{F}) &= (\operatorname{grad} h) \times \mathbf{F} + h \operatorname{curl}\mathbf{F} \end{aligned}$$

1.0.6 Example

The curl of the position vector field

$$\mathbf{r}(x,y,z) = x\,\mathbf{i} + y\,\mathbf{j} + z\,\mathbf{k}$$

is identically zero:

$$\operatorname{curl}\mathbf{r} = \mathbf{0}.$$

Therefore for any function $h(r)$ for $r = \sqrt{x^2+y^2+z^2}$, we have

$$\begin{aligned} \operatorname{curl}(h(r)\mathbf{r}) &= (\operatorname{grad} h(r)) \times \mathbf{r} + h(r)\operatorname{curl}\mathbf{r} \\ &= h'(r)\frac{\mathbf{r}}{r} \times \mathbf{r} + h(r)\mathbf{0} \\ &= \mathbf{0}. \end{aligned}$$

In particular, the curl of the solid angle vector field

$$\mathbf{A}(x,y,z) = \frac{1}{r^3}(x\,\mathbf{i} + y\,\mathbf{j} + z\,\mathbf{k})$$

is identically zero:

$$\operatorname{curl}\mathbf{A} = \mathbf{0}.$$

1.0.7 Basic Exercise

In the space, the **solid angle vector field** with respect to a point P is defined by

$$\mathbf{A}_P(X) := \frac{X-P}{|X-P|^3} \qquad (X \in \mathbb{R}^3 - \{P\}).$$

Show that

$$\operatorname{curl}\mathbf{A}_P = \mathbf{0}.$$

---

[6]The second equation is **Leibniz's rule**. It describes how to differentiate a product of two functions. In the above equation, if h is constant, then

$$\operatorname{curl}(h\mathbf{F}) = h\operatorname{curl}\mathbf{F}.$$

`Exercise`: In a rectangle with sides of lengths a and b, how much would the area increase if the sides increase by Δa and Δb respectively?

1.1 Irrotational Vector Fields

A vector field F is called an **irrotational vector field** if its curl is identically zero:

$$\operatorname{curl}\mathbf{F} = \mathbf{0}.$$

1.1.1 Basic Exercise

Show that the vector field $\mathbf{F}(x, y, z) = (yz, zx, xy)$ is irrotational.

The following theorem gives a way to get an irrotational vector field.

Theorem 1.1.2 *If a C^1 vector field $\mathbf{F}$ defined on an open set in $\mathbb{R}^3$ has a potential function, then $\mathbf{F}$ is irrotational.*[7]

Proof. If a vector field $\mathbf{F} = (f_1, f_2, f_3)$ has a potential function φ, we have

$$D_i f_j - D_j f_i = D_i(D_j\varphi) - D_j(D_i\varphi) = 0.$$

From this, we get $\operatorname{curl}\mathbf{F} = \mathbf{0}$. □

1.1.3 Basic Exercise

1. For the solid angle vector field $\mathbf{A}(x, y, z)$, show that

$$\mathbf{A} = \operatorname{grad}\left(-\frac{1}{r}\right)$$

and use this to explain that $\mathbf{A}$ is an irrotational vector field. Here r denotes the distance from the origin.

2. For the **angle element vector field**

$$\mathbf{a}(x, y, z) := -\frac{y}{x^2 + y^2}\,\mathbf{i} + \frac{x}{x^2 + y^2}\,\mathbf{j},$$

(1) show that $\mathbf{a}$ is an irrotational vector field. Also examine whether $\mathbf{a}$ has a potential function.

[7] Conversely, if you study differential equations a little bit(?), you can show that any irrotational vector field is **locally** conservative. Finding a potential function is the problem of integration. To know whether such a problem of integration has a solution, we may use differentiation.

(2) For an oriented closed curve $C \subset \mathbb{R}^3$ which doesn't intersect with the z-axis, show that the following integral gives an integer.[8]

$$\frac{1}{2\pi}\int_C \mathbf{a}\cdot d\mathbf{s}$$

(3) When a curve C is the intersection of the sphere $x^2+(y-3)^2+z^2=2^2$ and the cylinder $x^2+(y-2)^2=1$, find

$$\frac{1}{2\pi}\int_C \mathbf{a}\cdot d\mathbf{s}.$$

[8] This integer is the **winding number** of the curve C winding the z-axis counterclockwise.

Exercise: Chapter 18 Section 1

1. Find the curl of the following vector field.

   (1) $\mathbf{F}(x, y, z) = y\mathbf{i} + x\mathbf{j}$

   (2) $\mathbf{F}(x, y, z) = \left(x^2 + y^2 + z^2\right)(3, 4, 5)$

2. (**Electric Field**) By Coulomb's law, the electric field generated by a point charge q at a point P is given by[9]

$$\mathbf{E}(X) = \frac{q}{4\pi\epsilon_0}\mathbf{A}_P(X) = \frac{q}{4\pi\epsilon_0}\frac{X - P}{|X - P|^3} \qquad (X \neq P).$$

   In general, when the electric field generated by electric charges (P_1, q_1), (P_2, q_2), $\ldots (P_n, q_n)$ is given by

$$\mathbf{E} = \frac{1}{4\pi\epsilon_0}\left(q_1\mathbf{A}_{P_1} + q_2\mathbf{A}_{P_2} + \cdots + q_n\mathbf{A}_{P_n}\right),$$

   show that

$$\operatorname{curl}\mathbf{E} = \mathbf{0}.$$

   Also if the charge is continuously distributed on a region U and the "charge density" is given by $\rho(P)$ at each point P in U, then the electric field is[10]

$$\mathbf{E}(X) = \frac{1}{4\pi\epsilon_0}\iiint_U \rho(P)\,\mathbf{A}_P(X)\,dV(P).$$

   Show that the electric field is irrotational in $\mathbb{R}^3 - U$.

3. For a vector field

$$\mathbf{F}(x, y, z) = \sum_j f_j(x, y, z)\,\mathbf{e}_j$$

   defined near a point O in the space, the rotational force which rotates any point $P + \mathbf{x}$ around P by $\mathbf{F}$ is

$$\mathbf{x} \times \mathbf{F}(P + \mathbf{x}).$$

   Therefore the rotational force with which a sphere of radius r with the center at P rotates around P is

$$\mathbf{G}(r) := \iint_{|\mathbf{x}|=r} \mathbf{x} \times \mathbf{F}(P + \mathbf{x})\,dS.$$

---

[9] Here, 4π is the area of a unit sphere and ϵ_0 is a constant depending on units in use.

[10] The (definite) integral of a vector-valued function is obtained from integrating each component. In other words, If $\mathbf{F} = \sum_i f_i\,\mathbf{e}_i$, then

$$\int \mathbf{F} := \sum_i \left(\int f_i\right)\mathbf{e}_i.$$

If we let $\mathbf{x} = \sum_i x_i \mathbf{e}_i$, we get

$$\mathbf{x} \times \mathbf{F}(P+\mathbf{x}) = \sum_{i,j} x_i f_j(P+\mathbf{x})\,\mathbf{e}_i \times \mathbf{e}_j = \sum_{i,j,k} \epsilon_{ijk} x_i f_j(P+\mathbf{x})\,\mathbf{e}_k.$$

On the other hand, the Taylor expansion

$$f_j(P+\mathbf{x}) = f_j(P) + \sum_l x_l D_l f_j(P) + o(\mathbf{x})$$

gives us

$$\mathbf{x} \times \mathbf{F}(P+\mathbf{x}) = \sum_{i,j,k} \epsilon_{ijk} x_i \left(f_j(P) + \sum_l x_l D_l f_j(P) + o(\mathbf{x}) \right) \mathbf{e}_k.$$

Use this to show that

$$\mathbf{G}(r) = \frac{4\pi r^4}{3} \operatorname{curl} \mathbf{F}(P) + o(r^4).$$

From this equation, explain the meaning of the curl.

4. For vector fields $\mathbf{F}$ and $\mathbf{G}$, show that

$$\operatorname{div}(\mathbf{F} \times \mathbf{G}) = \operatorname{curl} \mathbf{F} \cdot \mathbf{G} - \mathbf{F} \cdot \operatorname{curl} \mathbf{G}.$$

2 Stokes' Theorem

In the fundamental theorem of calculus

$$\int_{[a,b]} f'(t)\,dt = f(b) - f(a) = \left[f\right]_{\partial[a,b]},$$

if we replace the domain $[a, b]$ by a curve C, we get the fundamental theorem of line integral

$$\int_C \operatorname{grad} f \cdot d\mathbf{s} = \left[f\right]_{\partial C}.$$

Similarly, in Green's theorem

$$\iint_D \operatorname{rot} \mathbf{F}(x, y)\,dx\,dy = \int_{\partial D} \mathbf{F} \cdot d\mathbf{s},$$

if we replace the domain $D\ (\subset \mathbb{R}^2)$ by a surface, we get Stokes' theorem.

Theorem 2.0.1 (Stokes' theorem) *For a surface $X : D \to \mathbb{R}^3$ defined on a region*[11] *$D \subset \mathbb{R}^2$ and a C^1 vector field $\mathbf{F}$ defined on the surface,*

$$\iint_X \operatorname{curl} \mathbf{F} \cdot d\mathbf{S} = \int_{\partial X} \mathbf{F} \cdot d\mathbf{s}.$$

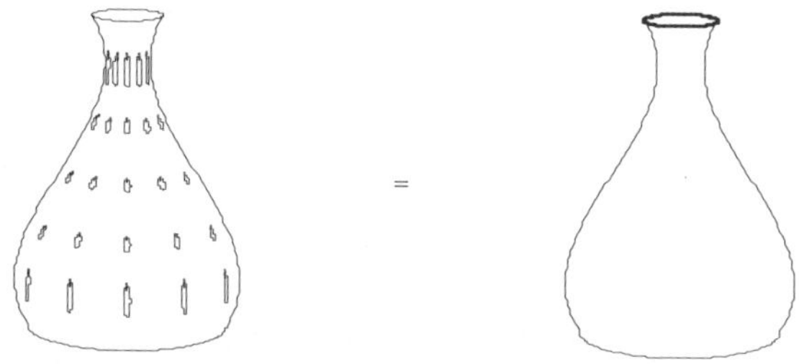

Picture: Stokes' theorem
– If we add up all circulations, only the value on the boundary is left.

[11] We assume that the region is a closed, bounded region whose boundary is a piecewise regular curve.

In the above equation, the **boundary**

$$\partial X : \partial D \to \mathbb{R}^3$$

of the surface

$$X : D \to \mathbb{R}^3$$

is a curve obtained by restricting the domain of X to the boundary ∂D of X.

We provide the proof of Stokes' theorem in the last section.[12]

Stokes' theorem is also true for a general oriented surface S without any concrete parametrization given. The orientation of the boundary ∂S is given in a similar way to give the orientation to the boundary of a plane region. That is, if we walk along the boundary in the direction of orientation, the surface S always lies on the left.[13]

Theorem 2.0.2 (Stokes' Theorem) *For a C^1 vector field $\mathbf{F}$ defined on an oriented surface S,*

$$\iint_S \operatorname{curl} \mathbf{F} \cdot d\mathbf{S} = \int_{\partial S} \mathbf{F} \cdot ds.$$

To prove the theorem, we may parameterize the surface locally and use "Stokes' theorem for a parametric surface".

2.0.3 Example

Suppose that the boundary of a parametric surface S is a simple closed curve.

---

[12]To avoid the proof of the general case, if we consider the case where U is a rectangular region and $\mathbf{F} = f\mathbf{e}_1$, we may prove an easy case.

[13]When we walk on an oriented surface, we walk in such a way that our head always points the standard normal direction of the surface. the orientation of the boundary of an oriented surface can be defined "mathematically" as follows: First, determine a unit vector $\mathbf{v}$ in the outward direction on the boundary of the oriented surface. Then the unit tangent vector $\mathbf{w}$ of the boundary is in the"right direction" if the pair $(\mathbf{v}, \mathbf{w})$ has the same orientation as the surface (that is, if $\mathbf{v} \times \mathbf{w}$ is the standard unit normal vector of the surface).

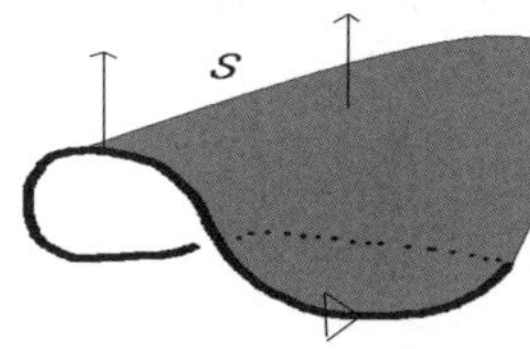

Since the vector field

$$\mathbf{F} = ye^z\,\mathbf{i} + xe^z\,\mathbf{j} + xye^z\,\mathbf{k}$$

is an irrotational vector field, that is, $\operatorname{curl}\mathbf{F} = 0$, we have

$$\int_{\partial S} \mathbf{F}\cdot d\mathbf{s} = \iint_S \operatorname{curl}\mathbf{F}\cdot d\mathbf{S} = 0.$$

2.0.4 Example

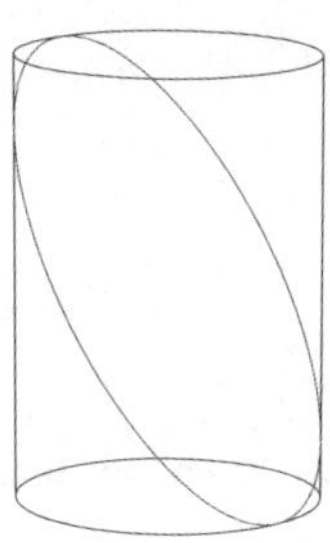

Let C be the intersection of the cylinder $x^2+y^2=1$ and the plane $x+y+z=1$. Find the line integral

$$\int_C -y^3\,dx + x^3\,dy - z^3\,dz. \tag{18.1}$$

(Here the orientation of the curve C is determined so that the orientation of the projection onto the xy-plane is counterclockwise.)

Solution. For the given vector field $\mathbf{F} := -y^3\,\mathbf{i} + x^3\,\mathbf{j} - z^3\,\mathbf{k}$,

$$\operatorname{curl}\mathbf{F} = \operatorname{grad}(-y^3)\times\mathbf{i} + \operatorname{grad}(x^3)\times\mathbf{j} + \operatorname{grad}(-z^3)\times\mathbf{k} = 3(x^2+y^2)\mathbf{k}.$$

The given curve C is the boundary of a surface

$$S \quad : \quad z = 1 - x - y$$

defined on a region $D := \{(x,y) \mid x^2+y^2 \le 1\}$. Therefore by Stokes' theorem, the above line integral (18.1) is equal to

$$\iint_S \operatorname{curl}\mathbf{F}\cdot d\mathbf{S} = \iint_D 3(x^2+y^2)\,dx\,dy.$$

If we transform it into polar coordinates, we have

$$\iint_D 3(x^2+y^2)\,dx\,dy = 3\int_0^{2\pi}\int_0^1 r^2\cdot r\,dr\,d\theta = \frac{3\pi}{2}.$$

The given problem can be solved without using Stokes' theorem. For example, if we parameterize C by

$$x = \cos t, \quad y = \sin t, \quad z = 1 - \cos t - \sin t \qquad (0 \le t \le 2\pi),$$

we have

$$\begin{aligned}
&\int_C -y^3\,dx + x^3\,dy - z^3\,dz \\
= &\int_0^{2\pi} \big((-\sin^3 t)(-\sin t) + (\cos^3 t)(\cos t) \\
&\qquad -(1-\cos t-\sin t)^3(\sin t - \cos t)\big)\;dt \\
= &\int_0^{2\pi} (\cos^4 t + \sin^4 t)\,dt - \int_0^{2\pi} (1-\cos t-\sin t)^3\,d(1-\cos t-\sin t) \\
= &\int_0^{2\pi} \left(\left(\frac{1+\cos 2t}{2}\right)^2 + \left(\frac{1-\cos 2t}{2}\right)^2\right)\,dt \;-\; \frac{1}{4}\big[(1-\cos t-\sin t)^4\big]_0^{2\pi} \\
= &\frac{1}{2}\int_0^{2\pi} \left(1+\cos^2 2t\right)\,dt \;+\; 0 \\
= &\frac{1}{2}(2\pi+\pi) = \frac{3\pi}{2}.
\end{aligned}$$

The following theorem is a simple application of Stokes' theorem.

Theorem 2.0.5 *Let S_1 and S_2 be oriented surfaces in an open region $U \subset \mathbb{R}^3$ with the same boundary and* $\mathbf{F}$ *be a vector field defined on U.*

(i) *If ∂S_1 and ∂S_2 have the same orientation, then*

$$\iint_{S_1} \operatorname{curl}\mathbf{F}\cdot d\mathbf{S} = \iint_{S_2} \operatorname{curl}\mathbf{F}\cdot d\mathbf{S}.$$

(ii) *If ∂S_1 and ∂S_2 have the opposite orientation, then*

$$\iint_{S_1} \operatorname{curl}\mathbf{F}\cdot d\mathbf{S} + \iint_{S_2} \operatorname{curl}\mathbf{F}\cdot d\mathbf{S} = 0.$$

If we think of a closed surface as a join of two surfaces, we get the following corollary.

Corollary 2.0.6 *If a vector field* $\mathbf{F}$ *is defined on a closed surface* S,[14]

$$\iint_S \operatorname{curl} \mathbf{F} \cdot d\mathbf{S} = 0.$$

2.1 Incompressible Vector Fields

A vector field $\mathbf{F}$ is called **incompressible** if it is divergence free:

$$\operatorname{div} \mathbf{F} = 0.$$

The following theorem gives us a way to get an incompressible field

Theorem 2.1.1 *If a vector field* $\mathbf{F}$ *in the three-dimensional space is the curl of some vector field, then* $\mathbf{F}$ *is incompressible.*[15]

Proof. If $\mathbf{F} = f_1\,\mathbf{e}_1 + f_2\,\mathbf{e}_2 + f_3\,\mathbf{e}_3$ is the curl of some vector field $\mathbf{G} = g_1\,\mathbf{e}_1 + g_2\,\mathbf{e}_2 + g_3\,\mathbf{e}_3$, then $f_k = \sum_{i,j} \frac{1}{2}\epsilon_{ijk}\left(D_i g_j - D_j g_i\right)$ and so

$$\sum_{k=1}^{3} D_k f_k = \sum_{i,j,k} \frac{1}{2}\epsilon_{ijk} D_k\left(D_i g_j - D_j g_i\right) = 0.$$

□

The converse of the above theorem is not always true. If the domain of the vector field has a hole, there is an incompressible field which is not the curl of any vector field. A typical example is the very **solid angle vector field**

$$\mathbf{A}(x, y, z) = \frac{1}{r^3}(x\,\mathbf{i} + y\,\mathbf{j} + z\,\mathbf{k}).$$

---

[14] As if the integral of a gradient vector field along a closed curve is zero.

[15] Conversely, if we study differential equations a little, an incompressible field is locally a curl of some vector field. Like this, an incompressible field can be thought of as the derivative of a "more fundamental" (?) vector field. As a conservative field has a potential function (locally) ...

We can see that this vector field can not be a curl from the following:

For the standard unit normal vector field $\mathbf{n}$ on the unit sphere S, $\mathbf{A} = \mathbf{n}$ and thus over the unit sphere S, we have[16]

$$\iint_S \mathbf{A} \cdot d\mathbf{S} = 4\pi.$$

If there exists a vector field $\mathbf{G}$ on $\mathbb{R}^3 - \{O\}$ such that $\mathbf{A} = \operatorname{curl} \mathbf{G}$, it follows from Corollary 2.0.6that

$$\iint_S \mathbf{A} \cdot d\mathbf{S} = \iint_S \operatorname{curl} \mathbf{G} \cdot d\mathbf{S} = 0,$$

which gives the contradiction.

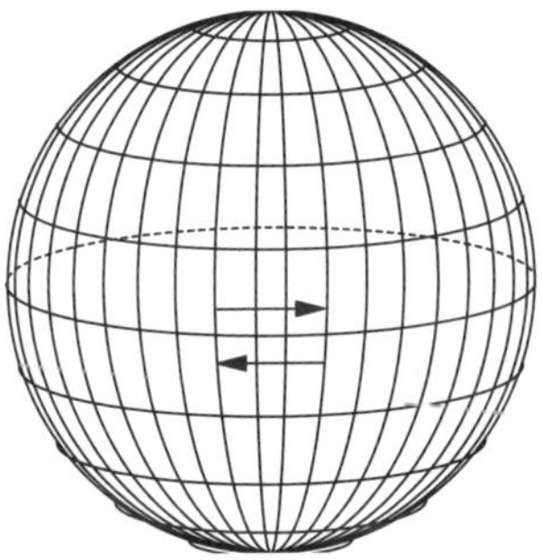

[16]In general, the integral of the solid angle vector field $\mathbf{A}$ over any surface R in $\mathbb{R}^3 - \{O\}$ is the area of the projection of R from the origin onto the unit sphere, which is called the **solid angle** of R.

Exercise: Chapter 18 Section 2

1. Verify Stokes' theorem for the upper hemisphere $z = \sqrt{1 - x^2 - y^2}$ and the vector field $\mathbf{F} = x\mathbf{i} + y\mathbf{j} + z\mathbf{k}$.

2. Verify Stokes' theorem for the triangle S with vertices $(1,0,0)$, $(0,1,0)$, $(0,0,1)$ and the vector field

$$\mathbf{F} = yz\,\mathbf{i} + zx\,\mathbf{j} + xy\,\mathbf{k}.$$

3. Show that if a vector field $\mathbf{F}$ is perpendicular to the boundary of the surface S, then $\iint_S \operatorname{curl}\mathbf{F} \cdot d\mathbf{S} = 0$.

4. Suppose that the surface S is the union of two surfaces

$$S_1 : x^2 + y^2 = 1, \quad 0 \le z \le 1$$

and

$$S_2 : x^2 + y^2 + (z-1)^2 = 1, \quad z \ge 1.$$

For the vector field

$$\mathbf{F} = (xz + yz^2 + x)\mathbf{i} + (xyz^3 + y)\mathbf{j} + x^2z^4\mathbf{k},$$

compute $\iint_S \operatorname{curl}\mathbf{F} \cdot d\mathbf{S}$. (Here, the orientation of S is determined so that $\mathbf{n} \cdot \mathbf{k} > 0$.)

5. For the vector field $\mathbf{F} = 3y\mathbf{i} - xy\mathbf{j} + yz^2\mathbf{k}$ and the surface

$$S : \quad 2z = x^2 + y^2, \quad z \le 2,$$

compute $\iint_S \operatorname{curl}\mathbf{F} \cdot d\mathbf{S}$. (Here, the orientation of S is determined so that $\mathbf{n} \cdot \mathbf{k} > 0$.)

6. Let D_r be the disk of radius r with the center at P which is perpendicular to a unit vector $\mathbf{n}$ in space. Show that there exists a point $P_* \in D_r$ such that

$$\frac{1}{\pi r^2} \iint_{D_r} \operatorname{curl}\mathbf{F} \cdot \mathbf{n}\, dV_2 = \operatorname{curl}\mathbf{F}(P_*) \cdot \mathbf{n}.$$

Use this to show that

$$\operatorname{curl}\mathbf{F}(P) \cdot \mathbf{n} = \lim_{r \to 0} \frac{1}{\pi r^2} \int_{\partial D_r} \mathbf{F} \cdot ds.$$

Also explain the meaning of this equation.

3 Proof of Stokes' Theorem, Derivatives of Vector Fields

3.1 Proof of Stokes' Theorem

Theorem 3.1.1 (Stokes' theorem) *For a surface $X : D \to \mathbb{R}^3$ defined on a region $D \subset \mathbb{R}^2$ and a $\mathcal{C}^1$ vector field $\mathbf{F}$ defined on the surface,*

$$\iint_X \operatorname{curl} \mathbf{F} \cdot d\mathbf{S} = \int_{\partial X} \mathbf{F} \cdot d\mathbf{s}.$$

Proof. First, for the surface

$$X(u, v) = (x(u, v), y(u, v), z(u, v)) \in \mathbb{R}^3, \qquad (u, v) \in D,$$

to parameterize its boundary $\partial X : \partial D \to \mathbb{R}^3$, consider one parametrization of the curve[17] ∂D

$$(u(t), v(t)) \in \partial D, \qquad a \leq t \leq b.$$

Then the boundary ∂X is parameterized by

$$X(u(t), v(t)), \qquad a \leq t \leq b.$$

Therefore the velocity vector of this curve is given by[18]

$$X_u \, u'(t) + X_v \, v'(t),$$

and we have

$$\int_{\partial X} \mathbf{F} \cdot d\mathbf{s} = \int_a^b \mathbf{F} \cdot (X_u u'(t) + X_v v'(t)) \, dt = \int_{\partial D} (\mathbf{F} \cdot X_u) \, du + (\mathbf{F} \cdot X_v) \, dv.$$

[17] In general, there may be several connected components of ∂D, but we can parameterize and consider each component.

[18] In this equation,

$$X_u = \left(\frac{\partial x}{\partial u}, \frac{\partial y}{\partial u}, \frac{\partial z}{\partial u} \right), \qquad X_v = \left(\frac{\partial x}{\partial v}, \frac{\partial y}{\partial v}, \frac{\partial z}{\partial v} \right).$$

On the other hand, since

$$\iint_S \operatorname{curl} \mathbf{F} \cdot d\mathbf{S} = \iint_D (\operatorname{curl} \mathbf{F}) \cdot (X_u \times X_v)\, du\, dv,$$

we are done if we show that

$$\iint_D (\operatorname{curl} \mathbf{F}) \cdot (X_u \times X_v)\, du\, dv = \int_{\partial D} (\mathbf{F} \cdot X_u)\, du + (\mathbf{F} \cdot X_v)\, dv.$$

Now, for the standard basis $\mathbf{e}_1, \mathbf{e}_2, \mathbf{e}_3$ in $\mathbb{R}^3$, if we let

$$\mathbf{F} = \sum_{i=1}^{3} f_i\, \mathbf{e}_i,$$

by Leibniz's rule, we get

$$\operatorname{curl} \mathbf{F} = \sum_{i=1}^{3} \nabla f_i \times \mathbf{e}_i,$$

and thus

$$(\operatorname{curl} \mathbf{F}) \cdot \big(X_u \times X_v\big) = \sum_{i=1}^{3} (\nabla f_i \times \mathbf{e}_i) \cdot (X_u \times X_v)\,.$$

So we have[19]

$$\begin{aligned}
& (\operatorname{curl} \mathbf{F}) \cdot \big(X_u \times X_v\big) \\
= \ & \sum_i (\nabla f_i \cdot X_u)\,(\mathbf{e}_i \cdot X_v) - \sum_i (\nabla f_i \cdot X_v)\,(\mathbf{e}_i \cdot X_u) \\
= \ & \left(\sum_i (\nabla f_i \cdot X_u)\, \mathbf{e}_i \right) \cdot X_v - \left(\sum_i (\nabla f_i \cdot X_v)\, \mathbf{e}_i \right) \cdot X_u \\
= \ & \left(\sum_i \frac{\partial f_i}{\partial u} \mathbf{e}_i \right) \cdot X_v - \left(\sum_i \frac{\partial f_i}{\partial v} \mathbf{e}_i \right) \cdot X_u \\
= \ & \mathbf{F}_u \cdot X_v - \mathbf{F}_v \cdot X_u = \frac{\partial\, (\mathbf{F} \cdot X_v)}{\partial u} - \frac{\partial\, (\mathbf{F} \cdot X_u)}{\partial v}.
\end{aligned}$$

---

[19]For arbitrary vectors $\mathbf{a}, \mathbf{b}, \mathbf{c}, \mathbf{d}$,

$$(\mathbf{a} \times \mathbf{b}) \cdot (\mathbf{c} \times \mathbf{d}) = \det \begin{pmatrix} \mathbf{a} \cdot \mathbf{c} & \mathbf{a} \cdot \mathbf{d} \\ \mathbf{b} \cdot \mathbf{c} & \mathbf{b} \cdot \mathbf{d} \end{pmatrix} = (\mathbf{a} \cdot \mathbf{c})(\mathbf{b} \cdot \mathbf{d}) - (\mathbf{a} \cdot \mathbf{d})(\mathbf{b} \cdot \mathbf{c}).$$

To make the proof easy, we can assume that X is a $\mathcal{C}^2$ surface and then there holds $X_{uv} = X_{vu}$.

Therefore

$$\iint_D (\operatorname{curl} \mathbf{F}) \cdot (X_u \times X_v)\, du\, dv = \iint_D \left(\frac{\partial (\mathbf{F} \cdot X_v)}{\partial u} - \frac{\partial (\mathbf{F} \cdot X_u)}{\partial v} \right) du\, dv$$

and by Green's theorem, the above integral is equal to

$$\int_{\partial D} (\mathbf{F} \cdot X_u)\, du + (\mathbf{F} \cdot X_v)\, dv.$$

So we complete the proof. □

3.2 Derivatives of Vector Fields

In the three-dimensional space $\mathbb{R}^3$, we can make the following diagram for differentiating functions or vector fields:

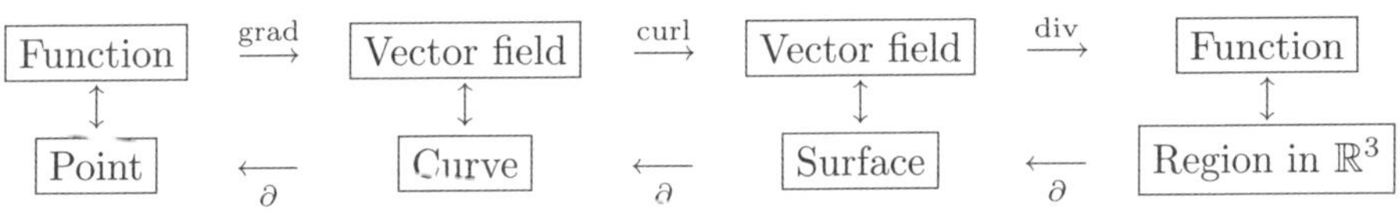

In this diagram, the first arrow $\xrightarrow{\text{grad}}$ is related to the fundamental theorem of line integral, the second arrow $\xrightarrow{\text{curl}}$ is related to Stokes' theorem, and the last arrow $\xrightarrow{\text{div}}$ is related to the divergence theorem.

So it is natural that the gradient vector field meets a curve, the curl of a vector field meets a surface, and the divergence of a vector field is integrated over a region in $\mathbb{R}^3$.

It is a "1-form", that is, a vector field that we integrate over a one-dimensional set like a curve, where we integrate the tangential component of the vector field.

Similarly, it is "2-form" that we integrate over a two-dimensional set like a surface, where we integrate the tangential component of the "2-form". However, for a surface in the three-dimensional space, the tangential component of a "2-form" is equal to the normal component of the vector field perpendicular to this "2-form". Because of that, a vector field in the three-dimensional space is integrated over a surface.

Another importance of the above diagram is that "the composition of two adjacent arrows becomes zero". In other words, a gradient vector field is irrotational, and a curl is incompressible.[20]

[20]The converse is locally true.

Chapter 18 Advanced Exercise

1. For any two mutually perpendicular unit vectors $\mathbf{v}_1, \mathbf{v}_2$ in $\mathbb{R}^3$, let

$$\mathbf{v}_3 := \mathbf{v}_1 \times \mathbf{v}_2.$$

For any vector field

$$\mathbf{F} := \sum_{i=1}^{3} f_i \mathbf{v}_i,$$

show that

$$\operatorname{curl} \mathbf{F} = \sum_{i=1}^{3} \nabla f_i \times \mathbf{v}_i = \sum_{i<j} \left(D_{\mathbf{v}_i} f_j - D_{\mathbf{v}_j} f_i\right) \mathbf{v}_i \times \mathbf{v}_j.$$

2. For vector fields $\mathbf{F} = \sum_i f_i \mathbf{e}_i$ and $\mathbf{G} = \sum_j g_j \mathbf{e}_j$, if we define

$$D_{\mathbf{F}} \mathbf{G} := \sum_j \left(D_{\mathbf{F}} g_j\right) \mathbf{e}_j = \sum_{i,j} \left(f_i D_i g_j\right) \mathbf{e}_j,$$

show the following.

$$\begin{aligned} \operatorname{grad}(\mathbf{F} \cdot \mathbf{G}) &= D_{\mathbf{F}} \mathbf{G} + D_{\mathbf{G}} \mathbf{F} + \mathbf{F} \times \operatorname{curl} \mathbf{G} + \mathbf{G} \times \operatorname{curl} \mathbf{F} \\ \operatorname{curl}(\mathbf{F} \times \mathbf{G}) &= D_{\mathbf{G}} \mathbf{F} - D_{\mathbf{F}} \mathbf{G} + (\operatorname{div} \mathbf{G})\mathbf{F} - (\operatorname{div} \mathbf{F})\mathbf{G} \end{aligned}$$

3. Show that when we denote the projection of a vector field $\mathbf{F}(x, y, z)$ onto a tangent plane T of a surface S by $\mathbf{F}_T$, the rotation of this plane vector field $\operatorname{rot} \mathbf{F}_T$ is equal to $\operatorname{curl} \mathbf{F} \cdot \mathbf{n}$. Here $\mathbf{n}$ is a unit vector perpendicular to T.

4. By Ampere's law, the integral of an electric field $\mathbf{B}$ along a closed curve C is proportional to the electric current through C:

$$\oint_C \mathbf{B} \cdot d\mathbf{s} = \mu_0 I.$$

If C is the boundary of the surface S and $\mathbf{J}$ is the **current density** (vector),[21] then

$$I = \iint_S \mathbf{J} \cdot d\mathbf{S},$$

and so Ampere's law becomes

$$\oint_{\partial S} \mathbf{B} \cdot d\mathbf{s} = \mu_0 \iint_S \mathbf{J} \cdot d\mathbf{S}.$$

From Stokes' theorem, derive

$$\operatorname{curl} \mathbf{B} = \mu_0 \mathbf{J}.$$

[21] Refer to Advanced Exercise in Chapter 17

5. **Maxwell's equations** are given by

$$\operatorname{div}\mathbf{E} = \rho/\epsilon_0, \qquad \operatorname{div}\mathbf{B} = 0$$
$$\operatorname{curl}\mathbf{E} = -\frac{\partial \mathbf{B}}{\partial t}, \qquad \operatorname{curl}\mathbf{B} = \epsilon_0\mu_0\frac{\partial \mathbf{E}}{\partial t} + \mu_0\mathbf{J}$$

for the electric field $\mathbf{E}$, magnetic field $\mathbf{B}$, charge density ρ and current density $\mathbf{J}$.

Faraday discovered that for a surface S,

$$\oint_{\partial S}\mathbf{E}\cdot d\mathbf{s} = -\frac{d}{dt}\iint_S \mathbf{B}\cdot d\mathbf{S}.$$

From this, derive the third Maxwell's equation

$$\operatorname{curl}\mathbf{E} = -\frac{\partial \mathbf{B}}{\partial t}.$$

Also from Lenz's law

$$\oint_{\partial S}\mathbf{B}\cdot d\mathbf{S} = \epsilon_0\mu_0\frac{d}{dt}\iint_S \mathbf{E}\cdot d\mathbf{S} + \mu_0\iint_S \mathbf{J}\cdot d\mathbf{S},$$

derive the last Maxwell's equation

$$\operatorname{curl}\mathbf{B} = \epsilon_0\mu_0\frac{\partial \mathbf{E}}{\partial t} + \mu_0\mathbf{J}.$$

6. For a vector field $\mathbf{F}$ defined on a simply connected region,

   (i) show that if $\mathbf{F}$ is irrotational, then $\oint_C \mathbf{F}\cdot d\mathbf{s} = 0$ for any closed curve in the region.

   (ii) show that if $\oint_C \mathbf{F}\cdot d\mathbf{s} = 0$ for any closed curve in the region, then $\mathbf{F}$ is irrotational.

Appendix

Appendix 1 MATH DICTIONARY

Here I collected what I mentioned in the text but had no chance to explain in detail.

Carroll: Lewis Carroll, 1832–1898. His real name is Charles Lutwidge Dodgson, and he taught mathematics at Christ Church, Oxford from 1854 to 1881. In 1865, children's books are mostly about "being a good and upright person" and "fun and imaginative" books were rare. Queen Victoria loved *Alice in Wonderland* so much that she requested copies of all Lewis Carroll's books. Among those books there was "Euclid and his Modern Rivals (1879)".

Chain rule: The derivative of the composition $(g \circ h)(x) := g(h(x))$ of functions $z = g(y)$ and $y = h(x)$ is

$$\boxed{(g \circ h)'(x) = g'(h(x))h'(x)}$$

The derivative is the "rate of expansion of the function". The rate of expansion of the composition of two functions is the product of rates of expansion of two functions. That is, it gives how much the length of the domain is stretched in the range.

Complete space: A complete space is a metric space in which any Cauchy sequence converges. To show that a sequence of points (x_i) in a metric space converges, we should pick an appropriate point x in the space and show that

$$\lim_{i \to \infty} \operatorname{dist}(x_i, x) = 0.$$

To prove the convergence, we need to know the limit x first. In a complete space, even though we don't know explicitly the limit where a sequence converges, as long as we can show that the sequence is Cauchy, we can get the convergence. A sequence (x_i) is called a Cauchy sequence when for any $\epsilon > 0$,

$$\sup \left\{ \operatorname{dist}(x_j, x_k) \mid j, k \geq i \right\} < \epsilon$$

if i is sufficiently large. Every convergent sequence is a Cauchy sequence.

Concave function: A real-valued function $f : I \to \mathbb{R}$ defined on an interval $I \subset \mathbb{R}$ is called concave if for any two points a, $b \in I$, the graph of f on the interval $[a, b]$ lies above the line segment connecting two end points of the graph.

A function $f(x)$ is concave if and only if $-f(x)$ is convex.

Convex function: A real-valued function $f : I \to \mathbb{R}$ defined on an interval $I \subset \mathbb{R}$ is called convex if for any two points a, $b \in I$, the graph of f on the interval $[a, b]$ lies under the line segment connecting two end points of the graph. In other words, $f : I \to \mathbb{R}$ is convex if

$$f((1-t)a + tb) \leq (1-t)f(a) + tf(b)$$

for any two points a, b in the interval I and any $t \in [0,1]$.

According to this definition, a constant function is convex. To exclude constant functions, we can define a **strictly convex function**.

When $f(x)$ is continuous, $f(x)$ is convex if and only if

$$f\left(\frac{a+b}{2}\right) \le \frac{f(a)+f(b)}{2} \qquad \forall\, a, b.$$

If $f(x)$ is differentiable and its derivative $f'(x)$ is increasing, then $f(x)$ is convex. If $f(x)$ is twice differentiable and $f''(x) > 0$, then $f(x)$ is convex.

Duality: In order that two sets A and B are dual, first there must exist a **natural** map

$$\langle\ |\ \rangle : A \times B \to \mathbb{R}, \qquad (a,b) \mapsto \langle a|b\rangle.$$

In other words, when an element a of A and an element b of B meet, a and b disappear and a real value $\langle a|b\rangle$ appears. Now if we let A^* and B^* be the space of all functions defined (allowed) on A and B respectively, such a **pairing** $\langle\ |\ \rangle$ derives maps

$$A \to B^*, \qquad a \mapsto \langle a|$$

and

$$B \to A^*, \qquad b \mapsto |b\rangle.$$

If these maps are isomorphic, $\langle\ |\ \rangle$ is called **non-degenerate**. A and B are called dual if there exists a natural non-degenerate pairing $\langle\ |\ \rangle$. Even though **Being natural** is a technical mathematical term used in **category theory**, it won't be impossible to explain this concept in calculus class.

Elementary function: An elementary function is a function built from a finite number of polynomials, rational functions, exponentials, logarithms, trigonometric functions and their inverses, through composition and combinations using the elementary operations (addition, subtraction, multiplication, division, and root extractions). Elementary functions whose antiderivatives are not elementary functions are as follows.

$$e^{ax^2},\ x^{2n}e^{ax^2},\ x^{-2n}e^{ax^2},\ x^{-n}e^{ax},\ x^{-1}e^{ax},\ x^{-1}e^{ax^2}$$

$$\sqrt{\log x},\ \frac{1}{\log x},\ \log(\log x),\ \frac{\log x}{x-a},\ \frac{\sin x}{x},\ \sqrt{\sin x},$$

$$\sqrt[3]{1+x^2},\ \sqrt{1+x^3},\ \frac{1}{\sqrt{1-x^4}},\ \sqrt{1+x^{-4}},\ \frac{1}{\sqrt{1-k^2\sin^2 x}},\ \sqrt{1-k^2\sin^2 x}.$$

Here n is a natural number, a is a nonzero real number, and $0 < k < 1$ [Marchisotto, Zakeri], [Birkhoff], [Risch].

Extreme value theorem: *"A continuous function defined on a bounded, closed set*[1] *in $\mathbb{R}^n$ attains its absolute maximum and minimum."* The rough proof of this theorem is as follows: First, prove[2]

> **Lemma** *A continuous function defined on a bounded, closed set is bounded.*

Now consider a continuous function f defined on a bounded, closed set U. Since this function is bounded by the above lemma,

$$m := \sup\{f(X) \mid X \in U\}$$

is finite.

If f doesn't attain the maximum, $f(X) \neq m$ for any $X \in U$ and so

$$g : U \to \mathbb{R}, \qquad X \mapsto \frac{1}{f(X) - m}$$

is continuous on U, and bounded by the lemma. However, from the definition of m, there exists a sequence $(X_i \in U)$ such that

$$\lim_{i \to \infty} f(X_i) = m,$$

and so g cannot be bounded. This contradiction comes from the assumption that f doesn't attain the maximum. Therefore f must have the maximum.

We can prove that f has the minimum in a similar way. Or if we consider $-f$ instead of f, finding the minimum becomes finding the maximum and so we are done.

Fields Medal: The Fields Medal is a prize founded at the behest of Canadian mathematician John Charles Fields (1863–1932). The Fields Medal is awarded every four years on the occasion of the International Congress of Mathematicians (ICM) to recognize outstanding mathematical achievement for existing work and for the promise of future achievement. The ICM was first held in the United States in 1893. At the second congress in Paris, France in 1900, David Hilbert (1862–1943) announced his famous list of 23 open problems in mathematics, now called Hilbert's problems, which served as a momentum for great development of 20th-century mathematics.

---

[1] In the Euclidean space, a bounded, closed set is a **compact set**.

[2] If a continuous function f defined on a bounded, closed set U is not bounded above, there exists a sequence $(X_i \in U)$ such that

$$f(X_i) > i \qquad (i = 1, 2, 3, \dots).$$

On the other hands, since U is a bounded, closed set, (X_i) has a subsequence which converges to a point X_∞ in U, and the difference between $f(X_\infty)$ and $f(X_i)$ gets larger eventually, which contradicts to the continuity of f. Therefore f is bounded above. Similarly, we can prove that f is bounded below.

Fixed point: A fixed point of a function $f : X \to X$ is a point such that

$$f(x) = x.$$

Fundamental theorem of algebra: It means Gauss' theorem, "Every polynomial with complex coefficients of degree≥ 1 has a complex root". Due to this theorem, we can say that we don't have to expand the field of numbers any further. In Descartes' day, the concept of negative numbers was not established yet. Descartes discovered "the way to find the number of real roots of polynomials", where he means a "positive root" by a real root [Anderson, Jackson, Sitharam]. There are several ways to prove the fundamental theorem of algebra, we will provide one proof using extreme value theorem as follows:

Suppose that a polynomial

$$f(z) = z^n + a_1 z^{n-1} + \cdots + a_{n-1} z + a_n$$

is given. If $|z|$ is very large, then

$$|f(z)| = |z|^n \left| 1 + \frac{a_1}{|z|} + \cdots + \frac{a_{n-1}}{|z|^{n-1}} + \frac{a_n}{|z|^n} \right| \geq \frac{1}{2}|z|^n.$$

So $\lim_{|z| \to \infty} |f(z)| = \infty$, the function $|f(z)|$ attains the minimum.[3] For convenience, we may assume that the minimum point of $|f(z)|$ is the origin.[4] Then $|a_n| = |f(0)|$ is the minimum value of $|f(z)|$. If $a_n \neq 0$, for $k < n$ satisfying

$$a_k \neq 0, \qquad k < i < n \Rightarrow a_i = 0,$$

we have

$$f(z) - a_n = z^n + a_1 z^{n-1} + \cdots + a_k z^{n-k} = z^{n-k}(z^k + a_1 z^{k-1} + \cdots + a_k).$$

In the above equation, if $|z|$ is very small, then $z^k + a_1 z^{k-1} + \cdots + a_k$ is a value near a_k, and if we multiply z^{n-k} to it, we get a value near $a_k z^{n-k}$. Now if we take $z = z_0$ properly, the "angle" between $f(z_0) - a_n$ and a_n is close to 180° and so we have

$$|f(z_0)| \leq |(f(z_0) - a_n) + a_n| < |a_n|,$$

which contradicts to the assumption that $|a_n|$ is the minimum value of $|f(z)|$.

Therefore $a_n = 0$ and $z = 0$ is a root of $f(z)$, and we are done.

[3] Here, the extreme value theorem is used.

[4] We may use parallel translation if necessary.

Fundamental theorem of calculus: It states that "For a continuous function f defined on an interval $[a, b]$, the function

$$x \mapsto \int_a^x f \qquad (x \in (a, b))$$

is differentiable, and its derivative is f." The proof is as follows.

If we fix a point $x \in (a, b)$, for $h \neq 0$ with small absolute value, we get

$$\frac{1}{h}\left(\int_a^{x+h} f - \int_a^x f\right) = \frac{1}{h}\int_x^{x+h} f.$$

Now if we let m, M be the minimum and maximum of f on the interval $[x, x+h]$ respectively, we have

$$f(m) \leq \frac{1}{h}\int_x^{x+h} f \leq f(M),$$

and so by intermediate value theorem, there exists $c \in [x, x+h]$ such that

$$\frac{1}{h}\int_x^{x+h} f = f(c).$$

Therefore as h approaches 0,

$$\lim_{h\to 0}\frac{1}{h}\int_x^{x+h} f = f(x)$$

and we have

$$\frac{d}{dx}\int_a^x f(t)\,dt = f(x).$$

□

For a function f defined on an interval, a function F whose derivative is f, that is, $F' = f$, is called an **antiderivative** of f. The antiderivative is unique up to constant, in other words, if F is an antiderivative of f, all other antiderivatives can be expressed as $F + \text{const}$. The fundamental theorem of calculus states that **Every continuous function has an antiderivative.** Often the antiderivative of f is denoted by

$$\int f \qquad \text{or} \qquad \int f(x)\,dx,$$

and called the **indefinite integral** of f.

Conversely, the following theorem tells how to integrate a continuous function from antiderivative.

If F is an antiderivative of a function $f : [a, b] \to \mathbb{R}$, then

$$\int_a^b f = F(b) - F(a) =: \big[F\big]_a^b.$$

This theorem can be proved as follows. The functions $\int_a^x f$ and $F(x) - F(a)$ have the same derivative f, and both functions converge to zero as $x \searrow a$. Therefore two functions coincide at any $x \in (a, b)$. Now as $x \nearrow b$, we get the conclusion we want.

The fundamental theorem of calculus expressed in the following form can be extended to the space of general dimension.

For a $\mathcal{C}^1$ function $F : [a, b] \to \mathbb{R}$,

$$\int_a^b F'(x)\, dx = F(b) - F(a).$$

The above equation can be understood as follows:

$$\int_a^b F'(x)\, dx = \lim \sum_i F'(x_i)(x_{i+1} - x_i) \approx \sum_i (F(x_{i+1}) - F(x_i)) = F(b) - F(a).$$

Golden ratio: In the oldest "Rhind Mathematical Papyrus" "the use of the golden ratio to build a pyramid" was described [김용운, 김용국, 數學序說, p. 218]. When the lengths of three sides of a right triangle form a geometric sequence, the ratio of its short side to the hypotenuse is the very golden ratio. The golden ratio has been represented by the Greek letter (phi), after ancient Greek sculptor Phidias, who is said to have employed it.

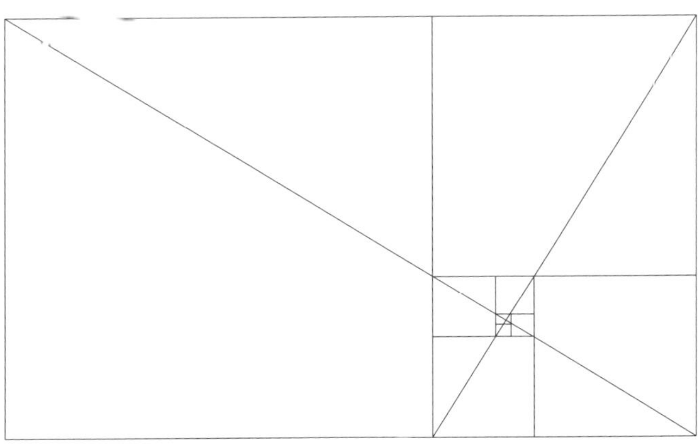

Picture: If we cut off a square from the golden rectangle, in which the ratio of the longer side to the shorter is the golden ratio , the remainder becomes a rectangle similar to the original golden rectangle. (Hidden Picture: Logarithmic Spiral)

The golden ratio appears in Parthenon, painting of Leonardo da Vinci, music of Hungarian composer Bela Bartók, grades given by Professor Kim, phone card, credit card, tobacco box and so on. Recently people think that the golden ratio is hidden in "reproduction" of living organisms. It appears in the ratio of sections of human finger, sunflowers or pineapples, horns and so many other examples. It is the average of 1 and $\sqrt{5}$. Its square ϕ^2 is 1 greater than ϕ, and its reciprocal $1/\phi$ is 1 less than ϕ. The **golden section** is approximately $(1 - \frac{1}{\phi}) \times 360° \approx 137.5°$.

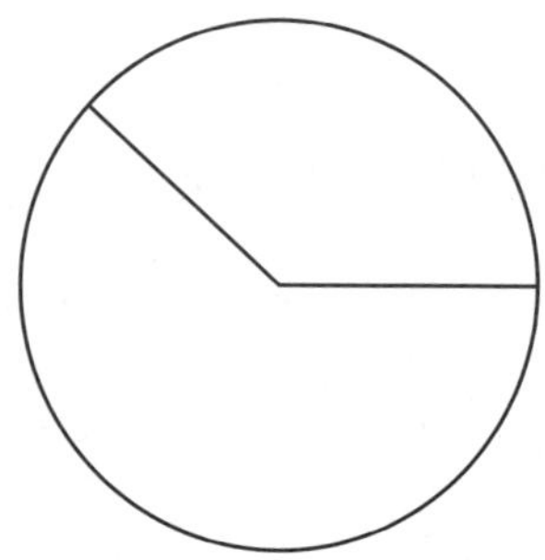

Golden section:

Intermediate value theorem: *Suppose that a continuous function defined on an interval attains A and B as its values. Then for any real number c between A and B, it attains c as its value.* For example, for a continuous function $f(x)$ defined on an interval $[1, 2]$, if $f(1) > 0$ and $f(2) < 0$, then there exists $x_* \in [1, 2]$ such that $f(x_*) = 0$. (Exercise: Suppose that Kildong was taller than Dongkil in middle school, but shorter than Dongkil in high school. Explain that there was a moment when Kildong and Dongkil are of the same height.)

Interval: There are four types of open intervals:

$$(a, b), \qquad (a, \infty), \qquad (-\infty, b), \qquad (-\infty, \infty).$$

Here a, b are real numbers and $a < b$. Besides them, there are the following types of intervals

$$[a, b], \qquad [a, \infty), \qquad (-\infty, b], \qquad [a, b), \qquad (a, b].$$

Among those nine types of intervals, $[a, b]$, $[a, \infty)$, $(-\infty, b]$, $(-\infty, \infty)$ are **closed intervals**, and $[a, b)$, $(a, b]$ are neither open nor closed.

Inversion: In the Euclidean space, for a point $P \neq O$, let's consider a point P' on the ray starting from O in the direction of P such that the product of lengths of line segments OP and OP' is the square of the radius of a sphere with the center at O. The transformation which corresponds P to P' is called an inversion. The inversion is defined similarly in the space of arbitrary dimension. The inversion was discovered first by Jakob Steiner (1796–1863) in 1824. [Boyer, p. 575]

Isotropic space: If a Riemannian manifold looks the same in all directions, it is called isotropic. A Riemannian manifold is isotropic if and only if any two geodesics with equal length are congruent, that is, two-point **homogeneous space**.

Leonardo da Vinci: Leonardo was born in 1452, at Anchiano near the town of Vinci, in the territory of Florence, and died in 1519. He was the illegitimate son of a notary, Ser Piero da Vinci, and Caterina, a peasant girl. An enormous range of interests and preoccupations on painting, drawing, music, anatomy, light and shadow, optics, physics, mechanics, hydraulics, canal construction, palace construction, mechanical engineering, civil engineering, weaponry, astronomy, geography, geology, botany and mathematics were recorded in voluminous notebooks. The journals are mostly written in mirror-image cursive. Among the paintings he drew, there were "The Last Supper" and "The Mona Lisa". He told his pupils that "wrinkles" should be considered most importantly in painting the cloth. "The flow of water" was one of the most mysterious object to him. While he designed chapels and niches, he found out that a "finite group of isometries" on a Euclidian plane is either a **cyclic group** or a **dihedral group** [Weyl]. The followings are in his journals.

- "Mechanics is the paradise of the mathematical sciences because by means of it one comes to the fruits of mathematics.."
- "Let no man who is not a Mathematician read the elements of my work."

The soccer ball used mostly nowadays in the world has 32 faces which consist of 12 pentagons and 20 hexagons. This shape appears in Leonardo's work.[5] Leonardo was deeply impressed by Archimedes' work.

Mean value theorem: Let m be the average rate of change of a differentiable function $f : [a, b] \to \mathbb{R}$:

$$m := \frac{f(b) - f(a)}{b - a}.$$

First we show that there exists a point $c \in (a, b)$ such that $m = f(c)$. For a linear function

$$l(x) := m(x - a) + f(a),$$

[5] In 1980s, chemists discovered a radioactive isotope of carbon C_{60} in high-temperature lab and its shape is the same as that of a "soccer ball". Chemists call this isotope Fullerenes after the American architect Richard Buckminster Fuller. Fuller was renowned for his geodesic domes, that can supports by itself without struts.

we have $l(a) = f(a)$, $l(b) = f(b)$ and so the function

$$F(x) := f(x) - l(x), \quad x \in [a, b]$$

attains zero as its values at a, b.

Consider the maximum and minimum points of the function $F(x)$ on the interval $[a, b]$.

If at least one of them is an interior point of the interval $[a, b]$, by the critical point theorem, the derivative at that point is zero. Therefore there exists a point $c \in (a, b)$ such that

$$0 = F'(c) = f'(c) - m. \tag{.1}$$

If the maximum and minimum points of $F(x)$ are both boundary points of the interval $[a, b]$, then $F(x)$ is identically zero, and so for any $c \in (a, b)$, (.1) holds.

Minimal surface: A surface in the Euclidean space is minimal if the area of the surface is locally minimal. The catenary or helix are examples of minimal surface. A surface is minimal if and only if the mean curvature is identically zero.

Stokes' theorem: In a broad sense, Stokes' theorem shows the **duality** of spaces and functions. It is like that we can guess trees from the land where they grow, and conversely we can guess the land from trees on it. If we describe it in detail, there are the **homology** theory and **cohomology** theory. In order to understand a space, the former emphasizes to consider how "subspaces" of the space, such as curves or surfaces, lie in the space , and the latter emphasizes to consider functions or vector fields, or more generally differential forms defined on the space.

Now, in integral calculus, the encounter of curves and vector fields is expressed as line integrals, the encounter of surfaces and vector fields is expressed as surface integrals. Moreover, in such an encounter, the following general principle, Stokes' theorem holds.

$$\int_S d\varphi = \int_{\partial S} \varphi$$

This is called the "fundamental theorem" in high school math, and expressed as

$$\int_{[a,b]} f'(x)\, dx = f(b) - f(a) =: [f]_a^b.$$

We can use Stokes' theorem to prove Brouwer's fixed point thoerem.

π: The ratio of a circle's circumference to its diameter is constant for every circle. This value is denoted by a Greek letter π.

For example, the circumference of a circle of the same diameter d meters as the equator of the earth and the circumference of a circle of diameter $d+1$ meters differ by π meters.

From ancient times to the present day there are endless efforts to compute the value of π. We can consider Archimedes as the first mathematician to know how to compute π. He knew that π lies between the perimeter of a regular polygon inscribed in a circle of radius 1 and the perimeter of a regular polygon inscribing a circle of radius 1, and used regular 96-sided polygons to discover $3+\frac{10}{71}<\pi<3+\frac{1}{7}$. The fact that π is not a rational number, that is, π is an irrational number was proved in 1768 by a Swiss, Johann Heinrich Lambert (1728–1777). In 1882, a German, F. Lindemann (1852–1939) used DeMoivre–Euler identity

$$e^{i\pi}+1=0$$

to prove that π is a **transcendental number**. A transcendental number is a number that is not the root of any integer polynomial. Therefore we know that squaring the circle, that is constructing a square with the same area as a given circle with only compass and straightedge, is impossible. (This is one of the three geometric problems of antiquity.) The symbol "π", which is the first letter of the Greek words $\pi\epsilon\rho\iota\varphi\epsilon\rho\epsilon\iota\alpha$ (periphery), was first used in 1647 by William Oughtred (1575–1660), but became in common use after Euler used it in 1737.

π is used to determine the capacity of newly devised high-speed computer chip, or for spies to make random numbers.

Appendix 2 References

[Kang, Lee] 강옥기, 이상구, *행렬이론의 발전과정과 현재*, 대한수학회 뉴스레터 **55** (1997), 20–27.

[Ko, Kim, Kim] 고영소, 김도한, 김홍종, 微積分學, 수정판, 서울대학교 출판부, 1993.

[Kim, Kim, Kye] 김성기, 김도한, 계승혁, *해석개론*, 서울대학교 출판부, 1995.

[Kim, Kim] 김용운, 김용국, *지성의 비극 — 수학정신의 탐구, 피타고라스에서 괴델까지*, 一志社, 1992.

———, 數學序說, 祐成文化社, 1993.

[Kim] 김홍종, *공돌리기 재주의 수학*, 과학동아 1997년 10월호.

[Lee] 이민섭, *도학연구*, 技文堂, 1999.

[Lee] 이일해 옮김, *힐버트 — 수학과 삶* (C. Reid 지음), 민음사, 1989.

[Lee, Kim] 이현구, 김홍종, *인문·사회계 미적분학*, 서울대학교 출판부, 1995.

[Lee] 이종우 편저, *기하학의 역사적 배경과 발달*, 京文社, 1997.

[Chung, Kim, Kim] 정재명, 김명환, 김홍종, *미적분학을 위한 벡터와 행렬*, 서울대학교 출판부, 1995.

[Chi] 지동표, *계산적 설계를 위한 곡선과 곡면론*, 서울대학교 출판부, 1990.

A. Aaboe, *Episodes from the early history of Mathematics*, Readings for Calculus, U. Dudley, Ed., The Math. Assoc. Amer., 1993.

V. Adamchik, S. Wagon, *A simple formula for* π, Amer. Math. Monthly **104** (1997), 852–855.

A. Aleksandrov, A. Kolmogorov, A. Lavrent'ev, ed., Mathematics, *Its Content, Methods, and Meaning*, The M.I.T. Press, 1969.

B. Anderson, J. Jackson and Sitharam, *Descartes' Rule of Signs Revisited*, Amer. Math. Monthly **105** (1998), 447–451.

G. Andrews, *The Theory of Partitions*, Addison-Wesley, 1976.

T. Apostol, *Mathematical Analysis*, Addison-Wesley Publishing Company, 1973.

T. Apostol, M. Mugles, D. Scott, A. Sterett, Jr., A. Watkins, *A century of Calculus*, Math. Assoc. Amer.

J. Barrow, *Theories of Everything*, Clarendon Press, Oxford, 1991.

M. Berger, *Geometry, I, II*, Springer-Verlag, 1987.

G. Birkhoff, *A source book in classical anaylsis*, Harvard Univ. Press, Cambridge, 1973.

G. Birkhoff, E.Kreyszig, *The Establishment of Functional Analysis*, Historia Mathematica **11** (1984), 258–321.

R. Boas, *A Primer of Real Functions*, The Math. Assoc. Amer., 1996.

W. E. Boyce, R. DiPrima, *Elementary Differential Equations and boundary value problems*, John Wiley and Sons, Inc., Fifth ed., 1992.

C. B. Boyer, *A History of Mathematics*, Princeton univ. Press, 1985.

J. Buhler, D. Eisenbud, R. Graham, C. Wright, *Juggling Drops and Descents*, Amer. Math. Monthly, **101** (1994), 507–519.

F. Cajori, *A History of Mathematical Notations*, Dover, 1993.

Soo Bong Chae, *Lebesgue Integration*, second edition, Springer-Verlag, 1995.

S.-s. Chern, *What Is Geometry?*, The Amer. Math. Monthly, (1990), 679–686.

______, *From Triangles to Manifolds*, Amer. Math. Monthly **86** (1979), 330–349.

J. H. Conway, R. K. Guy, *The book of Numbers*, Copernicus, 1996.

R. Courant, D. Hilbert, *Methods of Mathematical Physics I, II*, Interscience Publ., Inc., New York, 1953.

R. Courant, H. Robbins, *What is Mathematics?*, 2nd ed., Oxford Univ. Press, 1996. (Revised by I. Stewart)

Crowdis, Shelley, Wheeler, **Calculus** *for business, biology, and the social sciences*, Glencoe Press, 1976.

D. A. Danielson, *Vectors and Tensors in Engineering and Physics*, 2nd ed., Addison-Wesley Publ. Co., Inc., 1997.

H. Davis, A. Snider, *Introduction to Vector Analysis*, Wm. C. Brown Publishers, 1961, 1991.

K. Delvin, *Mathematics, The Science of Patterns*, Scientific American Library, 1994.

J. Dieudonné, *Mathematics — The Music of Reason*, Springer-Verlag, 1992.

C. T. J. Dodson, T. Poston, *Tensor Geometry — The Geometric Viewpoint and its Uses*, GTM 130, Springer-Verlag, 1977.

R. Driver, *Torricelli's law—an Ideal Example of an Elementary ODE*, Amer. Math. Monthly **105** (1998), 453–455.

Euclid, *The thirteen books of* **THE ELEMENTS**, Dover Publications, Inc., 1956.

H. Eves, *An Introduction to the History of Mathematics*, Sixth ed., the Sauders series, 1990. (이우영, 신항균 역, 경문사, 서울, 1996.)

______, *Great Moments in Mathematics*, The Math. Assoc. of Amer., 1983.

W. Faris, Review of Roland Omnés *The Interpretation of Quantum Mechanics*, Notices of Amer. Math. Soc. **43** (1996), 1328–1339.

T. Fay, *A Study in Step Size*, Mathematics Magazine **70** (1997), 116–117.

R. Feynman, *Surely You're Joking, Mr. Feynman*, Bantam Book, New York, 1985.

______, *Lectures on Physics*, vol. 1, Addison-Wesley, 1963.

H. Flanders, *Differential Forms with Applications to the Physical Sciences*, Dover, 1963, 1989.

A. Fomenko, *Mathematical Impressions*, Amer. Math. Soc., 1990.

M. Gardner, *Mathematical Magic Show*, Math. Assoc. Amer., 1965, 1990.

______, *Penrose tiles to trapdoor ciphers... and the return of Dr. Matrix*, The Math. Assoc. Amer., 1989, 1997.

Goldstein, Lay, Schneider, **Calculus** *and its applications*, brief edition, Prentice-Hall Inc., 1980.

J. V. Grabiner, *Was Newton's Calculus a Dead End? The Continental Influence of Maclaurin's* Treatise of Fluxions, The Amer. Math. Monthly **105** (1997), 393–410.

R. Graham, D. Knuth and O. Patashnik, *Concrete Mathematics — A Foundation for Computer Science*, Addison-Wesley Publ. Co., 1989.

J. Henle, *Non-nonstandard Analysis*, The Mathematical Intelligencer **21**, No. 1 (1999), 67-73.

L. Hoffmann, G. Bradley, **Calculus** *for Business, Economics, and the Social and Life Sciences*, McGraw-Hill Book Company, 1989.

H. Hopf, *Differential Geometry in the Large*, Lect. Notes in Math. **1000**, Springer-Verlag, 1983.

W. Y. Hsiang, *A concise introduction to Calculus*, World Scientific, 1995.

———, *The Geometry of Spheres*, Normat **1** (1994), 1–12.

G. Jennings, *Modern Geometry with Applications*, Springer, Unversitext, 1994.

J. Kappraff, *Connection, The geometric bridge between art and Science*, McGraw-Hill, Inc., 1991.

J. Kazdan, *Solving Equation, An Elegant Legacy*, Amer. Math. Monthly **105** (1998), 1–21.

H. Keisler, *Foundations of Infinitesimal Calculus*, Prindle, Weber & Schmidt, Inc., 1976.

O. Kellogg, *Foundations of Potential Theory*, Springer-Verlag, 1929, 1967.

V. Klee and S. Wagon, *Unsolved Problems*, Math. Assoc. Amer., 1991.

D. Knuth, *The TEXBook*, Addison-Wesley Publishing Co., 1986.

———, *Surreal numbers*, Addison-Wesley, 1974.

S. Lang, *Calculus of Several Variables*, Third Edition, Springer-Verlag, 1987.

D. Laugwitz, *On the historical development of infinitesimal mathematics*, The Amer. Math. Monthly **104** (1997), 447–455.

Maor, *To infinity and beyond*, Birkhüaser, 1987. (전대호 옮김, *무한, 그리고 그너머*, 사이언스북스, 1997.)

E. Marchisotto, G-A. Zakeri, *An Invitation to Integration in Finite Terms*, The College Math. J. **25** (1994), 295–308.

S. Mazur, S. Uhlam, *Sur les transformations isométriques d'espaces vectoriels normés*, C. R. Acad. Sci. paris **194** (1932), 946–948.

C. Misner, K. Thorne, J. Wheeler, *Gravitation*, Freeman, 1973.

A. Mizrahi, M. Sullivan, **Mathematics** *for Business and Social Sciences: An Applied Approch*, 1976.

E. Moise, *Elementary Geometry from an Advanced Standpoint*, Addison-Wesley, 1990.

R. A. Molin, *Prime-Producing Quadratics*, The Amer. Math. Monthly **104** (1997), 529–544.

T. Pappas, *The Joy of Mathematics*, 1989 : *More Joy of Mathematics*, 1991; Wide World Publishing/Tetra.

———, *Mathematical Scandals*, 1997; 고석구·이만근 옮김, *수학의 스캔들*, 10101, 1999.

H. Peitgen, H. Jürgens, D. Saupe, *Chaos and Fractals, New Frontiers of Science*, Springer-Verlag, 1992.

Z. Pogoda, L. Skołowski, *Does Mathematics Distinguish Certain Dimensions of Spaces?*, Amer. Math. Monthly **104** (1997), 860–869.

R. Risch, *The problem of integration in finite terms*, Translations of Amer. Math. Soc. **139** (1969), 167–189.

W. Rudin, *Principles of Mathematical Analysis*, Third Edition, McGraw-Hill, 1976.

H. M. Schey, *div, grad, curl and all that*, 2nd ed., W. W. Norton & Company, 1992.

M. M. Schiffer, L. Bowden, *The role of Mathematics in Science* Math. Assoc. Amer., 1984.

S. Schwartzman, *The Words of Mathematics: An Etymological Dictionary of Mathematical Terms Used in English*, Math. Assoc. Amer., 1994.

K. Smith, *Power Series from a Computational Point of View*, Springer-Verlag, 1987.

I. Stewart, "Nature's Numbers" (자연의 수학적 본성, 동아출판사, 1996)

D. W. Sumners, *The Role of Knot Theory in DNA Research*, Geometry and Topology, Marcel Dekker, Lect. Notes in Pure and Appl. Math., Vol. 105, 1987.

C. Taylor, *Exploring Music*, Institute of Physics Publishing, 1992.

G. Thomas, R. Finney, *Calculus*, 9th ed., Addison Wesley, 1996.

V. M. Tikhomirov, *Stories about Maxima and Minima*, Amer. Math. Soc. and Math. Assoc. Amer., 1990.

Truesdell, *Essays in the history of Mechanics*, Springer-Verlag, New York, 1968.

W. Weaver, *Lady Luck, The Theory of Probability*, Dover, 1963.

H. Weyl, *Symmetry*, Princeton University Press, 1952.

J. H. White, *Self-linking and the Gauss integral in higher dimensions*, Amer. J. Math. **91** (1969), 693–728.

S. Wolfram, *Mathematica*, 3rd ed., Addison-Wesley Publishing Company, 1996.

Appendix 3 Answers to Selected Exercises

Chapter 10

1.4 (1) is obtained by rotating the graph of the function $(\sin \pi r)/\pi r$ about $r = 0$.
(2) is obtained by rotating the graph of the function $2r^2e^{-r^2}$ about $r = 0$.

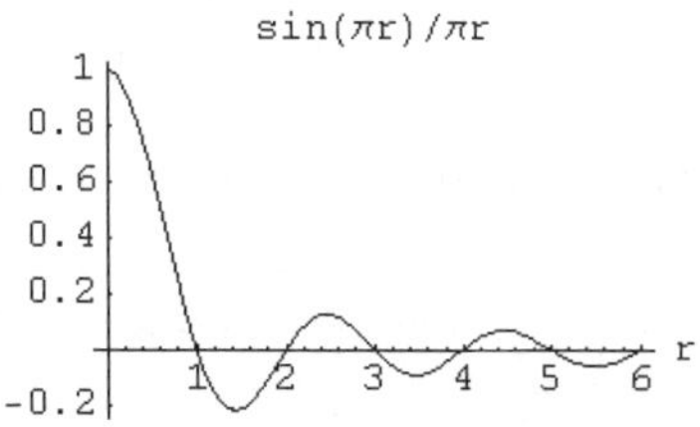

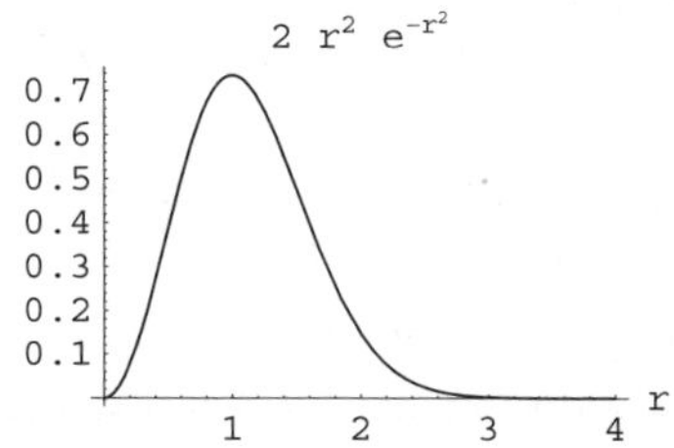

1.7 $(1,2,3)\cdot(1,1,1)/\sqrt{3} = 6/\sqrt{3} = 2\sqrt{3}$

2.1 (2) If we let $1/(x^2+y^2) = t$, the given problem is equivalent to $\lim_{t\to\infty} t/e^t$. Answer: 0

2.2 Since every function is continuous everywhere except at the origin, it suffices to investigate the continuity at the origin. (1) When $x = y^2$ and $(x,y) \neq (0,0)$, the value of the function is $1/2$. $\therefore$ Discontinuous.

(2) When $x \neq 0$ and $y = 0$,the value of the function is 1. $\therefore$ Discontinuous.

(3) Since $x^2y - xy^2 = xy(x-y)$ and $|xy| \leq \dfrac{1}{2}(x^2+y^2)$, $|f(x,y)| \leq \dfrac{1}{2}|x-y|$. Therefore the limit near the origin is 0. $\therefore$ Continuous.

2.4 $a = 0$, $b = 0$

2.6 If the domain is not closed or the domain is not bounded. (A closed set doesn't have to be an open set.)

3.1 $\text{grad} f(x,y,z) = (2xy^3, 3x^2y^2, 1)$ and thus $\text{grad} f(1,2,3) = (16,12,1)$.

3.2 (1) 52 (2) 3 (3) $2 + 1/e$ (4) $3\sqrt{2}/2$ (5) $3/2$

3.6 (1) $X/|X|^2$ (2) $-2e^{-|X|^2}X$ (3) $(-1+1/|X|)e^{-|X|}X$

3.9 $\dfrac{\partial A}{\partial W} = 0.007\cdot 0.425 W^{-0.575}H^{0.725} \approx 0.007\cdot 0.425\cdot 0.10\cdot 40.5 \approx 0.012$ and $\dfrac{\partial A}{\partial H} = 0.007\cdot 0.725 W^{0.425}H^{-0.275} \approx 0.007\cdot 0.725\cdot 5.4\cdot 0.25 \approx 0.007$. Therefore the body surface area of a person with weight 54 kg and height 165 cm is about $1.5\,m^2$, and if his weight increases by 1 kg then the body surface area increases by approximately $0.012\,m^2$ and if his height increases by 1 cm then the body surface area increases by approximately $0.007\,m^2$.

4.2 3, (1,0)

4.5 $\pm(2,1)/\sqrt{5}$

5.5 Since $f(x,y) = f(y,x)$, $\dfrac{\partial f}{\partial x}(1,0) = \dfrac{\partial f}{\partial y}(0,1)$. Answer: Both 2.

5.6 $D_1 f(x,y) = g(b(x,y))D_1 b(x,y) - g(a(x,y))D_1 a(x,y)$, $D_2 f(x,y) = g(b(x,y))D_2 b(x,y) - g(a(x,y))D_2 a(x,y)$

5.9 -3

5.10 $\frac{d}{dt} f(x,y) = \dfrac{\partial f}{\partial x}\dfrac{dx}{dt} + \dfrac{\partial f}{\partial y}\dfrac{dy}{dt} = 1\cdot 2 + (-1)\cdot 3 = -1$ (°C/time)

5.16 Since $\text{grad} f(X) = g'(|X|)\frac{X}{|X|}$, it is parallel to X.

5.17 Let $X(t)$ be a "spherical curve" connecting two points X_0, X_1 on the sphere with radius r and center at the origin. Then if we differentiate $|X(t)|^2 = r^2$, we get $X(t) \cdot X'(t) = 0$ and

$$\frac{d}{dt} f(X(t)) = \operatorname{grad} f(X(t)) \cdot X'(t) = g(X(t))X(t) \cdot X'(t) = 0.$$

Therefore $f(X(t))$ is a constant function and $f(X_0) = f(X_1)$.

6.4 $\frac{x_0 x}{a^2} + \frac{y_0 y}{b^2} - \frac{z_0 z}{c^2} = 1$, $\frac{x_0 x}{a^2} - \frac{y_0 y}{b^2} - \frac{z_0 z}{c^2} = 1$

Adv.Ex. 2. Yes. Yes. No

Adv.Ex. 3. At a point on the curve $y = x^2 - x^3$, (when the point is near the origin but not the origin)

$$f(x,y) = \frac{xy}{x^2 - y} = \frac{x(x^2 - x^3)}{x^2 - (x^2 - x^3)} = \frac{x^3 - x^4}{x^3} = 1 - x.$$

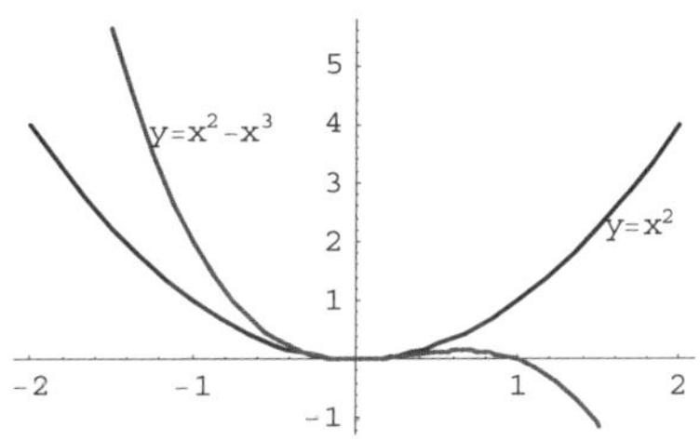

Therefore for a sequence of points which converges to the origin along this curve, the values of the function converge to $\lim_{x \to 0}(1 - x) = 1$. Since this limit is not equal to $f(0,0) = 0$, $f(x,y)$ is not continuous at the origin.

However we can see that the (x,y)-directional derivative exists at the origin as follows.

When $y = 0$, $f(x,0) = 0$ and so at the origin, the $(x,0)$-directional derivative is

$$\lim_{t \to 0} \frac{f(tx,0) - f(0,0)}{t} = 0.$$

When $y \neq 0$,

$$\lim_{t \to 0} \frac{f(tx,ty) - f(0,0)}{t} = \lim_{t \to 0} \frac{t^2 xy}{t(t^2x^2 - ty)} = \lim_{t \to 0} \frac{xy}{tx^2 - y} = \frac{xy}{-y} = -x.$$

However, the function $f(x,y)$ given in the problem, doesn't have the directional derivatives in the general direction at the "singular point" (that is, the point at which the denominator is zero). For example, at a singular point $(1,1)$, in order to get the (x,y)-directional derivative, there must exists the limit

$$\lim_{t \to 0} \frac{f(1+tx, 1+ty) - f(1,1)}{t} = \lim_{t \to 0} \frac{(1+tx)(1+ty)}{t((1+tx)^2 - (1+ty))},$$

however we know this limit doesn't exist.

Chapter 11

1.1 $3e^{x^7}/x - 2e^{x^5}/x + \int_{x^2}^{x^3} ye^{xy^2}\,dy = 3e^{x^7}/x - 2e^{x^5}/x + e^{x^7}/2x - e^{x^5}/2x = \left(7e^{x^7} - 5e^{x^5}\right)/2x$

1.3 $\pi/8$

1.4 (1) Since $\mathbf{n}(s)$ is a unit vector, $\mathbf{n}'(s)$ is perpendicular to $\mathbf{n}(s)$. Since $\mathbf{n}(s)$ is perpendicular to $\mathbf{t}(s)$, $\mathbf{n}'(s)$ is proportional to $\mathbf{t}(s)$. Here the proportional constant is $\mathbf{n}'(s)\cdot\mathbf{t}(s) = (\mathbf{n}(s)\cdot\mathbf{t}(s))' - \mathbf{n}(s)\cdot\mathbf{t}'(s) = 0 - (-\kappa(s)) = \kappa(s)$. Therefore $\mathbf{n}'(s) = \kappa(s)\mathbf{t}(s)$.

(2) Since

$$C_u'(s) = \mathbf{t}(s) + u\mathbf{n}'(s) = \big(1 + u\kappa(s)\big)\mathbf{t}(s),$$

if $|u| \ll 1$ then

$$L(u) = \int_{s_0}^{s_1} \big(1 + u\kappa(s)\big)\,ds.$$

If we differentiate it, by Leibniz's rule, we get

$$L'(u) = \int_{s_0}^{s_1} \kappa(s)\,ds.$$

2.2 From $D_1f(x,y) = y^2 + ye^{-x}(-\sin(x-y) + \cos(x-y))$, $D_2f(x,y) = 2xy + e^{-x}(\sin(x-y) - y\cos(x-y)$, we get ${D_1}^2f(x,y) = -2ye^{-x}\cos(x-y)$, $D_1D_2f(x,y) = 2y + e^{-x}(-\sin(x-y) + \cos(x-y) + y\cos(x-y) + y\sin(x-y))$, ${D_2}^2f(x,y) = 2x + e^{-x}(-\cos(x-y) - \cos(x-y) - y\sin(x-y)$.

2.4 Yes.

2.5 $4, 8$

2.10 $D_1f(x,y) = y/(1+xy)$, $D_2f(x,y) = x/(1+xy)$ and ${D_1}^2f(x,y) = -y^2/(1+xy)^2$, $D_1D_2f(x,y) = [(1+xy) - yx]/(1+xy)^2 = 1/(1+xy)^2$, ${D_2}^2f(x,y) = -x^2/(1+xy)^2$. Also ${D_1}^3f(x,y) = 2y^3/(1+xy)^3$, ${D_1}^2D_2f(x,y) = -2y/(1+xy)^2 + 2xy^2/(1+xy)^3 = -2y/(1+xy)^3$, $D_1{D_2}^2f(x,y) = -2x/(1+xy)^3$, ${D_2}^3f(x,y) = 2x^3/(1+xy)^3$ and thus $D^2_{(1,2)}f(1,1) = {D_1}^2f(1,1) + 4D_1D_2f(1,1) + 4{D_2}^2f(1,1) = -1/4 + 1 - 1 = -1/4$. Therefore it is concave in the direction of $(1,2)$ and $D^3_{(1,2)}f(1,1) = {D_1}^3f(1,1) + 6{D_1}^2D_2f(1,1) + 12D_1{D_2}^2f(1,1) + 8{D_2}^3f(1,1) = 2/8 - 12/8 - 24/8 + 16/8 = -9/4$.

3.1 (1) xy (2) x (3) $x^2 + y^2$ (4) $-\sin 1 + y\cos 1 + x^2\cos 1 + \frac{1}{2}y^2(\cos 1 + \sin 1)$ (5) xy (6) $x + xy + 2y^2 + x^2$ (7) $1 + x + y + (x+y)^2/2$ (8) $x - y + x^2/2 - y^2/2$ (9) $1 + x^2 + y^2$ (10) $y + xy$ (11) $1 + xy$ (12) $\sin 1 + xy\cos 1$

3.4 Since $|\sin(xy)| \le |xy|$, the limit exists and is 0.

3.5 0

3.6 $f(x,y,z) = \sum_{i,j,k} a_{ijk}x^iy^jz^k$. Since ${D_1}^i{D_2}^j{D_3}^kf(0,0,0) = i!j!k!a_{ijk}$, $Tf(x,y,z) = \sum_{n=0}^{\infty}\frac{1}{n!}(xD_1 + yD_2 + zD_3)^n f(0,0,0) = \sum_{n=0}^{\infty}\frac{1}{n!}\sum_{i+j+k=n}\frac{n!}{i!j!k!}x^iy^jz^k{D_1}^i{D_2}^j{D_3}^kf(0,0,0) = \sum_{i,j,k} a_{ijk}x^iy^jz^k = f(x,y,z)$.

3.7 $3+3(x-1)+3(y-1)+(x-1)^2+4(x-1)(y-1)+(y-1)^2+(x-1)^2(y-1)+(x-1)(y-1)^2$

3.9 This problem gives different answers depending on the way to set the function. For example, in (1), if we let

$$f(x,y) = x^y, \quad P = (1,1), \quad \mathbf{v} = (0.02, 0.01),$$

then

$$\text{grad}f(x,y) = (yx^{y-1}, x^y \log x), \quad D_{\mathbf{v}}f(P) = (1,0)\cdot(0.02, 0.01) = 0.02$$

and so

$$(1.02)^{1.01} = f(P+\mathbf{v}) \approx 1 + 0.02 = 1.02.$$

However if we let

$$g(x,y) = (1+2x^2)^y, \quad P = (0,1), \quad \mathbf{v} = (0.1, 0.01),$$

then

$$\text{grad}g(x,y) = (4xy(1+2x^2)^{y-1}, (1+2x^2)^y \log(1+2x^2)), \quad D_{\mathbf{v}}g(P) = (0,0)\cdot(0.1, 0.01) = 0$$

and so

$$(1.02)^{1.01} = g(P+\mathbf{v}) \approx 1 + 0 = 1.$$

As we can see from this example, $f(x,y)$ is a better function than $g(x,y)$. Now if we compute the second-order partial derivatives of $f(x,y)$ to estimate the remainder $f(x,y)$,

$$\frac{\partial^2 f}{\partial x^2} = y(y-1)x^{y-2}, \quad \frac{\partial^2 f}{\partial x \partial y} = x^{y-1} + yx^{y-1}\log x, \quad \frac{\partial^f}{\partial y^2} = x^y(\log x)^2$$

and in the interval $[(1,1),(1.02,1.01)$ $x^y \le 1.02^{1.01} \le 1.02^2 = 1 + 0.04 + 0.0004 \le 1.1$. From this, we get $M_2 \le 2.2$. Therefore

$$|R_1(\mathbf{v})| \le \frac{M_2}{2}(0.02+0.01)^2 = 1.1 \times 3^2 \times 10^{-4} \le 10^{-3}.$$

and so

$$1.02^{1.01} = 1 + 0.02 \pm 10^{-3}.$$

4.2 We may let $f(a,b) := \sum_i (y_i - ax_i - b)^2$ and find a point (a,b) such that $\dfrac{\partial f}{\partial a} = 0$, $\dfrac{\partial f}{\partial b} = 0$. (Or we may convert it into the form of perfect square.) $a = \dfrac{\overline{xy} - \overline{x}\,\overline{y}}{\overline{x^2} - (\overline{x})^2}$, $b = \dfrac{\overline{x^2}\,\overline{y} - \overline{x}\,\overline{xy}}{\overline{x^2} - (\overline{x})^2}$. Here we use the notations $\overline{x} = \sum x_i/n$, $\overline{y} = \sum y_i/n$, $\overline{xy} = \sum x_i y_i/n$, $\overline{x^2} = \sum x_i^2/n$. Here $\overline{y} = a\overline{x} + b$.

4.4 If P is a critical point, $\text{grad}f(P) = \frac{P-A}{|P-A|} + \frac{P-B}{|P-B|} + \frac{P-C}{|P-C|} = 0$. If we let $\mathbf{a} := \frac{P-A}{|P-A|}$, $\mathbf{b} := \frac{P-B}{|P-B|}$, $\mathbf{c} := \frac{P-C}{|P-C|}$, then they are all unit vectors which satisfy $\mathbf{a}+\mathbf{b}+\mathbf{c} = \mathbf{0}$. If we take the inner product of this equation and $\mathbf{a}$, we get

$$1 + \mathbf{a}\cdot\mathbf{b} + \mathbf{a}\cdot\mathbf{c} = 0.$$

Similarly, if we take the inner product of this equation and $\mathbf{b}$ or $\mathbf{c}$, then we get

$$\mathbf{a}\cdot\mathbf{b} + 1 + \mathbf{b}\cdot\mathbf{c} = 0, \qquad \mathbf{a}\cdot\mathbf{c} + \mathbf{b}\cdot\mathbf{c} + 1 = 0.$$

If we solve this system of equations, we get

$$\mathbf{a}\cdot\mathbf{b} = \mathbf{b}\cdot\mathbf{c} = \mathbf{c}\cdot\mathbf{a} = -\frac{1}{2},$$

which tells us that any two of those vectors make 120° angle.

5.3 (1) local maximum point: $\left(-\frac{1}{2}, 4\right)$ (2) saddle point: $(1,-1)$ (3) local minimum points: $\left(-3(-1)^k, k\pi\right)$, $k \in \mathbb{Z}$. saddle points: $\left(0, (k+\frac{1}{2})\pi\right)$, $k \in \mathbb{Z}$ (12) For this one, the critical points are $P := \pi(2n+1, 2m+1)$ and $Q_\pm := \pm\pi(2n+\frac{1}{3}, 2m+\frac{1}{3})$ $(m, n \in \mathbb{Z})$. Since the Hessian at the point P is the zero matrix, you cannot use the second derivative test. However if you consider the Taylor expansion $f(P+(x,y)) = \frac{1}{2}xy(x+y) + o((x^2+y^2)^2)$, P is a saddle point. The Hessian at $Q_\pm$ is $\mp\begin{pmatrix} \sqrt{3} & \sqrt{3}/2 \\ \sqrt{3}/2 & \sqrt{3} \end{pmatrix}$ and hence Q_+ is a local maximum point, Q_- is a local minimum point.

5.7 At a local maximum point, $\dfrac{\partial^2}{\partial x^2} \leq 0$ and $\dfrac{\partial^2}{\partial y^2} \leq 0$. Once we add these two inequalities, we get $\dfrac{\partial^2}{\partial x^2} + \dfrac{\partial^2}{\partial y^2} \leq 0$. We can deal with a local minimum point in a similar way.

6.3 (4) If we sketch the region, it is bounded and closed, and hence there exist a maximum point and a minimum point (and therefore extreme points). On the other hand, since $\text{grad} f(x,y) \neq \mathbf{0}$ inside the region, extreme points lie on the boundary. If we examine f on each part of boundary, $2x+y=0$, $x=0$, $y=0$, the the minimum value of f is 0, the maximum value of f is $f(\frac{5}{2}, 5) = 25/2$. There is no other extreme value. (6) The maximum value is 2 and the minimum value is -2. There is no other extreme value.

6.7 $\sqrt{3}/2$

6.8 $(2,2,4)$, $(-2,-2,4)$

6.10 (4) $3/\sqrt{6}$

6.11 The width, length, height are 10, 10, 20 (cm) respectively.

6.15 Notice that the constraint is a compact set and thus there exists the solution. the gradient of the function $f(x,y,z) = xy+yz+zx-2xyz$ is

$$\text{grad} f(x,y,z) = (y+z-2yz, z+x-2zx, x+y-2xy)$$

and thus if the extreme point (x,y,z) satisfies $x>0, y>0, z>0$, the gradient vector $\text{grad} f(x,y,z)$ must be parallel to $\text{grad}\,(x+y+z) = (1,1,1)$. Therefore we get

$$y+z-2yz = z+x-2zx = x+y-2xy.$$

From the first equation we get $0 = y-x-2z(y-x) = (1-2z)(y-x)$, that is, $z = \dfrac{1}{2}$ or $x=y$, from the second equation we get $0 = z-y-2x(z-y) = (1-2x)(z-y)$, that is, $x = \dfrac{1}{2}$ or $y=z$.

If $z = \dfrac{1}{2}$ and $x = \dfrac{1}{2}$, from the condition $x+y+z=1$ we get $y=0$ and

$$f\left(\frac{1}{2}, 0, \frac{1}{2}\right) = \frac{1}{4}.$$

If $z = \dfrac{1}{2}$ and $x \neq \dfrac{1}{2}$, then $y=z$ and from the condition $x+y+z=1$ we get $x=0$, $y = z = \dfrac{1}{2}$ and

$$f\left(0, \frac{1}{2}, \frac{1}{2}\right) = \frac{1}{4}.$$

If $z \neq \dfrac{1}{2}$ and $x = \dfrac{1}{2}$, then $x=y$ and from the condition $x+y+z=1$ we get $x = y = \dfrac{1}{2}$, $z=0$ and

$$f\left(\frac{1}{2}, \frac{1}{2}, 0\right) = \frac{1}{4}.$$

If $z \neq \frac{1}{2}$ and $x \neq \frac{1}{2}$, then $x = y = z$ and from the condition $x + y + z = 1$ we get $x = y = z = 1/3$ and

$$f\left(\frac{1}{3}, \frac{1}{3}, \frac{1}{3}\right) = \frac{7}{27}.$$

If the extreme value is attained when $z = 0$[6], the function to minimize is

$$f(x, y) = xy$$

and the constraint becomes $x + y = 1$, $x \geq 0$, $y \geq 0$. The minimum value is 0 and the maximum value is $1/4$.

Therefore the maximum and minimum value we want to find are $\frac{7}{27}$ and 0, respectively.
6.16 Answer: 5
Adv.Ex. 3. For a function

$$u(x, y) := \frac{1}{4}x^4 + \frac{1}{6}y^6 + x^2 y - 3ye^{\sin x} + 7x - 5y,$$

if a point (x, y) moves far away from the origin, the value of $u(x, y)$ becomes very large. Therefore $u(x, y)$ has the minimum value and so there exists a point (x_0, y_0) such that

$$\operatorname{grad} u(x_0, y_0) = \mathbf{0}.$$

This point is a solution of the given system. [Kazdan].

Chapter 12

1.1 On the sphere of radius ρ, the area of the region

$$\varphi_0 \leq \varphi \leq \varphi_0 + \Delta\varphi, \qquad \theta_0 \leq \theta \leq \theta_0 + \Delta\theta \qquad (*)$$

is

$$\rho^2(\cos\varphi_0 - \cos(\varphi_0 + \Delta\varphi))\Delta\theta.$$

Therefore the volume of a region which satisfies the equation (*) and

$$\rho_0 \leq \rho \leq \rho_0 + \Delta\rho$$

is

$$\begin{aligned} \Delta V &= \frac{1}{3}\left((\rho_0 + \Delta\rho)^3 - \rho_0^3\right)\left(\cos\varphi_0 - \cos(\varphi_0 + \Delta\varphi)\right)\Delta\theta \\ &= \left(\rho_0^2\Delta\rho + o(\Delta\rho)\right)\left(\sin\varphi_0 \cdot \Delta\varphi + o(\Delta\varphi)\right)\Delta\theta \end{aligned}$$

and so

$$\lim \frac{\Delta V}{\Delta\rho \cdot \Delta\varphi \cdot \Delta\theta} = \rho_0^2 \sin\varphi_0.$$

1.6 The Jacobian determinant is identically zero. The image of $(u, v) = F(x, y)$ is a curve $v = \sin u$ and so the rate of change of area is zero.

---

[6]The given function $f(x, y, z)$ is symmetric.

Chapter 13

1.1 If a point (x, y) satisfies $(x-a)^2+y^2 = a^2$, then $\big((x,y)-(a,0)\big)\cdot\mathbf{F}(x,y) = (x-a, y)\cdot(-2xy, x^2-y^2) = -(x-a)2xy+y(x^2-y^2) = -y(x^2+y^2-2ax) = -y\big((x-a)^2+y^2-a^2\big) = 0$. So $\mathbf{F}$ is perpendicular to the radius of the circle, and thus we get the conclusion we want.

2.1 (1) 4π (2) -2π (3) 9

2.2 Answer: (1) $\pi/2$ (2) $\pi/2$ (3) $-3\pi/2$

2.3 (1) (a) 3 (b) -1

4.1 (1) 69/4 (2) 6/5 (3) 105/2 (4) 315 (5) -9

Chapter 14

3.1 (1) 16 (2) $4224/5 - 160\sqrt{10} - 72\sqrt{6}$ (3) 0 (4) 0

3.2 $3(2e-5)$

3.3 $400(4/e-1)$

3.5 Let B be the center of the base R, and R_x be the cross section cut by the plane parallel to the base with the distance x from the vertex. Then the area of the cross section R_x is $(x/h)^2A$ and the center is $\frac{x}{h}B + \left(1-\frac{x}{h}\right)P$. Therefore the center of the cone is

$$\frac{1}{\frac{1}{3}Ah}\int_0^h (x/h)^2 A\left(\frac{x}{h}B + \left(1-\frac{x}{h}\right)P\right)dx = 3\int_0^1 \left(t^3B + t^2(1-t)P\right)dt = \frac{3}{4}B + \frac{1}{4}P.$$

Thus we are done.

3.9 (1) 7 (2) 11/5 (3) 4/15 (4) π^2 (5) 5/6 (6) $\pi/4$ (7) $\frac{3\pi}{4}\log 5$ (8) $-\log\sqrt{2}$ (9) 1/3 (10) $\sqrt{2}/3 - 1/6$ (11) $2(2\sqrt{2}-1)/15$ (12) $(2\sqrt{2}-1)/15$ (13) 6 (14) $(e-2)/2$ (15) 1 (16) $2e-4$ (17) $\frac{5}{4}\log 5 - 1$

3.10 (3) 8

3.14 When we pick the x-axis and cut the rock perpendicular to the x-axis, let $g(x)$ be the area of cross section, and $f(x)$ be the area of a particular mineral. Here we assume that $f(x), g(x)$ are both continuous functions. Then there exists a point x_0 such that

$$\int f(x)\,dx \;:\; \int g(x)\,dx = f(x_0) : g(x_0).$$

4.6 The pictures of orthogonal projections onto the xy-plane of cylinders $x^2+z^2 \le 1$, $y^2+z^2 \le 1$ and $x^2+y^2 \le 1$ are as follows.

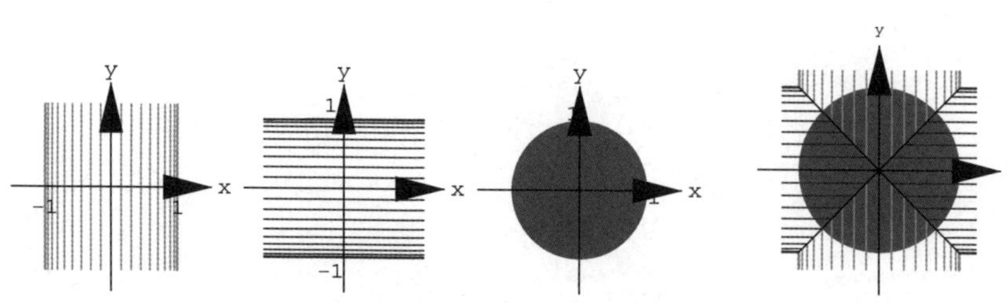

Therefore to know the volume of their intersection it suffices to know the volume of the sector-shaped part. Thus if we use cylindrical coordinates, the volume we want is $16\displaystyle\int_0^{\pi/4}\int_0^1\int_0^{\sqrt{1-r^2\cos^2\theta}} r\,dz\,dr\,d\theta =$

$$16\int_0^{\pi/4}\int_0^1 r\sqrt{1-r^2\cos^2\theta}\,dr\,d\theta = 16\int_0^{\pi/4}\int_1^{\sin\theta} -\frac{t^2dt}{\cos^2\theta}\,d\theta = \frac{16}{3}\int_0^{\pi/4}\frac{1-\sin^3\theta}{\cos^2\theta}\,d\theta =$$
$$\frac{16}{3}\int_0^{\pi/4}\sec^2\theta - \frac{1-\cos^2\theta}{\cos^2\theta}\sin\theta\,d\theta = \frac{16}{3}\left[\tan\theta - \frac{1}{\cos\theta} - \cos\theta\right]_0^{\pi/4} = 16-8\sqrt{2}.$$

4.11 (5) $(0, 0, 3/8)$
5.1 $I_x = I_y = 400\mu/3$, $I_z = 16\mu/3$
5.2 $M = 5/2$, $(\bar{x}, \bar{y}, \bar{z}) = (\frac{8}{15}, \frac{8}{15}, \frac{8}{15})$, $I_x = I_y = I_z = 11/6$
Adv.Ex. 5. We can get it by replacing T by $\sqrt{2a}T$.

Chapter 15

1.5 $+$
1.6 $\text{div}\mathbf{F}(\mathbf{x}) = k$
2.3 When $|X| > r_0$, $\text{grad}\,\varphi(X) = 2\pi r_0\,\text{grad}\log|X|$ and thus

$$\varphi(X) - 2\pi r_0 \log|X| = \text{const}.$$

Now all you have to do is to derive

$$\text{const} = \lim_{X\to\infty}(\varphi(X) - 2\pi r_0\log|X|) = \lim_{X\to\infty}\int_{|P|=r_0}(\log|X-P| - \log|X|)\,ds = 0.$$

5.3 $3\pi a^2$
5.4 $3\pi/8$
5.5 $3/2$

Chapter 16

3.6 (1) 16π (2) 16π (3) 20π

Chapter 17

1.2 $12\pi/5$

Chapter 18

2.4 0
2.5 -12π

Index